Pioneers of Ecological Restoration

*Publication of this book was made possible, in part,
by a gift from Dr. Kato Perlman in memory of her beloved husband,
Professor David Perlman.*

The University of Wisconsin Press
1930 Monroe Street, 3rd Floor
Madison, Wisconsin 53711-2059
uwpress.wisc.edu

3 Henrietta Street
London WC2E 8LU, England
eurospanbookstore.com

Library of Congress Cataloging-in-Publication Data
Court, Franklin E., 1939–
Pioneers of ecological restoration : the people and legacy of the University of Wisconsin
Arboretum / Franklin E. Court.
p. cm. — (Wisconsin land and life)
Includes bibliographical references and index.
ISBN 978-0-299-28664-4 (pbk. : alk. paper) — ISBN 978-0-299-28663-7 (e-book)
1. University of Wisconsin—Madison. Arboretum—History. 2. Arboretums—
Wisconsin—Madison—History. I. Title. II. Series: Wisconsin land and life.
QK480.U52W533   2012
580'.7377583—dc23
2011041958

Pioneers of Ecological Restoration

∾

The People and Legacy of the University of Wisconsin Arboretum

Franklin E. Court

THE UNIVERSITY OF WISCONSIN PRESS

WISCONSIN LAND AND LIFE

Arnold Alanen
Series Editor

for the Arboretum staff, guides, rangers, and volunteers

To have and to hold . . . the Arnold Arboretum . . . for one
thousand years . . . yearly rental of one dollar . . . and so on from
time to time forever.

INDENTURE BETWEEN HARVARD COLLEGE
AND THE CITY OF BOSTON, December 30, 1882

It would seem to be peculiarly appropriate for our Universities
and Colleges to secure upon the grounds by which they are surrounded, at least one good specimen of each tree and shrub that
grows naturally in Wisconsin; and I will venture to predict that
the University or College that shall first surround itself with such
an "Arboretum" will first secure the patronage and good opinion
of the people, and will thus outstrip those institutions that show
a lack of taste and refinement by omitting to plant trees.

INCREASE A. LAPHAM, "The Forest Trees of Wisconsin," 1853

When we build, let us build forever. Let it be not for present
delight, not for the present use alone; let it be such work as our
descendants will thank us for.

JOHN RUSKIN, "Lamp of Memory," 1849

Contents

List of Illustrations   ix

Preface   xi

Acknowledgments   xvii

List of Abbreviations   xix

1   The Beginnings: John Nolen, Michael Olbrich, Paul E. Stark,
     and the Prairie School of Landscape Design   3

2   1930–1932: Acquiring the Land, Building the Road   18

3   1932–1934: Pre-Dedication Years, Edward M. Gilbert,
     Aldo Leopold, G. William Longenecker, the Politics
     of Governance   45

4   1934–1935: Two Directors, Dedication, Camp Madison, and
     the CCC   65

5   1935: "Go Make a Prairie," Fassett's Planting Experiments,
     Sperry's Fieldwork   95

6   Late 1930s: Frustrations, "Golden Years," Efforts to
     Increase Funding   110

7   The 1940s: Curtis Years, Postwar Challenges, Leopold's Death   126

8   The 1950s: Outside Threats, "the Beltline Cometh"   159

9   The 1960s: Last Lost City Lots, Building a Reputation,
     Shifting Focus   180

10   The 1970s: World Famous Prairies, Conflicts in Leadership,
     New Appointments and a Visitor Center   214

11   The 1980s and Beyond: A Restoration Ecology Legacy,
     "Earth Partnerships," Reflections   237

     Notes   265
     Bibliography   287
     Index   299

Illustrations

John Nolen   5

Michael Balthazar Olbrich   6

Paul E. Stark   10

Madison Parks Foundation map showing six land parcels   22

Colonel Joseph W. "Bud" Jackson   33

Map showing the Island at the east end of Lake Wingra   37

Louis Gardner   43

Edward M. Gilbert   48

Aldo Leopold   56

Albert F. Gallistel   59

G. William Longenecker   69

CCC enrollees from a Camp Madison barracks   89

CCC enrollees in front of Camp Madison supply warehouse   91

Norman C. Fassett   96

Ted Sperry   105

CCC enrollees fighting the mess hall fire   114

John T. Curtis   127

Robert McCabe and Jim Hale banding pheasants   144

Grant Cottam   184

Rosemary Fleming   202

Katharine T. Bradley   226

Virginia "Gina" Kline 230
William R. Jordan III 238
Gregory Armstrong 244
Kevin McSweeney 259

Preface

It is no exaggeration to say that the reputation of the University of Wisconsin's Arboretum in Madison is closely intertwined with the evolving global history of restoration ecology. Yet, like so many other consciously preserved and cultivated wonders of nature, the Arboretum's early history was largely local in scope and its mission only partially defined. Thus, the early chapters of this book concentrate in some detail on substantive, if sometimes lackluster, local issues such as the first land purchases, university governance concerns, staffing challenges, decisions about what to plant and where to plant, and continuously frustrating efforts throughout the 1930s and 1940s to gain adequate funding and to identify precisely what the primary mission of the Arboretum's future should be and who should direct it.

As the Arboretum evolves, however, from the postwar 1940s and 1950s into the 1970s and 1980s, its mission, though more diverse, is also more focused. Its international seed distribution program, its trailblazing ties to national and international environmentalism and to emerging ecological movements and trends, particularly in the field of land restoration, also begin to gain more visibility. By the 1990s, the Arboretum was nationally and internationally recognized as a laboratory for scientists and naturalists, as a place for environmental learning and research in the production of ecological systems rich in native species, and also as an international resource for the promotion of instruction in land restoration techniques intended for the general public.

The University of Wisconsin's "Forest Preserve Arboretum and Wild Life Refuge" was officially founded—under that title—by the University Board of Regents on April 26, 1932. Born in the shadow of John Muir and the American Conservation Movement and framed by the visions of Boston's John Nolen, pioneering landscape architect and city planning specialist, and Madison's

Michael Olbrich and Paul E. Stark, both nature enthusiasts and dedicated builders of city parks, the University of Wisconsin Arboretum had been part of an effort, beginning with the first land purchases in the 1920s, to preserve and protect the land bordering Madison's four intercity lakes. The main object was to guarantee public access to the lakes and protection from exploitive development. When the Arboretum opened, it comprised six very ecologically different land parcels, each acquired piecemeal with a novel purchase history of its own. All in all, the Arboretum's grounds in 1932 consisted of 245 acres bordering the western and southwestern shores of Lake Wingra. By 1980 the grounds would include over 1,200 acres.

After a somewhat fuzzy start in which the new Arboretum's mission was conceived as something between a park and a botanical collection of trees and shrubs, the Arboretum's mission expanded to include not only the preservation and protection of land but also a new direction, very much in keeping with the regional Chicago "prairie style" of planting and landscape design promoted by Ossian Cole Simonds and Jens Jensen, both staunch proponents of natural Midwest landscapes and native plant gardens. It was a vision promoting the merits of "natural design" first articulated in the context of the UW Arboretum, in 1928, by Norman C. Fassett, UW botany professor, and six years later in June 1934 by Aldo Leopold, director of Arboretum research, in an address at the Arboretum's dedication, in which he encouraged the planting of a tall grass prairie reminiscent of Wisconsin in the 1840s in order to attract and to provide habitat for game birds. The hope, into the 1930s and 1940s, was that this Arboretum, unlike most other arboretums, both in the United States and Europe, would function less as a "museum of trees and shrubs," in the words of Ossian Simonds, and more "naturally" as a protected wildlife refuge and experimental forest preserve.

During those early years, from the mid-1930s into the 1940s, the Arboretum pioneers—G. William Longenecker, Aldo Leopold, Colonel Joseph W. "Bud" Jackson, Edward M. Gilbert, Albert F. Gallistel, Maurice McCaffrey, and other local civic and UW supporters—envisioned themselves as promoters of "wildness" in a potential wildlife refuge, a contained game sanctuary. They saw themselves involved in a "grand experiment" primarily in wildlife ecology, habitat restoration, and forest experimentation. Concerns over the success or failure of the experiment frequently played out against the historically interesting but often frustrating backdrop of Depression Era politics. Obtaining the help of the Civilian Conservation Corps (CCC) along with National Park Service (NPS) staff that included Ted Sperry, an experienced prairie ecologist, was viewed as a solid measure of success in the great experiment. With the 1935

arrival of the CCC in what was then called Camp Madison, the realization
of Fassett's and Leopold's visions of restored "wildness"—forests, savannas,
ponds, marshes, and a native tall grass prairie—would be aided by dependable
CCC laborers and the availability of heavy equipment, making the reclamation task possible.

But after the 1930s and 1940s, experience rather than wishful thinking ultimately defined the Arboretum's mission. Periodic gifts of money and the
development of the Monroe Street corridor in the early 1940s expanded the
Arboretum's size and enhanced its developing civic reputation as a dandy place
to visit and to commune with nature on warm sunny afternoons. By the time
of Leopold's death in April 1948, it was clear, and I think Leopold realized
it, that the Arboretum would never be what he and most of the Arboretum
founders wanted it to be—a protected wildlife refuge and sanctuary for game
birds. For one thing, it would never be able to acquire adequate funding, at
least not on the federal level. But to the relief of many, including Leopold, it
would not be a public park either, in spite of challenges to its pristine naturalness and its restored ecosystems. The national desire to sacrifice wildness to
government-supported recreational demands was becoming more of a threat
as the number of visitors began to increase through the 1940s and into the
1950s, and the Arboretum was forced to confront the realities of urban expansion. By the early 1950s the city of Madison was encroaching. Its population
topped a hundred thousand. Neighborhood dogs and cats were roaming freely
on Arboretum grounds; bird trapping had to be suspended; and the construction of the Beltline highway bisecting Arboretum property was underway.
Yet, interest in the Arboretum as an experimental environmental lab, both
nationally and internationally, also increased considerably as the Arboretum
developed an international identity. The staff reported receiving inquiries relative to Arboretum matters from as far away as Yugoslavia. John T. Curtis, the
new director of research, initiated a seed exchange program with arboretums
and botanical gardens worldwide, an effort that did much to expand the Arboretum's global identity into the 1960s. Curtis's signature contribution to
the exchange was the conviction that although routine seed offerings were a
mainstay, a more desirable exchange should involve plants native to the Midwest of which the Arboretum now had a bounty. The mailing list included
twenty-three institutions in the United States and 132 in foreign countries.

By the 1960s and 1970s, the Arboretum had evolved considerably. Gone
were the matters integrally and primarily associated with the desire for a federal partnership. Even the official letterhead had dropped the "Experimental
Forest Preserve" label. The Arboretum was now known as the University of

Wisconsin Arboretum and Wildlife Refuge. Soon even the "Wildlife Refuge" appellation was dropped. An increasing emphasis on research forced the Arboretum Committee to look outside the university for funding. Also, the Arboretum's increasing vulnerability to outside intrusions and damaging pollutants became more and more apparent.

Arboretum management decided that one solution to the daily intrusions and environmental threats was education. The Arboretum's survival depended on the public's understanding of what the Arboretum was, what it was not, and what was and was not accessible to them. Into the 1970s and 1980s, educating the public about the Arboretum's mission became vitally important, as did a shared civic consternation over threats to the survival of Arboretum plant communities. A new focus on outreach and education characterized this transitional period and carried the Arboretum into the new millennium.

By the 1970s and 1980s, the Arboretum also had developed an international reputation as a leader in prairie restoration, as the place where prairie ecologists Ted Sperry, John Curtis, and Henry Greene had set the precedent for prairie restoration, leaving behind a legacy of information on how to do it. Curtis's work was recognized worldwide, and Greene's prairie and the fascinating story of his single-handed effort to plant it became legendary. By the 1980s, owing largely to the hiring of William R. Jordan III as the Arboretum's first public service coordinator, the Arboretum emerged as the definitive center for the global study and propagation of restoration ecology. By the 1990s, the Arboretum's commitment to the recovery and restoration of plant and animal ecosystems was increasingly viewed as the logical extension of the "land ethic" promoted by the first director of research, Aldo Leopold. Owing also to Jordan's efforts, the Arboretum began the sponsorship of the journal *Ecological Restoration*, a peer-reviewed quarterly that published articles on the restoration of ecological communities and landscapes worldwide.

By the end of the century, the Arboretum's growing reputation as a good place to visit for nature-based educational tours, for professional instruction in land reclamation, and for nature-based programs designed for children and adults necessitated staff increases that enlisted a cadre of talented and dedicated naturalists. The Arboretum's reputation as a good place for viewing seasonal changes, for spotting a red-tailed hawk, for photographing a flower, or for taking quiet hikes through the prairies or the woods also began to gain well-earned national and international attention. "A world-famous outdoor laboratory of nature" was how one news service referred to it. The Arboretum, in the eyes of many, had evolved into something larger than life, something "immortal," other-worldly, a "wilderness" with "exotic birds"; a wonderful place in which to walk, to sit down, to "watch"—a place, where, if one sat quietly enough, as radio and television broadcaster and entertainer Arthur Godfrey suggested, after a late 1960s visit, "the wildlife starts to come around."

What the Arboretum continues to stand for as we enter the second decade of the twenty-first century is a living testament to what G. K. Chesterton in 1913 in "The Victorian Age in Literature" observes about the synchronic nature of the phenomenon we call "historical development." Real development, as Chesterton notes, "is not leaving things behind as on a road, but drawing life from them, as from a root." "Progress" or "improvement" is often mistaken to mean leaving our homes behind us. But, as Chesterton advises, real improvement means exalting the towers of our homes and extending the gardens.

One is reminded of Leopold's plea for respect for the land and the promotion of efforts to assist its recovery, advice as alive and relevant today as it was in April 1932 when the University of Wisconsin Board of Regents accepted the deeds to six land parcels, first acquired by Olbrich and Stark, on Lake Wingra's southern shore, and a small group of civic and university visionaries commenced the monumental task of turning 245 acres of tired ground into a flourishing Arboretum.

Acknowledgments

At the top of the list of acknowledgments, special thanks must go to University of Wisconsin–Madison senior scientist emerita Kato Perlman, a longtime Arboretum supporter, whose generous gift supporting the publication of this history will help to keep its cost in a range that will make it affordable to the general reader, because it was conceived and written with the general reader in mind.

For assistance obtaining photos and illustrations, I wish to thank Vicki Tobias of the University of Wisconsin–Madison Archives, and Lisa Marine and Simone Munson, Wisconsin Historical Society. I also extend warmest thanks for months of help during the early research phase to David Null, Bernie Schermetzler, and Cathy Jacob of the UW–Madison Archives. I am also indebted to the library staffs at UW–Madison's Memorial Library, the Wisconsin Historical Society, and the Newberry Library in Chicago. Thanks also to the UW Archives, the Wisconsin Historical Society, and the Arboretum for permission to reprint photos and to quote from tapes and printed documents.

I wish particularly to thank Paul Zedler for his sound professional advice on the early draft, Mrill Ingram for her critical reading of early chapters, and Judy Kingsbury for alerting me to pertinent online reference material. Kathy Miner continuously shared her love of Arboretum history with me and proved always to be a helpful and encouraging listener. Molly Fifield Murray and I compared notes frequently on what we knew and were still learning about Arboretum history as the book progressed. My thanks to William R. Jordan III for the informative phone conversations, for the many email exchanges, for sharing details of his Arboretum days, and for establishing a storehouse of historical documents, slides, and tapes at the McKay Visitor Center. Thanks also to Robert Lange, UW Oral History Program, who invited me to share the benefits of Gregory Armstrong's candid insights into his productive years as Arboretum director during a series of interviews. Thanks also for shared information from Sue Bridson on the guide program and Ken Wood on the Lost City. I'm also grateful to Kevin McSweeney, current Arboretum director, for his encouragement and the assurance that a new history of the Arboretum was overdue and very welcome, and to Donna Paulnock, associate dean of the UW Graduate School, for her interest and support.

For their insights into family history and the Arboretum past, I wish to acknowledge the help of Lynne Archbald, Cinda Johnson (E. M. Gilbert's granddaughter), Jerry Schwarzmeier, Do Cary, and Caroline Greenwald. I am also indebted to the Arboretum staff, guides, rangers, and volunteers who regularly asked for updates on the book's progress. In particular, I wish to thank Cheryl Bauer-Armstrong, Susie Berg, Paul Borolsky, Peggy Brown, Susan Carpenter, Sara Christopherson, Vijoya Dasgupta, Marian Farrior, Jim Fitzgibbon, Steve Glass, Susan Kilmer, Kristin Lamers, Sylvia Marek, Chris Reyes, Shelley Stillman, Stephanie Williams, Levi Wood, and Fawn Youngbear-Tibbets. I also wish to thank the Arboretum's Pat Brown for help with the photos and for access to the media clip file and Carole McEvoy, Arboretum librarian, for assistance with the Arboretum tape file and photo collection. I also wish to acknowledge my faithful friends Sophie Lein and Ben DiGeifran for our many conversations about what is history and what isn't.

Finally, one does not get to the publication stage with a book of this dimension without the help of careful and concerned editors. My thanks to Gwen Walker, acquisitions editor, UW Press, and her assistant, Angela Bagwell, for their professional guidance in the early stages, and to Jane M. Curran, my copyeditor, and Adam Mehring, my project editor, for their helpful insights and suggestions during the final stages.

My biggest thank-you of all, however, is reserved for my wife and partner, Abigail Loomis, for her continuing encouragement and her astute advice on portions of the text.

Abbreviations

AA	Arboretum Archives, McKay Visitor Center
AAC	Arboretum Advisory Committee
AC	Arboretum Committee
ACM	Arboretum Committee Minutes
AC TAPE	Arboretum Collection Tape, McKay Visitor Center
BBS	U.S. Bureau of Biological Survey
C&NW	Chicago and North Western (railway)
CCC	Civilian Conservation Corps
CWA	Civil Works Administration
EPF	Earth Partnership for Families Program
EPS	Earth Partnership for Schools Program
FERA	Federal Emergency Relief Administration
FOA	Friends of the Arboretum
IMS	Institute of Museum Sciences
MATC	Madison Area Technical College
NSF	National Science Foundation
NPS	National Park Service
TRC	Technical Research Committee
UBC	University Bay Committee
USDA	U.S. Departmental of Agriculture
USFS	U.S. Forest Service
UW	University of Wisconsin–Madison
WARF	Wisconsin Alumni Research Foundation
WCD	Wisconsin Conservation Department
WERA	Wisconsin Emergency Relief Administration
WSHC	Wisconsin State Highway Commission

Pioneers of Ecological Restoration

❧

CHAPTER I

The Beginnings

JOHN NOLEN, MICHAEL OLBRICH, PAUL E. STARK, AND THE PRAIRIE SCHOOL OF LANDSCAPE DESIGN

On January 30, 1923, John Nolen, nationally known and celebrated landscape architect and city planner, mailed a letter from his office at Harvard Square, Cambridge, Massachusetts, to Michael B. Olbrich, in Madison, Wisconsin, in care of the law firm of Aylward, Davies, Olbrich, Brown and Siebecker. Nolen had Madison on his mind that January day. He was interested now in how the city planning efforts he had recommended in 1911 had fared, and he wrote to Olbrich, whom he had never met and never would, to get an assessment.

John Nolen Comes to Madison

Nolen's relationship with Madison had begun in 1908, when John Olin, president of the Madison Park and Pleasure Drive Association, a club dating from the 1890s that promoted scenic carriage and automobile drives for weekend excursionists and that also functioned as the city's unofficial parks department, invited Nolen to visit the city. Olin was a prominent Madison civic leader, an influential attorney, and a dedicated builder of parks. By 1908, Olin had been president of the Pleasure Drive Association for sixteen years. During his tenure as president, the association had raised nearly $215,000 toward the purchase of land. Thus, a strong nucleus supporting a park system already existed by 1908 when Nolen arrived in town.

Nolen, who had studied landscape architecture with Frederick Law Olmsted Jr. at Harvard, earning his Master's degree at age thirty-six in 1905, was, by 1908, well on the road to a successful city planning career. Olmsted, who had designed Harvard's Arnold Arboretum, instilled in Nolen a strong appreciation for natural landscape designs that linked Nolen directly with the

American Conservation Movement developing nationally at the time. Nolen
also had learned from Olmsted and other progressive city planners, especially
Ebenezer Howard, during visits to Europe in order to study Howard's "Garden
City" experiments, the benefits of designing urban centers around preserved
or restored greenbelt areas.[1]

Initially, when Nolen came to Madison, he was viewed as a candidate for
the recently vacated position of Madison parks superintendent, but the salary was low. Olin, in an effort to keep Nolen in Madison, suggested that the
University of Wisconsin–Madison create a chair in landscape architecture for
him. University funding was limited, however, and the Board of Regents was
unwilling to act on the possibility at the time. Olin hoped that Nolen would
ultimately choose to live in Madison. Nolen found the possibility appealing
but was hesitant to give up a potentially promising career in the East.

In spring 1908, Nolen settled with the city of Madison for a three-year contract and annual salary of $3,000 for temporary consulting and design work.
He worked directly with the "Madison fifty," a committee of fifty citizens,
including Olin, who oversaw his work. Nolen also convinced his Madison
backers that residing in Madison was unnecessary, that he could provide them
with a satisfactory design for the city's future while still maintaining his Harvard Square offices.[2] During spring and summer 1908, he made frequent visits
to Madison. He was originally asked to design a park system, but by 1909
he had managed to convince the "Madison fifty" that the city would benefit
more from a broader plan, one that included not only parks but also practical
recommendations for the downtown business area, residential neighborhoods,
the lakes, and the university. In a report first made available in 1909 and subsequently published in 1911 under the title *Madison: A Model City*, Nolen laid
out his grand plan. He addressed the need for wider streets, more trees, more
plazas, and the regulation of the height and style of downtown buildings; and
he suggested connecting a park extension that linked the State Capitol with an
ambitious Lake Monona esplanade two blocks away, a foreshadowing of the
current Monona Terrace that now borders the north end of Lake Monona,
designed originally in 1938 by Frank Lloyd Wright and opened to the public
in 1997.

The presence of the UW–Madison so close to the city center also captured Nolen's imagination. He envisioned the university manifesting itself in
Madison's future in two ways: first, "in the development of the grounds and
buildings of the University itself" and, second, in the establishment of forms,
forces, and expressions of science, art, and culture that would "differentiate
Madison from all other cities of the State." He was also impressed with open

John Nolen, landscape architect and city planner, first cited the need for an arboretum in *Madison: A Model City* (1911). (Wisconsin Historical Society, WHi-12506)

university land stretching for over a mile westward along the shores of Lake Mendota. The university was large with potential, Nolen reported, but in need of professional advice from a landscape architect. He thought a "State University devoted largely to horticultural and agricultural interests" should have at least a 20-acre botanical garden, a water garden, an aquarium, and an arboretum. He cited, as his model arboretum, Harvard's 200-acre Arnold Arboretum, designed by Olmsted. He also recommended a forest preserve for the UW of 1,000 to 2,000 acres, a summer engineering camp on Lake Mendota's shores, and a university pleasure garden.[3] His report, received with enthusiasm, formed the basis of Madison's developmental planning designs for decades to come.

Michael Balthazar Olbrich

When Nolen wrote that letter to Olbrich on January 30, 1923, he was aware of Olbrich's emerging importance as a force behind Madison's civic

Michael Balthazar Olbrich, prominent Madison attorney and member of the UW Board of Regents, acquired the first parcels of land for the UW Arboretum in the 1920s. (Wisconsin Historical Society, WHi-84345)

beautification effort. The effort, especially Madison's greenbelt system, was precisely what Nolen had recommended twelve years earlier. Olbrich had read assiduously Nolen's 1911 monograph, *Madison: A Model City*, and referred to it as his "civic bible."[4] Wisconsin Supreme Court judge Marvin Rosenberry, in 1933, recalled Olbrich introducing him to Nolen's monograph. Rosenberry also recalled Olbrich carrying with him daily a map taken from Nolen's book.[5] As early as 1923, Olbrich had developed a reputation as a visionary with a particular talent for acquiring land with which to build city parks. The more interested he became in creating parks, the more interested he also became in purchasing land for those parks along the shores of Madison's four intercity lakes. He realized that Madison was expanding and that building parks on the lakeshores not only would protect the land but also would guarantee unlimited public access to the shorelines.

Olbrich was an unpretentious person who liked simple things and the outdoors. He grew lady's slippers and other native wildflowers and prairie plants in his backyard, and he enjoyed long walks in Madison's parks and forests. A native Illinoisan, he was born in 1881 on a farm in McHenry County, a few miles northwest of Chicago. He moved to Madison in 1898 to attend the University of Wisconsin, graduating in 1905 from the UW law school.[6] He was also involved in a variety of Madison nature societies and recreational clubs. In the tradition of Wisconsin's John Muir, he believed that nature's beauties were to be shared, that availability to natural wonders was not just the privilege of the affluent few, and that the preservation of nature benefited all humankind, especially city dwellers hungry for spiritual nourishment. As early as 1916, he had begun to raise money. By 1921, he had raised enough to purchase a large portion of land along the eastern shore of Madison's Lake Monona, land that now includes Olbrich Botanical Gardens, a world-famous 16-acre cornucopia of outdoor garden varieties, and an expansive park, also named after him, that extends in a northerly line parallel with the gardens, curving west around the Lake Monona shoreline to the Yahara River.

Olbrich's usual business strategy for his land purchases involved raising money, purchasing the land, and then offering it at cost for public use as a park to a civic authority. In the case of the Lake Monona project, for instance, that authority was the city of Madison. In the case of the University of Wisconsin Arboretum, the authority was the University Board of Regents acting for the state of Wisconsin. In 1919, he offered the Lake Monona land to the Madison City Council. He had paid out over $40,000 for the purchases and at least wanted to break even. The council was disinterested at first. The prospect of investing in the Lake Monona Park, much of which was subject to flooding, was controversial. Finally, after two years of negotiation, on July 22, 1921, the city purchased the land for $50,000. It was a bargain for both. The city of Madison would have a large new park on the southeastern shore of Lake Monona, and Olbrich had $50,000 with which to begin purchasing more of Madison's lakeshore land for parks.[7]

For his next major project, Olbrich looked westward toward the western and southwestern shores of Lake Wingra, a small lake slightly southwest of the downtown Madison area and about a half mile south of the UW campus. His new land acquisitions, beginning in 1922, would be the first steps toward the creation of possibly another park, this time on the shores of Lake Wingra. But the project required more than $50,000. So, in 1922, he created the Madison Parks Foundation with a capital stock investment of $100,000. The foundation's purpose was "the acquisition of land" for parks in and about the city of Madison, with a view toward the subsequent development and sale of the land to the city or other appropriate public agencies. By 1925, Olbrich had enlisted 450 subscribers to stock investments in the Madison Parks Foundation. It was also in 1925 that Olbrich was appointed to the University Board of Regents and, as fate would have it, was provided with an opportunity to travel to Boston. The Boston visit would prove decisive.

Olbrich Goes to Boston

On March 23, 1925, Olbrich posted a letter to UW professor Edward T. Owen, a stockholder in the Madison Parks Foundation and a professor of French language and literature. Olbrich told Owen that he now had close to $50,000 in stock subscribed. He added: "the purpose of the articles is very general and . . . we are not committed to moving in any particular direction. . . . Where we will go when this is completed will depend upon the immediate exigency of the situation."[8] The note of indecision in Olbrich's letter suggests that he was open to the possibility of using the Madison Parks Foundation money for something other than just another municipal park.

In summer 1925, Olbrich traveled to Boston. According to his good friend Professor Edward M. Gilbert of the UW Botany Department, Olbrich unexpectedly went to Boston on the pretext of attending a convention. While there, however, recalling John Nolen's high praise for Harvard's Arnold Arboretum, he visited it and was most impressed with what he saw, and as Gilbert recorded, returned to Madison with an armload of arboretum literature. Eight years later, in 1933, Colonel Joseph W. "Bud" Jackson, another Madison civic luminary, in a November 23 letter to supporters of an arboretum in Madison, corroborated the date as the first time he was ever aware of Olbrich's interest in creating an arboretum instead of a park. It happened sometime in 1925, Jackson wrote. In the same letter, he added that Mrs. Olbrich also had told him that "Mike's interest in the Arboretum" began in 1925. She remembered because it was the same year he was appointed to the Board of Regents.[9]

Olbrich had known Ed Gilbert since Gilbert's days as an undergraduate. Now, in late summer 1925, after his return from Boston and the Arnold Arboretum, Olbrich sought out Gilbert for his botanical expertise and his experience with subjects related to the creation of gardens and arboretums. Olbrich was now seriously considering the possibility of building an arboretum, one similar to but bigger than the Arnold Arboretum, one that also had the potential to become both a wildlife sanctuary and an experimental forest preserve promoting the survival of indigenous, native plants. To that end, Gilbert recalled, shortly before his retirement from the university in 1946, how he and Olbrich in summer 1925 "thus began a work that has taken much of my time for more than fifteen years. There were morning sessions while my colleagues were still asleep; there were sessions in Mr. Olbrich's office, often way into the night; there were afternoon and week-end trips to study areas around Madison, and I shall never forget the pleasure enjoyed when in 1927 Mr. Olbrich had accumulated $64,000 and went before the Regents with an offer of the first tract of land for the proposed Arboretum. When the Regents accepted the offer, Mr. Olbrich insisted that I be made Chairman of the Arboretum Committee."[10] And, indeed, when, in 1933, the president of the university, Glenn Frank, looked for a professor to chair the first Arboretum Committee, he turned to Ed Gilbert, Olbrich's friend, who at the time was chairing the Botany Department.

The Nolen-Olbrich Correspondence

Two years after his trip to Boston, in December 1927, Olbrich sent John Nolen an important letter in which he asked for professional advice on nothing less than the promotion of an actual "arboretum project." It would be called the Wingra Project. At the time, Olbrich was concerned about a recommendation from Paul E. Stark, a Madison realtor who worked closely with Olbrich on the Lake Wingra land purchases. Stark thought that a road around Lake Wingra should supersede efforts to create an arboretum. Olbrich was dismayed. The road was not uppermost in his mind. "Ultimately," he had written in 1925, "I hope to see a drive continue on toward the city . . . and finally clear about the lake. But I am trying to take each step as we get to it."[11] Stark, however, felt an urgency about completing the road. Colonel Bud Jackson, whose efforts to create an arboretum would be outmatched only by Olbrich's, recalled later a conversation that he had with Stark in December 1945 shortly before Stark died. They were discussing how the University of Wisconsin Arboretum actually had gotten underway. Stark told him "that he thought he was probably the first one to make a definite proposal which he submitted to Mike Olbrich in 1927." Stark was referring to his proposal to Olbrich that they first build a road around the east and south shore of Lake Wingra and then, subsequently, acquire "about 800 acres for the arboretum." Stark told Jackson that Olbrich had cut him short and, without hesitating, came right back at him "with a suggestion that it be made 2,000 acres."[12]

Stark was not alone in 1927 in his desire to have a road constructed before an arboretum. For years, a road around Lake Wingra had been the hope of many Madisonians, particularly those involved with the Madison Park and Pleasure Drive Association and those who owned property in the Lake Forest development on the southeast shore of Lake Wingra. For the Lake Forest property owners, a passable road around the lake from the west, replacing the old mud-rutted farm road, would, for the first time, provide westerly access to

Paul E. Stark of Stark Realty worked closely with Olbrich in the 1920s and later with Joseph "Bud" Jackson to ensure that the proposed Arboretum would become a reality. (photo courtesy of Phillip Stark)

their property, much of it still for sale. The road, eventually to be called Arboretum Drive and later McCaffrey Drive, would continue as a major object of contention right into the 1980s. In fact, after the Arboretum land acquisition, the subject of most importance in the early history of the University Arboretum was the disposition of that "road around the lake," which took years to construct owing to a muddle of problems that clearly in 1927 neither Stark and his supporters nor Olbrich could have anticipated.[13]

So Olbrich, somewhat dismayed, wrote to John Nolen, seeking advice on what should take precedence: the road or the Arboretum. Olbrich's letter, dated December 2, 1927, provides a clear assessment of his commitment to the hope for an arboretum promoting, in the true "prairie style," the preservation of native plants, wildflowers, and wildlife. It merits quoting at length.

My dear Mr. Nolen:

I take the liberty of writing you . . . on a subject that has been very much on my mind for the past four or five years. You will recall that opposite page 150 in

Madison: A Model City your map shows the major city park surrounding Lake Wingra. On page 70 . . . you suggest the University's need of a botanical garden, water garden and aquarium, etc. . . .

For two years I have been seeking to create a combination enterprise based on your suggestions and patterned somewhat after the Arnold Arboretum by taking over the lake shore area as a University project—a living laboratory in everything that has to do with wild life . . . furred, feathered, or finned, together with shrubs, wild flowers, etc.

. . . I am [now] confronted with the suggestion made by one of our public spirited and progressive realtors. . . . He is pressing me to acquiesce in his program of an *immediate* good arising from the construction of a road in the vicinity of the lake . . . as against the remote and nebulous possibility of creating an arboretum. . . . I finally told him this afternoon that I would write you a letter asking for your offhand reaction to his suggestion.

Would in your judgment, the construction of a road . . . be incompatible with the creation of the combined forest park and arboretum . . . ? That is, it is the old problem of striking the right proportion between beauty and utility, between park and road. . . .

If it will not unduly burden you, I should like to have your general notion upon this situation.[14]

Whether or not Nolen responded is uncertain. No letter survives. Likely, however, Nolen, true to character, did respond. His high respect for Olbrich, his interest in preserving Lake Wingra wildlife (he referred to Wingra as "a veritable gem" in his 1909 report), and his desire to see a University Arboretum built would have compelled him to respond. Olbrich, more determined than ever after his plea to Nolen and undaunted by Stark's persistent desire for the road around the lake, forged ahead with his plans for first acquiring land on the south shore of Lake Wingra for the Arboretum. The completion of the road could—and did—wait.

On January 3, 1928, a month after posting his letter to Nolen, Olbrich wrote to Zona Gale, American novelist and playwright, who was also a fellow member of the UW Board of Regents. She and Olbrich shared a deep interest in city planning and in native gardening. In his letter, he noted that he had received an encouraging letter from Nolen. He referred to him now, in glowing terms, as "the godfather of . . . Greater Madison." He expressed gratitude for Nolen's advice and told her that he was going to invite him to visit Madison within the next few months. He hoped, should Nolen accept the invitation, that a lecture at the university for an honorarium could be arranged. He wanted Gale's help with the board in swaying opinion in favor of the honorarium. Olbrich also had on his mind at the time the prospective size of the Arboretum. Stark's 800 acres were small change; now he really was thinking seriously of 2,000 acres. He would also ask Nolen for his advice on the matter of size.

Unfortunately, the hoped-for 1928 visit to Madison by Nolen never materialized, nor is there extant record of any correspondence between Nolen and Olbrich after the letter of January 3, 1928. But the desire to have Nolen return to Madison for a visit persisted. Nolen had made a lasting impression on many of Madison's most influential civic leaders during his 1908 to 1910 visits. And *Madison: A Model City* accounted for much of the continued attention after 1911. It provided a vision for Madison's civic planners. And as Stuart D. Levitan recently observed, "Nolen's plan did not fade away. Instead, it lodged itself deep in the city's consciousness—an invitation to dream of Madison not as it is, but as it could become."[15]

And so the record is indisputable. By 1925, Olbrich clearly had made up his mind. He wanted more than a park or pleasure drive on the southwest shore of Lake Wingra. He wanted an arboretum, a big one. In 1927, he approached the Board of Regents with an offer to provide the land. On December 7, 1927, the board passed the following resolution: "That the unpledged balance in the Tripp Estate, approximately $83,000, be appropriated to aid in the purchase of land adjoining Lake Wingra and the Nakoma Golf Course, for a Forest Preserve Arboretum and Wild Life Refuge."[16]

The Board of Regents resolution proposing the new Arboretum received scant attention, however. The *Capital Times* newspaper on January 1, 1928, published a small four paragraph article on a $100,000 Madison City Council appropriation that assured the "establishment of a 600 or 700 acre arboretum, or wild life sanctuary, on the shores of Lake Wingra."[17] The city money was appropriated for the purchase from the Madison Parks Foundation of property on the south shore of Lake Monona between the Yahara River and Elmside Boulevard to be designated as a municipal park. It came at a time when Olbrich needed additional funding to keep what he now called "the arboretum proposition" afloat. The Regents had pledged the $83,000, but the deal called for Olbrich and the Madison Parks Foundation to come up with a matching $83,000 in support of the venture before the Regents would accept the deeds to the land making it officially university property. In April 1928, Olbrich could report to Professor Gilbert that with $100,000 in city money, the project should "make pretty definite progress this summer." Yet, as a caution, he also appealed to Gilbert to help raise subscriptions for additional funds to cover interest payments. He asked Gilbert to appeal to the university staff.[18] On the

campus in 1928, however, news of the future possibility of a University Arboretum sparked minimal interest. Nevertheless, Botany Professor Norman C.
Fassett, who would figure largely in the Arboretum's early development, particularly in valuable early experiments with native prairie plants and natural
landscape designs, did catch the attention of the *Capital Times* in June 1928 by
having included as a requirement for one of his spring semester botany classes
the identification of plants growing in the proposed Arboretum. By 1928, the
university still had not officially taken title to the property. Undaunted, Fassett's students combed the area and identified over 250 varieties of land plants
and shrubs, approximately 30 kinds of trees, and more than 20 kinds of water
plants on land that four years later would actually belong to the university.

Appropriately, the *Capital Times* article also cited the importance of the
Arnold Arboretum in Olbrich's decision to create an arboretum in Madison.
In response, Fassett was quoted as observing that although "Wisconsin cannot
expect to make an Arboretum on the scale of the Arnold Arboretum with its
fine collection of Asiatic plants," the Wisconsin Arboretum "has a chance to
make one as unique, and as valuable scientifically, *of native Wisconsin plants.*"[19]
That was June 1928. Fassett's observation, an early twentieth-century echo of
Increase Lapham's 1853 observation on the value of having a Wisconsin Arboretum replete with restored native Wisconsin plants, was prophetic.[20] Six years
after Fassett's 1928 *Capital Times* statement on the merits of native Midwest
plants and natural Midwest landscapes, Aldo Leopold, director of Arboretum
research, in his address at the June 17, 1934, dedication of the Arboretum,
would echo the theme, expressing similar sentiments about the need to recon-
struct with native plants a habitat, a sample of what Wisconsin looked like in
the middle of the nineteenth century. By 1936, the link between Arboretum
planning and the restoration of plants native to Wisconsin would emerge as a
defining future direction for the new Arboretum and its staff.

The *Capital Times* news article on Fassett's course underestimated the size of
the proposed Arboretum as 100 acres; in fact, the proposed acreage in 1928 was
closer to 200 acres. Fassett's spring 1928 botany course, however, was the first
UW course to incorporate research at the Arboretum as a requirement. The
fact that it was Fassett's botany course that required the research was appropriate, given the close ties that Fassett had, especially with Arboretum efforts at
prairie restoration, for the rest of his distinguished UW career.

Olbrich's 1928 Rotary Address

In March 1928, Colonel Bud Jackson sent Olbrich a letter offering to help him
create an Arboretum. He wrote, "I well realize the fact that you have had to

play pretty much of a lone part in bringing into reality the beautiful but practical picture you have painted for the future of this community. That ought
not to be so, Mike, but at the same time the very fact that you have led the
procession and carried the load should give you that much more personal satisfaction in the accomplishment." Jackson expressed the hope that his children
should live to see an Arboretum built on the stretch of woods and uplands on
Wingra's south shore. He added, "I don't know just what I can do to help you
with that project, but if I can do anything at any time I wish you would feel
free to tell me."[21] Little did Jackson know at the time how prophetic his words
would be and the role he personally would play after Olbrich's unexpected
death in 1929 in the realization of Olbrich's and Nolen's dream.

Two months after writing his letter offering help, Jackson was sitting in
the audience at an open meeting of the Madison Rotary Club on May 17,
1928, at which Olbrich delivered a historical address. The public was told in
an advance notice that Olbrich was going to paint the "picture" of what the
proposed Arboretum would mean to the city of Madison. And the picture that
Olbrich was going to paint was based largely on Nolen's influence, Olbrich's
1925 visit to the Arnold Arboretum, and his more recent visit, in early May
1928, to the Morton Arboretum in northern Illinois. Only three days before
his Rotary address, Olbrich had received a letter from Joy Morton, who was
away when Olbrich, in the company of Judge Marvin B. Rosenberry, visited
the Morton Arboretum. Olbrich went there seeking advice and answers to
questions. In his letter of 14 May, Joy Morton attempted to provide Olbrich
with some answers. He told Olbrich that the general plan for the Morton
Arboretum had been suggested by Charles S. Sargent, who, until his death in
1926, had managed the Arnold Arboretum. Morton also told him that at Sargent's suggestion, he had engaged Ossian Cole Simonds, the famous Chicago
landscape architect and proponent of the prairie style of planting, as his landscape designer. Sargent considered Simonds the "best man in the country" for
landscape design. He suggested that Olbrich contact him. Olbrich, of course,
was familiar with Simonds' work. Actually, in 1919, Simonds had drafted the
proposal for the development of the Lake Monona park project. And so, urged
on by Morton's letter, Olbrich contacted Simonds. By May 16, 1928, he could
report to Morton that Simonds was coming to Madison, and that he would
spend the next day, the day of Olbrich's Rotary address, "going over" the details of the new Arboretum project.[22]

Olbrich's Rotary Club address on the value of the proposed Arboretum for
Madison was well attended. Colonel Jackson, of course, was there. Also, sitting in the audience was Aldo Leopold, who had come to Madison in 1924 to

work in Madison's Forest Service Products laboratory. In 1928, Leopold would resign from the U.S. Forest Service and leave the Forest Products Laboratory to work as a private game survey consultant. Leopold was also aware in 1928 that the Arboretum—the Wingra Project, as it was known around town—was still in the planning stage. He appears to have taken no active role at the time in promoting the project. But the subject appealed to him. He also had been privy to a UW College of Agriculture inquiry into the possibility of establishing an experimental forest preserve in connection with the U.S. Forest Products Laboratory. He was also becoming increasingly interested in the link between conservation and game management, and he recognized the potential benefit that could come from a federally funded migratory game bird refuge in Madison. He wanted to hear what Olbrich had to say.

Olbrich began his metaphorical picture of an arboretum that evening by quoting Simonds, who was in attendance: "An Arboretum . . . a museum of trees and shrubs . . . should be more than a museum. It should be a work of art, showing to advantage the hills and valleys . . . tall trees, retaining large open areas so that the foliage, the sky lines, and the reflections in water can be seen to advantage. In short, it should be a beautiful place affecting one like a beautiful painting." Olbrich followed up the quotation with the observation that Nolen in his 1911 report on Madison had advanced similar sentiments about the value of a University Arboretum based on natural design and native gardens and had advised that it should be located close to the UW campus. Since property to the west of the campus that Nolen had recommended was no longer available, Olbrich suggested to the audience that Madison look south—that the city was fortunate to have an "almost untouched" area of "suitable magnitude and of unique and beautiful setting equally within the scope of Mr. Nolen's plan," an area of "six or seven hundred acres . . . surrounding Lake Wingra which lies a scant half mile South of the University Campus." Fortunately, he added, there was at the time nothing in the way of private development that should interfere with the creation of an Arboretum as ambitious as anyone could hope for. The location had potential that, if realized, would rank the University of Wisconsin Arboretum with Harvard's Arnold Arboretum, the Morton Arboretum, Washington University's Shaw Gardens in St. Louis, and even the magnificent arboretum in the Royal Botanical Gardens at Kew. In fact, Olbrich continued, specialists in the field of landscape architecture that he had contacted, including Ralph E. Griswold, the celebrated landscape architect and historian, had concluded that the Wingra Project exceeded in its possibilities "those of any of the institutions mentioned," because, as Griswold had observed during a recent visit to

Madison, there was no site in this country, in his estimation, that offered as much potential "for combining scenic beauty with cultural and educational development."[23]

A University Arboretum in Madison, like the Arnold Arboretum, Olbrich continued, would mean much to Wisconsin's university but even more to Wisconsin's citizenry. As he put it:

> It would cater to the needs of all, from the scientist to the child. It would mean the most . . . to the children of Madison and of this community. How could they be more effectively educated in the love of the beautiful than by the creation and preservation of a wonderland of natural beauty here at the Capital of the state. . . . But the appeal of such an institution would be not merely to child-hood and the specialist. . . . More than any other thing that could be established by . . . the University, it would appeal to the average man . . . This Arboretum or park will bring back into the lives of all confronted by a dismal industrial tangle, whose forces we so little comprehend, something of the grace and beauty which nature intended all to share.[24]

And then Olbrich capped his presentation by citing the famous line from Henry David Thoreau's essay on "Walking": "in wildness is the preservation of the world." He went on with the quotation from Thoreau: "Hope and the future for me are not in lawns and cultivated fields, not in towns and cities, but in the impervious and quaking swamps. . . . A township where one primitive forest waves above while another primitive forest rots below—such a town is fitted to raise not only corn and potatoes, but poets and philosophers for the coming ages." Olbrich added, as a conclusion: "to the subtle ministry of such a place will come not merely the working man alone, but all those whose souls are sickened and surfeited with city life."

The audience must have left the hall that May evening in 1928 satisfied that Olbrich had made his case. Leopold sent Olbrich a letter four days later asking for the source of the "wildness" quotation from Thoreau. Olbrich sent him the citation and also asked Leopold if he wished to make a donation to the project, to which Leopold replied by thanking Olbrich for the citation but informing him that although he was "not in a position to help out financially," he would be willing to help out with work on the Arboretum.[25] At the time, of course, Leopold had no idea just how much work he eventually would do for the Arboretum.

Jackson, after the speech, was reported to have been "set on fire" by Olbrich's plea.[26] It was a fire that ignited something special in Jackson's soul that smoldered for two years. On October 9, 1929, Olbrich committed suicide at the age of forty-eight for unknown reasons.[27] Following Olbrich's death, however, Jackson eventually would realize that Olbrich's work needed to be finished and that he was the one to do it.

CHAPTER 2

∾

1930–1932

ACQUIRING THE LAND, BUILDING THE ROAD

After Michael Olbrich's death in October 1929, efforts to build the Arboretum stalled. The nation was entering the Great Depression. Investing UW funds in the as yet unrealized Arboretum was not high priority. Olbrich's monumental proposal of December 7, 1927, approved by the Board of Regents, would remain in abeyance for five years. The Regents had passed the 1927 resolution pledging $83,000 toward the creation of a "Forest Preserve Arboretum and Wild Life Refuge," but the deeds were never transferred, and neither the board nor the university had taken any action to acquire any of the land parcels, a total, by 1930, of six parcels, or 245 acres.

In 1930, Paul E. Stark, in an effort to salvage the project, recommended measures to be taken immediately. They included first appointing a committee charged with the renewed sale of bonds and the task of working out a plan for completing the Wingra Project that would satisfy both the city of Madison and the UW Regents. By 1931, Bud Jackson, encouraged by Stark, agreed to take up Olbrich's mantle in an attempt to revive interest in the plan. By November 1931, acting on Stark's recommendations, members of the Madison Parks Foundation reinstated an Arboretum planning committee that was called the Citizens Committee and was led by Jackson and Stark. Jackson, never one to waste time, arranged for a meeting of the new Citizens Committee for the evening of November 20, 1931. Jackson wanted the committee to consider the feasibility of applying for federal assistance—Jackson's persistent solution to the continuing lack of funds—in the hope of reactivating the now almost defunct Arboretum project. These were early Depression years, Herbert Hoover was still in the White House, and Jackson was keen on the possibility of convincing the federal government to take over the proposed Arboretum as a federal project—as, in fact, a wilderness refuge and experimental forest

18

preserve. He and others on the Citizens Committee envisioned a potential partnership between the Arboretum project and Madison's USDA Forest Products Laboratory.

Maurice McCaffrey, Board Secretary

On November 20, 1931, the day the Citizens Committee met, William J. P. Aberg, secretary-treasurer of the Madison Parks Foundation, sent a letter to Alvin E. Gillette, secretary of the Madison Association of Commerce. Aberg offered his opinion on Jackson's hope for federal assistance. Keep in mind this was 1931. The nation was deep in the Great Depression, and Roosevelt and the New Deal were still two years away. Aberg acknowledged that he could see no way possible for obtaining federal funds from Hoover's administration or for the Arboretum to be financed as a federal project. His words were prophetic.

Aberg's assessment of the challenges facing the Citizens Committee in 1931 was politically astute, especially as it related to the university, particularly to the lack of high-level assistance coming from the UW administration. His assessment was based largely on his experience as secretary-treasurer of the Madison Parks Foundation during Olbrich's tenure. Aberg was aware of Olbrich's clout with the Regents. The reality was that Olbrich, as a member of the board, had been the driving force behind *all* the actions taken by the Regents in support of the creation of the Arboretum from 1925 until his death in 1929. Since Olbrich's death, however, the Regents had done nothing to procure the Arboretum. Implicit in Aberg's assessment was the suggestion that someone with university influence was needed who could pressure the board and the university into finally acting, "someone closely . . . affiliated with the University," he observed, who would have connections with "the Board of Regents and also the College of Agriculture," someone who also "had a vital interest in this project, and who would be willing to devote the time." That particular someone, fortunately, was waiting in the wings. He was already an influential board figure. During the Olbrich years, he had been relatively out of the picture but would now emerge as a forceful political spokesperson for the creation of the Arboretum. That person was Maurice Erve McCaffrey, who, at the time, was the secretary of the Board of Regents.

McCaffrey believed in the Arboretum project, and as UW president E. B. Fred observed in his address at the July 1953 dedication of McCaffrey Drive, McCaffrey had been the university's "business genius," and "of all that he did for the University nothing gave him more pleasure than the lead he took in the acquisition and development of the Arboretum." Any Arboretum business that related to the Regents or to UW management funding passed through McCaffrey's hands. As a staunch advocate of the Arboretum, he was duly honored in 1953 by having the Arboretum road named after him.[1] In 1931 the Citizens Committee needed support from the Regents. They would also need the assistance of as many UW faculty members as possible. McCaffrey had influence with both the board and the faculty. Fittingly, Jackson turned to McCaffrey. "As you may possibly have heard," he told McCaffrey in December 1931, "a group of Madison men are reviving interest in the Wingra Arboretum project." Jackson asked McCaffrey to send him an extract of the original resolution passed by the board on December 7, 1927, and to send along with it "any comments which may occur . . . with reference to the entire proposal."[2] Jackson was asking for help; McCaffrey would now be the "man on the inside"—the Citizens Committee's link to the board and to the university.

In the early months of 1932, Jackson, with McCaffrey's assistance, arranged for a decisive meeting between the Citizens Committee and a select committee of members from the Board of Regents on March 5, 1932. The only item of business on the agenda was the feasibility of convening at some future date the full Board of Regents to discuss the possibility of finally conveying the deeds to the university for the six land parcels under the terms of the December 7, 1927, resolution. The deeds were still in the hands of the Madison Parks Foundation. The March meeting, significantly, was held in McCaffrey's office. The board members at the meeting, after a lengthy discussion, scheduled a future meeting with the full board. Jackson would acknowledge afterward McCaffrey's persistent role in efforts to rejuvenate the stalled Wingra Project. As secretary of the board, McCaffrey, Jackson noted, had "cooperated to the fullest and . . . had a genuine heart interest all the way through." By the end of the meeting, the Citizens Committee also had drafted a proposal, a lengthy amendment to the original 1927 resolution, which would be voted upon by the full board at a meeting arranged for April 26, 1932.

The Decision on April 26, 1932

Finally, after five years, the matter of Olbrich's moribund 1927 resolution would be revived. On April 26, after a lengthy debate, the Board of Regents rescinded the old resolution in favor of the amended proposal, a milestone of decisive importance in the history of the UW Arboretum. The Regents officially accepted the deeds to the six original parcels of land, which Olbrich, Stark, Jackson, and other committed members of the Madison Parks Foundation had so diligently worked to acquire from 1922, when the Parks Foundation was formed, to 1932. Ten years of effort had finally paid off. Passing this motion marked the official creation of the University of Wisconsin's "Forest Preserve Arboretum and Wild Life Refuge."

The Six Land Parcels

The March 5 proposal from the Madison Parks Foundation also outlined in detail the terms by which each of the six land parcels would be transferred to the Regents. To facilitate action on the transfer of each deed, the Parks Foundation circulated a document that explained the terms and details of the six transfers. Included with the document was a colored map that easily enabled the board members to identify the size and location of each parcel. Each parcel had also been assigned a letter of the alphabet that corresponded to that parcel's position on the map. The letters ranged from F to K and were considered in descending order (see Parks Foundation map):

Parcel K: a 15-acre Wingra Marsh tract bordering Lake Wingra's southwestern shore.

Parcel J: Spring Trail Park, 1 ½ acres located on the east side of Nakoma Road directly across the street from the Old Spring Tavern.

Parcel I: 10 acres on the west side of parcel K, bordering parcel J on the southwest corner, and extending almost to Monroe Street on the north.

Parcel H: 30 acres abutting parcel J on the northwest corner, parcel I on the north, parcels K and G on the east, and the Nakoma Country Club on the south.

Parcel G: 140 acres containing the farm owned by Charles Nelson and abutting parcel K on the north, parcel H on the west, Lake Forest property on the east, and parcel F on the north and east.

Parcel F: 50 acres extending west along the south shoreline of Lake Wingra to parcel G, east along the shoreline to the Lake Forest property line, and south to parcel G and a portion of the Lake Forest property.

Following is a brief account of the acquisition and the significance in terms of both local history and, in one instance, local folklore, of each of the six original parcels that ultimately were joined on April 26, 1932, to form the original UW Arboretum grounds.

Parcel K, the Marston Estate

The history of efforts to acquire this parcel of land went back at least twenty years to 1912 and to Leonard Gay's land purchases along the western shoreline of Lake Wingra. Among his 1912 purchases, Gay, an enterprising Madison

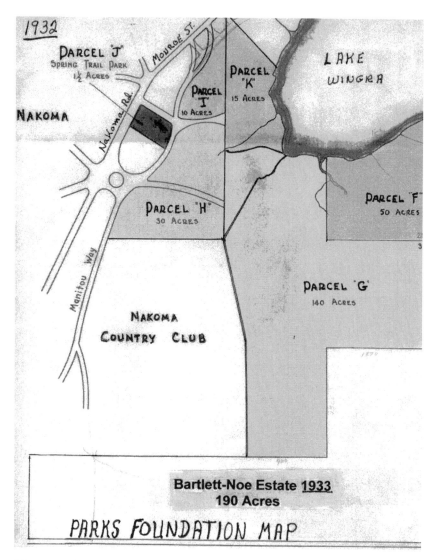

The original Madison Parks Foundation map showing the location of the six parcels of land acquired by the UW in 1932. The Bartlett-Noe property was acquired in 1933. (Arboretum Archives)

realtor, bought the Marston Estate, a parcel of land dating from 1840, bordered roughly by the Ice House, an old Madison landmark on the northwestern shore of Lake Wingra near Knickerbocker Street, and the Plough Inn, another Madison landmark dating from 1853, located northwest on Monroe Road, which in 1912 was a rural extension of Monroe Street.[3] On the west side of Monroe Road, the Marston land extended west to an old Illinois Central Railroad line that is currently used as a bike path.

In 1939, Bud Jackson observed, regarding the Marston land gift, that in 1922 Olbrich "engineered a trade, whereby [Stark's] Madison Realty Company and the University" exchanged "lots and about 30 acres of Marston estate marsh land." Jackson commented that this exchange "was the real start of the acquisition of land for the Arboretum." Of the 30 acres of acquired marsh land, 15 acres constituted parcel K, which Olbrich set aside for the Arboretum and which today marks the northern and westernmost section of the Arboretum's Wingra Marsh.[4] Throughout the 1920s, Leonard Gay, Olbrich's friend and business associate, would be instrumental in Arboretum land dealings. He was also one of the partners, along with Chandler B. "Bernie" Chapman, in the Lake Forest Land Company, which owned 840 acres in the southeast Lake Wingra basin, land that over a thirty-year period of time gained fame and notoriety as the legendary Lost City, land that also, from the early 1930s to 1972, would be deeded to the university in pieces.

Parcel J, Spring Trail Park

Parcel J was originally promised to Olbrich as a gift from Stark's Madison Realty Company. In turn, in 1927, Olbrich promised it to the university. The history of the parcel dates from 1911, when the western Madison suburb known as Nakoma, an Ojibway term for "I honor my oath," was being planned by the University Land Company. The area was known at the time as Gorham Heights. The Gorham family had owned land there as recently as 1899. Seventy-seven lots in the Gorham Heights land development were platted on 40 acres of Town of Madison farmland that spread across the gentle hills that overlooked the Old Spring Tavern, originally a stop on the Madison-Monroe stagecoach line heading into southwest Wisconsin. The tavern, which faced the lake on what is now the extreme north end of Nakoma Road, was built in 1854 by Charles C. Morgan and James W. Gorham. A large spring behind the tavern accounted for its name. In 1860, Morgan sold the tavern to Gorham, who closed it after volunteering for service in the Civil War. In 1895 the tavern became a private residence, which it still is. In 1914 Madison Realty bought the Gorham Heights land along with 46 nearby acres.[5]

When Madison Realty first purchased the land in 1914, the realtors were delighted to find just opposite the Old Spring Tavern a pond and a variety of beautiful springs in a small park called Spring Trail Park. Both the old tavern and the springs, Stark thought, added considerably to the neighborhood's charm. Hence, Madison Realty decided early on that the springs, the pond, and Spring Trail Park should be preserved for the public. Madison Realty then "created some pools, filled in low ground, partially surrounded the one and a half acres contained in the park with a stone wall [rumored to have been designed by Frank Lloyd Wright], constructed paths, and [did] a great deal of planting."[6] The landscape design was largely supervised by Franz A. Aust, a UW professor of landscape design in the Department of Horticulture.

Madison Realty initially had intended to present Spring Trail Park as a gift to the city of Madison, but after the company board learned that Olbrich had already decided to add to his plan for an Arboretum the 30 acres adjoining the park on the east (parcel H), they thought he would welcome the opportunity to acquire the little park for the Arboretum. Stark believed that the park eventually would be an excellent UW laboratory for aquatic plant experiments.[7]

Parcel I

The history of the purchase of parcel I began, like parcel J, in the early to mid-1920s and also involved Paul Stark and Madison Realty. On December 7, 1928, Stark told Olbrich that Madison Realty was willing to sell to the Madison Parks Foundation the land lying between Monroe Street and the 30 acres that had previously been conveyed to Olbrich (parcel H). Madison Realty would also give him two years to come up with the purchase money. Of course, Stark didn't realize in December 1928 that in less than a year Olbrich would be dead. Hence, no contract actually deeding the land to the Madison Parks Foundation was ever signed or delivered. Nevertheless, Stark and Olbrich had a gentleman's agreement, and Stark honored the arrangement. In 1932, parcel I was deeded to the university.[8]

The Marion Dunn Prairie and holding pond, created in 1983, are currently within the boundaries of parcel I. The tall grass prairie was named after Marion Dunn, a dedicated Arboretum volunteer. The pond at the center of the prairie, built by the city of Madison, was designed by Arboretum ecologist Virginia Kline.

Parcel H, Gorham Farm

Parcel H, consisting of 30 acres on the north side of the Nakoma Country Club, just southeast of Spring Trail Park, like parcel J, originally was owned by the Gorham family. Within the boundaries of parcel H today are the

Arboretum's Viburnum Garden, which contains more than 80 species and varieties of viburnums and also 110 species and varieties of arborvitae. The E. Ray Stevens Memorial Aquatic Gardens and Pond are also located there. The pond, completed by the CCC and dedicated on November 1, 1936, was named after Judge Edmond Ray Stevens, a member of the Wisconsin Supreme Court and a staunch early supporter of the Arboretum project prior to his death in 1930.

Bud Jackson was particularly pleased with the Stevens Memorial. He told Judge Marvin Rosenberry in October 1936 that the long-term plan was to have "all of the various marsh and bog flowers of the entire area" gathered in the gardens, and he predicted that the memorial would become a "show place" for both the Arboretum and the city of Madison. He also pointed out that the memorial was purposely designed as the forerunner in a plan, then underway, for the development of a memorial to Michael Olbrich, planned for the Arboretum's west entrance.[9]

Parcel G, Nelson Farm: A Fraudulent Egg Business, a Motive for Murder, and a Piece of Madison Folklore

As early as 1922, Charles N. Brown, the former secretary of the Madison Park and Pleasure Drive Association, suggested to Olbrich that, as soon as possible, the Madison Parks Foundation should bid on the Nelson farm, 140 acres of land west of Lake Forest and south of Lake Wingra, with a portion of its northeastern border fronting the lake shoreline. Since 1901, the land had belonged to Charles Nelson, who was unsuccessful in his attempt to farm it. By 1922, Nelson was in financial trouble. The farm was unproductive; he was rumored to have made bad investments.[10] Brown feared, as did Olbrich, that Nelson, if pressed, would sell the land to the first bidder. They were especially concerned about the potential disposition of the northern lakeshore portion of the property. If sold indiscriminately, it could jeopardize the dream of the Madison Park and Pleasure Drive Association of someday having a suitable road circling Lake Wingra. With the money in hand, purchasing the Nelson land and keeping it in abeyance until the university or the city decided what to do with it seemed simple enough. In any case, the land would be safe from developers. As of 1922, remember, there was still no clear idea about creating an arboretum on the south shore of Lake Wingra. Olbrich and other interested civic officials still referred to the land buying effort as the Wingra Project. The primary objective was to buy up land in order to protect and preserve the Wingra shoreline. What transpired in regard to efforts by Olbrich or Brown to purchase the Nelson property between 1922 and 1927 is unclear. What is clear is that it would take ten full years (well after Olbrich's death) and an annoying sequence of bizarre problems before the deed would finally be delivered to the UW Regents in April 1932.

The full story of the acquisition of the Nelson property involves an odd but fascinating turn of events that reads like an aside but that is actually directly connected both to Olbrich's ongoing but frustrating efforts to purchase the land and to an enthralling part of Madison folklore that necessitates a retelling every time the subject of the now legendary Nelson farm and Nelson family comes up.

At some point in 1927, Olbrich asked the Madison Parks Foundation board to let him use the $40,000 invested in 1922 after the sale of the Lake Monona property to buy the Nelson farm. The Nelson property would be the center-piece, the "nucleus for the Wingra Park Arboretum and Wild Life Refuge."[11] At the time, Charles Nelson, in his seventies, still owned the property. It's also likely that foreclosure was imminent, as Nelson's fortunes were diminishing. To make matters worse, on March 6, 1927, his thirty-five-year-old son, Harvey, was pursued by police on a double homicide charge and committed suicide in a deserted garage in Lake Forest. The murder victims had been neighbors named Henderson who owned a chicken farm slightly southwest of the Nel-son farmhouse. On the morning of March 5, 1927, Harvey Nelson allegedly drove to the Henderson farm and shot to death Allen Henderson, the farm owner, and his twenty-year-old son, Walter. As to a motive for the murders, all the newspaper accounts could provide were reports that Harvey Nelson "nursed a grievance," that he and Allen Henderson had an argument, and that Nelson had a history of mental problems, possibly stemming from his WW I service. Charles Nelson, when asked about a motive for the slayings, acknowl-edged his son's enmity for the Hendersons, which Charles explained away simply by saying, "They've slurred us for the past two years."[12] By late 1927 the Nelson farm was still unsold. Judge Rosenberry called for "immediate action." Olbrich agreed and set out determinedly at the time to acquire the 140 acres.

On March 7, 1928, Olbrich wrote to Carlyle Wurster, one of the Madison Parks Foundation investors. He warned of a threatening option that had been taken on the Nelson farm: "Mr. Heilman who represents the group which has taken an option on the Nelson farm called . . . to say that he was arranging a conference of all the attorneys for Friday afternoon at three o'clock." Olbrich urged quick action.[13] On May 7th, Stark hurriedly sent a letter to Olbrich in which he offered the following advice on the Nelson farm purchase. Stark was aware that a banker from Mazomanie, Wisconsin, named J. P. "Bill" Hudson held the mortgage to the Nelson property. Stark told Olbrich that in order to transfer the mortgage, he should offer Hudson $8,000 in cash or, as an alternative, what Stark called "the Black Earth farm," which was a reference to a farm in Black Earth, Wisconsin, known as the Ward farm. This offer to Hudson, Stark advised, should be limited. Nelson should not be given an opportunity to work out a counterproposition. Should Hudson refuse to take either the cash or the land, Olbrich should buy out the bank's first mortgage. Stark added that Madison Realty believed that Hudson was almost certain to accept either the cash or the Ward farm.[14]

In spite of Olbrich's persistent efforts to acquire the property, however, the title to the Nelson farm remained oddly out of reach right into the early months of 1929. Something was amiss. And at this point, the story truly takes a bizarre turn. On March 6, 1929, McCaffrey sent Olbrich a letter in which he included "the abstract of title to the Charles Nelson farm, in three pieces"; the abstract, McCaffrey noted, was last certified on June 27, 1928. McCaffrey added that Olbrich should already have in his possession a copy of the "opinion" written by Assistant Attorney General Adeline J. Meyer, dated three months earlier, December 11, 1928. McCaffrey offered no hint in his letter as to what the "opinion" addressed.

Opinions of the Attorney General of the State of Wisconsin for 1928 does not include an opinion signed by Adeline Meyer, but it does include an opinion, dated December 12, 1928, signed by Glenn D. Roberts, the district attorney for Dane County. The opinion pertains to a matter directly involving the Nelson farm that sheds some light on a possible motive for the enmity between the Nelsons and the Hendersons that possibly accounts for the double murders in 1927. The opinion also helps to explain why Olbrich was having no success acquiring the property.

The DA's opinion is a direct response to a written query, an "anonymous citizen complaint," about a possible violation of a legal statute prohibiting fraudulent advertising. The query concerned an anonymous dealer operating in Madison who sold "eggs in a carton along the top of which, in prominent letters," appeared the claim that the contents were "high vitamin eggs." On the bottom of the carton appeared the acknowledgment that the eggs were "From Nelson's Certified Egg Farm." The dealer, the complaint notes, actually had "no egg farm of his own but goes out among the farmers and picks up enough eggs to meet his trade." "None of the farmers" supplying the eggs, the complaint continues, "have certified egg farms."

Now how would the anonymous author of the complaint know with any certainty that the eggs were not from certified egg farms, and why would the author care, unless perhaps he or she was a nearby chicken farmer also in the business of selling eggs? Keep in mind that the Hendersons actually *were* chicken farmers and in all probability were also in the egg-selling business. There is nothing that suggests that the Nelsons were chicken farmers, but they obviously were running a bogus egg business. Highly likely is the possibility, since the two families lived so close together, that the Hendersons had warned the Nelsons to stop cutting into their market with competitive eggs sold under false pretenses. "They've slurred us for the past two years," was Charles Nelson's only recorded response, remember, when asked why his son held a "grievance" against the Hendersons.

"Certified egg farms" in the 1920s, after the discovery of vitamins A and D, were egg farms where the vitamins were added to the chicken feed. The DA's legal opinion concludes, given "the statement of facts" provided by the anonymous complaint, that the egg seller was clearly violating the state statute. The eggs were not from a registered "certified egg farm," and because the dealer got twenty cents a dozen above the market price owing to his "high vitamin" advertising, he was deceiving his customers and, therefore, "was guilty of fraudulent advertising."[15] The complaint and the judgment against Nelson's Certified Egg Farm obviously compounded the already Byzantine matter of property ownership. How could one take title to property under investigation? The matter of the sale and ownership of the Nelson farm continued into the spring of 1929. On March 23, 1929, Charles Nelson sent Olbrich a labored handwritten letter in pencil in which he wrote: "Dear Mr. Olbrich: I am going to try to write but little, and ask but one question—can I have the whole farm . . . can ask more later if there is any need of them." He added cryptically, "I think I woun't like the whole of it—but sall not be greatly disappointed if I cannot have it all . . . but will do the best I can on a little of it—and perhaps it will be better for me that way. Respectfully, Chas. Nelson."

A month later, on April 24, 1929, Olbrich sent a letter to Nelson asking him "to drop in sometime so that we can make service with *a summons and complaint* in an action to quiet the title to your property" (italics mine). Olbrich appears to have received no reply, prompting him on May 16, 1929, to send yet another letter addressed to Nelson at General Delivery, Madison, Wisconsin, in which he wrote: "My dear Mr. Nelson: I have been trying to reach you in various ways by letter. Should you get this at the post office, I wish you would make it a point to drop over and see me. Yours very truly, M. B. Olbrich."[16]

Where was Nelson? Why did Olbrich want to speak with him? What was keeping the title from being cleared? My guess is that it was litigation, and that Nelson was restrained under an indictment, likely for fraudulent advertising, in which case the farm (Nelson's Certified Egg Farm), where the advertising originated and where the bogus "high vitamin eggs" were packaged, would be integral to prosecution, if circumstances came to a prosecution.

On August 14, 1929, Olbrich received a letter from C. L. Femrite, Dane County treasurer, in which Femrite requested that Olbrich take over and pay the tax certificates on the Nelson farm. At some point after August 1929, the Nelson farm was finally purchased by the Madison Parks Foundation for $45,000 on a foreclosure sale. On April 26, 1932, five years after Olbrich's original proposition and three years after his untimely October 1929 death, the Board of Regents finally took title to parcel G, the 140-acre Nelson property.

In January 1933, the Regents had the title, but problems associated with the troublesome Nelson farm purchase just would not go away. On January 30, four years after Olbrich's death, the Regents recommended that the university serve notice of vacation of premises "upon the present tenant of the Nelson Farm." The tenant, unknown, was given until March 1 to leave. The recommendation also noted that the Regents should begin looking for an "experienced person to carry out various projects." The "experienced person" would be a live-in caretaker. The Regents had plans for the Nelson farmhouse and for the Nelson barns.

The Madison Parks Foundation paid $54,000 for parcel G in 1929. In 1930, the appraisal value was $77,000. In the quest for Arboretum land, the UW had gotten a bargain. The Nelson farmhouse remained standing, and when the CCC moved into the Arboretum in 1935, the building, known simply as "the old farm-house," served as headquarters for the National Park Service. The Nelson barn was renovated in 1934, the bottom floor turned into a mess hall, the top floor into a recreation room. The barn accidentally burned to the ground in March 1937; fortunately no one was hurt. In October 1939, the Nelson farmhouse was razed. The water lines and stone foundation were removed, and the property graded.[17]

The acquisition of Nelson's 140-acre property, parcel G, was essential to the creation of the Arboretum. As Olbrich had observed, it was "the nucleus," the centerpiece, of the land holdings. Its boundaries now include the southernmost section of the Wingra Marsh, the Overlook Prairie, the Arboretum Visitor Center and service area, the northernmost portion of the Curtis Prairie, and most of Gallistel Woods, including effigy mounds, relics of a historic Native American culture that date from the ninth century A.D. Without parcel G, the road running west to east along the southern Lake Wingra shoreline might never have been constructed. Once parcel G was in the hands of the Madison Parks Foundation, Jackson, Stark, and the others could bargain with Chandler B. "Bernie" Chapman of the Lake Forest Company for parcel F, the

50-acre plot on the east border of the northern third of parcel G. Owning parcel F meant that the road along the southern Wingra shoreline was almost a certainty. Stark, all along, had wanted that road, but Lake Forest's Bernie Chapman wanted it even more.

Parcel F, Lake Forest Land

In the 1920s, parcel F, 50 acres, was owned by the Lake Forest Land Company. It was bordered on the north by the south shore of Lake Wingra, on the west by parcel G, and on the south by parcel G and a small westerly section of Lake Forest. Today, its boundaries include the Wingra Woods, the Wingra Woods Native American effigy mounds, and the Big Spring, one of the largest of the springs feeding water year round to Lake Wingra. Parcel F was given to the Regents in April 1932, deed free and clear of encumbrances. But the Madison Parks Foundation arrangement with the Lake Forest Company for the sale of parcel F involved two major stipulations. The foundation would pay $25,000, $10,000 up front in either cash or securities and the balance to be made up by the Lake Forest Company's selection of any one of three parcels of land that were part of the University's remaining Tripp Estate endowment.

A second condition attached to the deeding of parcel F to the Regents involved the agreement that "at such a time as a road" was brought "to the Lake Forest property from the east to parcel 'F', the University" would guarantee "a right-of-way easement through parcel 'F' westerly to Manitou Way." Apparently, in April 1932, the time of the transfer of the six land parcels to the university, plans for the Arboretum Road called for a road crossing from east to west and then running north of the Nakoma Country Club to Manitou Way.

The final location of the Arboretum Road went south of the Nakoma Country Club and west to an exit on what is now Seminole Highway but was known as Bryant Street in 1932. The Arboretum Road leading south through parcel G (the Nelson farm property) and then west to the Seminole Highway exit guaranteed that the road as it curved south would provide access to a larger portion of the Arboretum property than would a road exiting north and west to Manitou Way. The advantage, a timely one, was made possible by the rather unexpected gift in 1933 of 190 acres of land that was part of the Bartlett-Noe estate.

Acquiring the Bartlett-Noe Property

The possibility of purchasing the Bartlett-Noe land surfaced in 1929. On April 27 of that year, six months before he died, Olbrich told George W. Mead, a member of the Board of Regents, that he thought the Bartlett-Noe property, about 200 acres, could be bought for $70,000, a price that Olbrich judged most reasonable. The farm, actually 190 acres, had been in the Bartlett family since 1860. The farm was located south of both the Nakoma Country Club and the Nelson property (see the Parks Foundation map). The eastern quarter of the property extends north and is bordered on the north and the west by the Nelson property and on the east by Lake Forest. The property became known as the Bartlett-Noe estate when Jessie, Seth Bartlett's daughter, married Walter C. Noe, a Madison businessman. Currently, within the boundaries of the Bartlett-Noe acquisition are most of the 60 acres that comprise the Curtis Prairie, the first of the Arboretum's restored prairies; the Leopold Pines, a 21-acre pinetum with red and white pines, some red maple, and some white birch; and Noe Woods, 41 acres of white and black oaks, some 150 years old. In the extreme northeast portion of the parcel, sandwiched between parcel G and what was Lake Forest land, are the Teal Pond and Teal Pond wetlands.

On March 31, 1933, Bud Jackson appeared before the Executive Committee of the Board of Regents and presented a communication from Jessie B. Noe dated March 10, 1933, in which she offered the farm to the Regents to be added to the University Arboretum. The Executive Committee voted to take the matter under advisement. On April 27, 1933, the Board of Regents accepted the deed at a purchase price of $47,500, to be paid over a ten-year period at 4 percent interest per annum. The price was much less than the $70,000 that Olbrich, in 1929, had thought reasonable. The university had gotten another bargain.[18]

The acquisition of the Bartlett-Noe property meant that by April 1933 the Arboretum consisted of seven parcels of land amounting to 435 acres. The acquisition also meant that it was now possible to extend a road through Arboretum property far enough south and west to a convenient exit. But this was 1933. In August 1934, the plan was still to extend the road west to Manitou Way.[19] Eventually, the UW Arboretum Committee would realize that a road cut through the Bartlett-Noe property, exiting west on Seminole Highway, was the wiser choice. But the disposition of the eastern third of the Arboretum road would remain uncertain through 1934 and 1935.

Like the Nelson farm property, there is also a touch of local folklore connected with the Noe Woods located on the west end of the Bartlett-Noe property. In the nineteenth century it was called Bartlett's Woods, and locals believed it was haunted. Farmers in the area reported hearing the sounds of an axe blade repeatedly chopping into a tree over one summer-fall period. Yet, there was no evidence of tree chopping: no axe marks on trees, no chips on the

ground. And no one at the time seemed to have had the nerve to venture into
the woods at night to see who was wielding the axe. Finally, as the story goes,
the mysterious chopping ceased. Neither the chopper's identity nor the source
of the sound was ever discovered.[20]

The Jackson-Nolen Connection

Throughout 1932 and the early months of 1933, Bud Jackson and various UW
representatives were consumed with ironing out difficulties standing in the
way of simply getting the Arboretum project off the ground. There were questions and concerns about the feasibility of the conveyance of the property
deeds to the UW Regents. Decisions about the roles of the state and the city
of Madison in Arboretum planning needed to be resolved. Once the university took control of the land, administrative committees had to be appointed,
meetings scheduled, and some plan for governance devised. As late as spring
1933, however, almost a year after the Regents had accepted the deeds, there
was still no strategic plan for the future. Arboretum supporters were interested
in soliciting as much advice and feedback as possible. Jackson, in late summer
or early fall 1933, wrote to John Nolen and asked for advice. Nolen sent Jackson copies of relevant publications on his urban planning efforts in Roanoke,
Virginia, and Little Rock, Arkansas. Nolen, being unfamiliar with Jackson,
also asked Jackson to tell him something about himself. The request gave Jackson the opportunity to provide Nolen with a brief summary of his eventful
life, a welcome effort for Colonel Jackson, who admittedly relished opportunities to talk about his colorful past. In a letter dated September 7, 1933, Jackson
thanked Nolen for the copies of the Roanoke and Little Rock publications
and then noted that though both were interesting, the most helpful publication on urban planning for Madison, in his estimation, was Nolen's own 1911
monograph on Madison. Jackson was a button-pusher when it came to dealing personally and professionally with people. He knew the right things to say
and he knew how to endear himself to people. He had a wonderful sense of
humor, and during the 1930s and 1940s he would almost single-handedly raise
more money for the Arboretum, particularly money for land purchases, than
any other figure involved with the project, either professional or professorial.

Jackson told Nolen in his September 7 letter that he was born in Madison
in 1878 and that in 1899 he had traveled west, where he spent twenty-one
years working as a cattle rancher in North Dakota. He also told Nolen that
he was in the Great War, returning afterward to live in Madison close to Lake
Wingra and the location that had now been chosen for the Arboretum. What

Colonel Joseph W. "Bud" Jackson, with cane and fedora, on one of his frequent walks through the Arboretum. (photo by G. Wm. Longenecker, Arboretum Photo Collection)

he did not tell Nolen, perhaps out of modesty, was that he had come from a very distinguished family, that his father had been a Civil War surgeon and the founder of Madison's Jackson Clinic. Nor did he tell Nolen that he served with General Pershing during the Great War, that he earned the rank of Colonel, and that he organized in France a mounted cavalry unit that was, very likely, one of the last active cavalry units in the U.S. Army. Surely, Nolen was amused with what he did learn about Jackson in that September letter. And, surely, he was to learn more about the colorful Colonel Jackson, as he was known around town, as he and the colonel grew closer.

When the correspondence between Nolen and Olbrich had ended in 1928, all that Nolen knew about the Wingra Project was that it was in progress. Now in September 1933, he wanted to know more. The request enabled Jackson to provide some details: since April 1932, the Arboretum had become a reality and now consisted of 450 acres.[21] He bragged that the Arboretum land had been acquired almost without funds, and that he and other supporters realized that before going much further they ought to have the benefit of "some thoroughly competent consultant" to produce a large-scale, general plan. And, of course, he hoped Nolen could provide it. Jackson also made it clear, however, that the Arboretum had limited finances. Nevertheless, he hoped that Nolen would help; if Nolen should be interested, Jackson could send him information on the Arboretum that would enable him to see how his old "dream" was "becoming a reality." Jackson's wheedling, of course, was pretty transparent. He was "buttering up" Nolen by appealing to Nolen's generosity and to his stake in Madison's history. He wanted Nolen to come to Madison, to see firsthand what the Arboretum looked like. A series of letters ensued. At some point in the correspondence, Nolen did raise the possibility of a visit. Jackson proposed a winter gathering of Madison service clubs for which outside speakers were generally solicited. Payment would cover travel expenses and include a modest honorarium. To sweeten the offer, Jackson also held out the possibility of arranging for Nolen to give a lecture at the University of Wisconsin.[22]

Jackson was a tenacious fund-raiser and salesman, not to be denied. If he wanted something he worked very hard to get it. He once told William McKay, owner of the McKay Nursery, that begging for money and help was not easy. "Sometimes," he told McKay, "you can get them with honey, and sometimes with vinegar, but once in a while you've got to knock them down, drag them out, but honey is the best . . . when it works."[23] On December 8, 1933, Nolen told Jackson of his continuing interest in Arboretum developments. He also advised Jackson on newly legislated government funds that could be used to finance planning and consulting projects. Nolen acknowledged that a speaking engagement before the service clubs appealed to him, as did the possibility of a university lecture. Finally, Jackson had done what Olbrich couldn't do. He had convinced Nolen to commit to a Madison visit. Jackson wrote back on December 14, with the disappointing news that the service clubs lacked the money for an outside speaker, but that he was going to approach UW president Glenn Frank with the suggestion that Nolen be brought to the campus for a lecture or two sometime after January 1934.

Jackson was determined to have Nolen visit. On April 23, 1934, he wrote to Nolen informing him of the official dedication of what he now referred to as the "University of Wisconsin Arboretum and Wild Life Refuge," which had been scheduled for Sunday morning, June 17. The plan was for a 9:30 breakfast in the remodeled Nelson barn, followed by a program at which Nolen would be the main speaker. "I most earnestly trust," Jackson remarked, that "you may find it possible to accept." He added that he was also convinced that the city of Madison wanted to engage his services for a 1934 follow-up on his 1911 report. Mayor James R. Law, an architect himself, was most interested in having Nolen speak to Madison's Association of Commerce and other interested city figures about "fitting Madison . . . for the New Era." And so Nolen's return to Madison in June 1934 would keep him busy and was to be two-tiered: a presentation at the dedication of the Arboretum combined with an appearance before Madison's Association of Commerce to address the future of the city.

The Matter of "the Road"

At some point in 1933, Paul Stark recalled the conversation he had with Olbrich six years earlier, in 1927, over the merits of completing that long-sought-after road around Lake Wingra. He recalled proposing a plan that involved acquiring all of the lake frontage property, including the 50 acres that were labeled parcel F when the deeds to the six land parcels were accepted on April 26, 1932 (see the Parks Foundation map). Bernie Chapman, who owned Lake Forest Company, had agreed in 1927 to sell the 50 acres at a loss to the Madison Parks Foundation, provided that the foundation or the city of Madison build a road for him westward from Lake Forest. The road would connect Capitol Avenue on the east, the main Lake Forest road extending eastward to Mills Street, with Monroe Street on the west. The result would be a road that would finally run east to west through Lake Forest property and through the entire extent of Arboretum property.

At the time, Chapman wanted the road because he knew it would enhance the sale of the Lake Forest high land on the southeast shore of Lake Wingra, the small suburban area now known as Forest Park. The high land was free of the major drainage concerns that plagued the doomed lower lots. And Chapman and the Lake Forest Company desperately needed additional capital in 1927 to keep the company going, since it was barely solvent and had been struggling since the company's collapse in 1922. For three years, from 1927 to 1930, negotiations pertaining to the matter of building the road were stalled. Chapman was impatient. In 1930, Stark told Jackson that finally the matter of the road had to be confronted because Chapman had decided that unless the road was guaranteed before summer 1930, he would sell the lake frontage land, including parcel F, in order to pay expenses. In a letter to Jackson, Stark noted that he was bringing the matter to the attention of Jackson because, as he put it, "I feel that no project of greater importance to Madison as a whole has been brought to the attention of its citizens in recent years."

Acquiring the Lake Forest land along the Lake Wingra shoreline meant that practically the entire lakeshore would finally be under public control, an objective that went back to the days of John Olin and the Madison Park and Pleasure Drive Association and one that had been a primary motivating force behind Olbrich's efforts to guarantee a road. Olbrich's plan, however, did not include, as Stark was now suggesting in 1930, *all* of the 50 acres in parcel F. Olbrich's earlier plan called only for the possibility of acquiring a strip of land on the southern lakeshore, 350 to 400 feet wide, upon which to build a road that would parallel the shore.

Building the Road, Acquiring the Island and the Gardner Marsh

After the 50-acre parcel F had been deeded to the university on April 26, 1932, Chapman and the Lake Forest Company assumed that work on the construction of the Arboretum road providing easement through parcel F westerly to Nakoma would begin by the fall. Chapman, impatient, wanted his road. The insolvent Lake Forest Company needed money, but the university was not acting fast enough for him. Work on the road did not commence that fall 1932 as he had hoped. Frustrated, Chapman turned to the city of Madison. In November 1932 he offered the city a 90-acre plot of Lake Forest land, known as the Island. The Island is located at the extreme east end of the lake, bounded by Murphy's Creek (now Wingra Creek) on the east, the Lake Wingra shoreline on the west and the north, and on the south by the channel to the lake that had been dug by the Lake Forest Company in the early 1920s (see the map showing the Island).

The Island is still literally an "island" surrounded by water. Chapman had tried unsuccessfully for years to use it as a bargaining chip for getting his coveted road built. Now he offered to deed the Island to the city if city officials could guarantee that the road would be built. As it turned out, the city was not interested in procuring the Island because it saw no advantage in building a road for Chapman. Ironically, as Chapman soon learned, the university was willing to try to complete the road if it could get some kind of governmental assistance. Chapman would have to be patient, but patience was not a virtue that Chapman cultivated. The UW in November 1932 was not acting on the matter of the road because it was laboring through the process of trying to put together some form of workable governance for the new Arboretum. By early December 1932, the university had appointed an Executive Committee to oversee developments in the Arboretum. On December 1, 1932, Edward M. Gilbert (of the Botany Department), who had been chosen to chair the new Arboretum Committee, wrote to Jackson praising him for his outline of

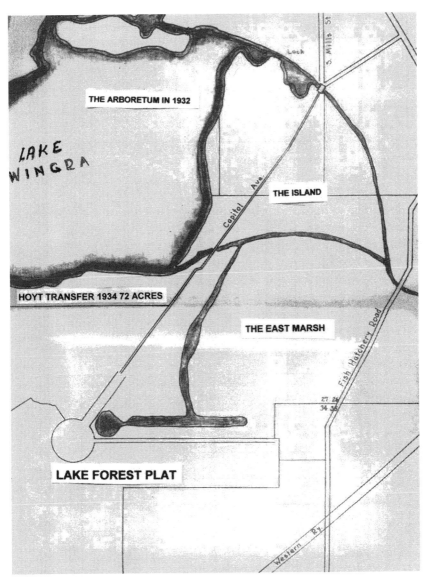

A map showing the Island at the east end of Lake Wingra. (illustration by author)

projects, one of which was the construction of the controversial road. Within two weeks, by December 15, 1932, the newly formed Arboretum Committee (AC) had taken up the burdensome matter. Gilbert wrote to College of Agriculture dean Harry L. Russell on the 15th, explaining that members of the Arboretum Committee were planning a visit to the shore area in order to draft a "preliminary survey" that would identify a possible route for the road, which, as Gilbert cautioned in a reference to the deal for parcel F, "we may have to build as per contract with the Lake Forest Company." On January 30, 1933, McCaffrey presented to the Board of Regent's Executive Committee a map of Arboretum property and adjoining lands, showing the location of the road chosen by the AC. A motion to approve the location was passed, but that was the only action the Regents took. No funds were appropriated. The Regents were not paying for Chapman's road.

Gilbert, speaking for the AC, expressed concern over the lack of funding. He also realized that at stake was the Island, the coveted 90 acres of wetland that would be added to the Arboretum's acreage once the university built Chapman's road. Since the Regents were not providing money for the project, the hope was that the road work could be done as a Dane County Work Relief Project.[24] So work on the road, which Chapman had hoped would have started in fall 1932, was delayed into the early months of 1933 when relief laborers would be available.

Work on the road moved at a snail's pace through spring and summer 1933. Transient laborers, hired for the job, were unreliable. Many of them were hoboes, "on the bum," riding the rails from one charity shelter to another. The major construction task at hand in 1933 was to provide landfill for the road bed. By November 1933, a section of the land was filled, beginning at the bridge crossing Murphy's Creek and extending westward through the Island to the Lake Wingra shoreline. The fill continued along the shoreline to the point where it intercepted Capitol Avenue and then extended westerly for a quarter of a mile in a direct line to the woods at the base of what was known at the time as Twenhofel Hill, the high bluff area in the Forest Park housing area south of the current Arboretum Road where homes were built, including the first home built in 1922 by UW professor W. H. Twenhofel (Geology). By 1933, Twenhofel, incidentally, was serving on the AC.

Work on the fill was slow and seemed to irritate everyone involved in the project. On November 24, 1933, Jackson, acting for the AC, asked M. H. Hovey, the city of Madison director for public works projects, that the Arboretum Drive project be given precedence and that more relief workers be assigned. Jackson, sympathetic with Chapman's frustration, was concerned that the road construction was stalled. It was now the end of 1933, and the road bed was not even filled in completely.

The Hoyt Transfer

By December 1933, the Board of Directors of the Lake Forest Company decided that perhaps if they transferred to the Regents ownership of a parcel of Lake Forest land along the lakeshore that linked Capitol Avenue in the east with parcel F in the west, the Arboretum staff would have more incentive for completing the road. The Lake Forest Board suggested that Frank W. Hoyt, a Madison banker and a member of the Madison Parks Commission and of the Lake Forest Board, be given the land. Hoyt, in turn, would agree to cover the interest on some notes, pay the taxes, and, similar to the land parcels transferred in April 1932, transfer the property to the Regents. On December 2, 1933, at a special meeting of the Lake Forest directors, the transfer was approved. Hoyt paid the taxes—$1008.44—and subsequently took possession of all of what was called, at the time, Wingra Beach.[25] The "Hoyt transfer" amounted to 72 acres, which, on January 21, 1934, officially became UW property. The university now owned enough land to complete the road from Mills Street in the east to a westerly entrance bordering somewhere on the suburb of Nakoma. Building the road seemed simple enough, but the subsequent history tells a different story.

By March 1934, the road fill still had not been completed. Chapman was upset and ill. Jackson tried to console him on March 22 with the assurance that as soon as the Federal Emergency Relief Administration (FERA) program began in early April 1934, the first priority for the Dane County Highway Committee would be completing the fill. He also assured Chapman that the county and the city of Madison had no problems working together on the project, "that the City could rightly complete the fill, grade, etc. out as far along the lake shore as the city limits." Furthermore, a very confident Jackson added, "the city limits could be stretched a bit if necessary." There had been a controversy over the fill between City of Madison and Town of Madison officials when it was discovered that part of the road was in the township. But enough dirt would be available, and Jackson assured the long-suffering Chapman that "Carl Felton of the County Board, and Chairman of the Madison Township . . . will push it as a Township Road," if necessary. Furthermore, John Icke, the city engineer, was now on record as assuring the contractors that he had located enough dirt to finish the job.[26]

Chapman was ill, and the construction delay, particularly the lakeshore portion, worsened his condition. Jackson urged him to stop fretting. He also took him to task for his earlier failure to explain exactly how he wanted the width and height of the grade, especially where the road branched west from Capitol Avenue to Twenhofel Hill. Jackson assured Chapman that he had support, that his friends agreed that he was "unduly disturbed" over the matter, and that he should take heart because, as Jackson noted, "there is good reason to believe the job will be completed in ship shape satisfactory manner in all reasonable time this Spring."[27] Jackson was referring to spring 1934, but it was not to be.

The Arboretum dedication on June 17, 1934, came and went, and there was still no noticeable progress on the road. Five days before the dedication, on June 12, 1934, Jackson reminded Chapman that John Nolen was coming to town for the dedication and that Nolen might have some helpful advice. He told Chapman to take heart, adding, "we are working hammer and tongs" to complete the project.[28] Yet, another year passed with little noticeable progress. In April 1935, R. A. Smith, FERA work secretary for Dane County, informed Jackson that there was 600 feet of road left to complete but not enough laborers to get it done. By the end of August 1935, FERA, soon to be phased out, stopped all work on the road completely.[29]

In mid-August 1935, a still optimistic Jackson told Chapman, who was at the end of his rope, that the CCC had moved into Camp Madison and that their "first work project would be the completion of Arboretum Drive." But Jackson also cautioned that there might be a delay since the government had notified the university that the CCC could not and would not "do any work on land . . . *not in public ownership*" (italics mine). Jackson warned Chapman that the government policy would have a direct bearing now on the old matter of ownership of the Island, the 90 acres at the extreme northeast end of what was still, in 1935, Lake Forest land (see the map of the Island). The upshot of Jackson's caution was to serve notice on the bewildered Chapman that work on the portion of the road that passed through the Island could not and would not begin until Chapman delivered the deed to the Island to the Regents. Chapman was stunned. The university was poised to realize another bargain. "I feel confident," Jackson told Chapman, offering little comfort, that he should give the disposition of the deed his "immediate consideration." Of course, the AC was thrilled with the prospect of gaining another 90 acres. Jackson was the cat who had eaten the canary. What Chapman felt and continued to feel can only be imagined. But Chapman now had no choice in the matter. If he wanted his road, he would have to deed the Island to the Regents. Chapman, however, went into a funk and refused to act. In September 1935, Jackson, still hopeful, told him that stalling was futile, that by law the CCC could not work on the road until the deed to the Island passed to the university. Chapman, ignoring Jackson's advice, refused to give it up. Also, complicating matters, the AC had learned that Chapman and the Lake Forest Company were now interested in selling the entire 190-acre East Marsh.

In July 1935, Jackson, once again convinced of the need for federal financial assistance, approached the U.S. Bureau of Biological Survey (BBS) with a request for financial aid in purchasing the 190-acre East Marsh sandwiched between the 90-acre Island in the north and Carver Street to the south (see the map of the Island). The Lake Forest Land Company was asking $80 an acre for the marsh. The BBS office examined the marsh and offered $15 an acre. Chapman was appalled and refused to sell. In approaching the BBS office, incidentally, Jackson was acting for Aldo Leopold, in particular, who had been hired in 1933 as the new UW professor of game management and who wanted East Marsh as "his outdoor laboratory." Jackson told the National Park Service (NPS) on July 18, 1935, in an effort to convince the NPS to provide funding, that, in the judgment of Arboretum management, "the amount of money required to buy the 190 acres—$15,000—will be . . . tripled in the expenditure of labor, materials, plantings, etc. that will be done by the CCC program."[30] The NPS told Jackson that the Arboretum would have the support of the NPS but no financial help. In a final August 1935 appeal to J. N. "Ding" Darling, director of the BBS, Jackson argued that Leopold, the new Arboretum research director, was "creating what we all believe will be one of the really outstanding laboratories in the country. . . He has been extremely anxious to acquire [the] 190 acres . . . and has worked out a plan for developing it that is a masterpiece." Then Jackson laid on the honey, adding, "as you well know, Aldo is the last man on earth to ask for something that might have the slightest semblance of appearing to be for his benefit, but he is so eager to acquire this tract because he is so conscious of its possibilities that I am taking this responsibility of urging its fullest consideration. I do that because I have such confidence in his ability to show practical, worthwhile results."[31]

Jackson's efforts to convince the BBS to help with the funding of Leopold's "outdoor laboratory" were commendable, but the BBS offer remained at $15 an acre. And Chapman, insulted, would not sell. This was mid-August, and once again Chapman had set a deadline for the purchase: October 20, 1935. The Lake Forest Company needed money to cover its debts. The October deadline gave Jackson and the AC only two months to raise the money. Jackson and the committee made an earnest effort but again were unsuccessful.

Chapman extended the deadline to December 20, but the money still was not there. On December 12, 1935, eight days before the deadline, Louis Gardner, a Nakoma businessman and owner of the Gardner Baking Company in Madison, donated $15,000 to the AC for the purchase of East Marsh. "Santa Claus," as Jackson referred to Gardner, had provided a Christmas gift that would finally secure, as Jackson noted, "Leopold's refuge, demonstration area and outdoor biological research laboratory." Leopold, Jackson noted, now had "300 acres in a block for his migratory game refuge." Jackson, of course, was including the 90-acre Island in his tabulation. Acquiring it was a sure bet, he thought. But his hopes were premature. Chapman, down but by no means out, was still refusing to give up the deed to the Island. He told Jackson that he regretted the deal for the East Marsh because "the price was too low." Angry, Chapman refused to accept the $15,000 and also refused to relinquish the two deeds. He would hold on to them until March 1936—three months. Through January and February 1936, Jackson tried repeatedly to assuage Chapman, to convince him to accept the $15,000, and to transfer the deeds to the university.

Jackson had a way of softening Chapman. They were friends, and Jackson was a super salesman. He tried to convince Chapman that, given the expenditure of CCC hours and government material, he was actually getting the equivalent of $100,000 worth of work toward completion of his road. In turn, he was giving up 90 acres of marsh land that he had been trying unsuccessfully to sell. He was not losing money, Jackson insisted. He was getting a bargain. As late as March 5, 1936, however, Chapman was still stalling. There would be no work on the road; Leopold would not have his game bird sanctuary. On March 5 Jackson unburdened his frustration to Stark, telling him that Chapman had "jeopardized the entire project." The AC "has been forced to the conclusion that there is some yet unexplained reason why Chapman has quibbled. . . . As far as I am personally concerned . . . this 90 acres has been the most difficult of all the Arboretum problems. Although we have completed [the road] 90 to 95 percent, he has made it impossible to put the finishing touches." Jackson had given up: "I have reached the end of my patience with it and would like to wash my hands from it entirely."[32]

Jackson was finished appealing to Chapman. Shortly after March 5, however, someone or something must have gotten to Chapman. Perhaps it was Jackson after all; perhaps it was Stark. Nevertheless, Chapman finally agreed to take the money and give up the deeds. On March 11, 1936, a quitclaim deed to the 90-acre Island and a warranty deed to the 190-acre East Marsh were accepted by the Regents. Secretary McCaffrey delivered $15,000 to the Lake Forest Company. From the time of the purchase, the 190-acre marsh would be known as Gardner Marsh, an expression of gratitude for Louis

Louis Gardner, Arboretum benefactor, in 1979, age ninety-two. (photo by
Norman Lenburg, UW Information Services, Arboretum Photo Collection)

Gardner's generosity. Gardner, incidentally, would continue in his support of
the Arboretum, especially in later years by providing funds for the Friends of
the Arboretum. He would live to be ninety-two, dying on December 14, 1979.
Jackson could boast in 1936 that the UW Arboretum, thanks to Gardner, had
over 800 acres, and he was bent on acquiring more. He had his eye on the
Grady Tract, 200 acres to the southwest of the Arboretum.

Work to finish the Arboretum road began again in earnest with CCC assistance in summer 1936. By the end of 1936, a two-lane graveled road was
completed. In the summer of 1937, the road was blacktopped. There would
be ongoing problems with the road—heavy traffic, repairs, and vexing questions of jurisdiction—well into the end of the century. But after 1936, one
could finally drive a vehicle through the Arboretum from east to west and west
to east.

Gardner Marsh was a welcome addition to the Arboretum, but it was also
damaged, a victim of history. Early in the twentieth century the Madison Park
and Pleasure Drive Association dredged Murphy's Creek to create a waterway

for boats. The dredging, however, lowered the water level, resulting in a damaging rise in the marsh land level. In 1916 the Lake Forest Land Company dredged the marsh again in order to create "Venetian" canals in the doomed Lake Forest housing development. The dredging took another toll. Damaged portions were overrun by invasive plants, especially cattails. In the 1930s, Leopold worked hard to turn Gardner Marsh into a bird sanctuary, initially planting trees and brush, mostly tamarack and river birch, along the road in order to create a screen. The marsh soil, predominantly peat, provided fertile beds for large numbers of small white and yellow lady's slippers and clumps of rare orchid hybrids as well as pink phlox and yellow-eyed grass. There was also an area that contained rare Juncus that Norman Fassett located. Canals, left over from the aborted Lake Forest land development, were cleared and redredged in order to create a scattering of protected islands for nesting birds. A shorebird beach was planned at a spot where the water level was lowered in August or September, exposing mud flats without weeds. In spring the beach was flooded to create a "flooded pasture" condition. Thick brush around the marsh provided protection, and efforts were begun to encourage the return of wood ducks, among other shore birds that once inhabited the area, but as the years passed the efforts produced limited success.[33]

Nancy Sachse, in *A Thousand Ages*, the first book-length history of the UW Arboretum, noted that the piecemeal nature of the Arboretum's land acquisitions resulted in a curious and ironic advantage. It produced an Arboretum singular in its variety of terrain and plant communities, all midwestern, all indicative of a regional natural heritage, including "open marsh, shrub, savanna, conifer swamp and forest, grassland, deciduous forest, horticultural areas, and experimental plots," yet with enough variety for centuries of research and life study.[34] She certainly was right.

CHAPTER 3

1932–1934

PRE-DEDICATION YEARS, EDWARD M. GILBERT, ALDO LEOPOLD, G. WILLIAM LONGENECKER, THE POLITICS OF GOVERNANCE

The first Arboretum land parcels were officially acquired by the university on April 26, 1932. No one on campus, however, had ever built an arboretum. Nor did anyone emerge with a clear idea of how an arboretum should be governed or precisely what direction it should take. Harvard's Arnold Arboretum, Olbrich's inspiration, was acknowledged as the original model for the UW Arboretum. But by 1932 the UW Arboretum already had exceeded the 200-acre Arnold Arboretum in size, and it contained a far more varied distribution of land, wildlife, and native plant communities. There was a general agreement among the Arboretum founders that the facility should aspire to be more than a park or a botanical garden, and at Olbrich's urging, the Board of Regents in December 1927 had designated the land as a "Forest Preserve Arboretum and Wild Life Refuge." If, as Ossian Cole Simonds maintained, arboretums—usually "museums of trees and shrubs"—should be more than museums, "affecting one like a beautiful painting," this arboretum would be weighted down from the start living up to a designation that called for much more than the artistic propagation of trees and shrubs.[1] It would be even more heavily weighted down during the early months with confusion over who was in control and what its fundamental charge should be. Since April 1932 the land had belonged to the university, but the university was not acting.

A brief front-page notice published on April 27, 1932, informed readers of the *Wisconsin State Journal* that "a natural forest reserve and wild life refuge of several hundred acres" had been officially approved by the UW Board of Regents and was now planned for Lake Wingra's south shore. A far more detailed and informative account, running to ten paragraphs, with a headline reading "Announce Establishment of Wisconsin Arboretum," turned up in an uncataloged file of news clippings in the Arboretum archival collection. The news clipping—unsigned, undated, and of unknown origin—appeared in print only a few days after the official establishment of the Arboretum on April 26. The news story assured readers that this Arboretum, once underway, will be devoted not only to the development of trees and shrubs but also "to the solution of reforestation problems and the propagation of wild life."[2] No source for the quote was cited in the article, but coincidentally Aldo Leopold, when interviewed for a February 1934 article titled "The Arboretum" in the *Wisconsin Alumni Magazine* two years later used the same language, verbatim, to explain the Arboretum's purpose. Leopold also observed that the Arboretum would provide "the University and the State [with] an excellent opportunity for experimentation in reforestation and the propagation of wild life."[3] So exactly who was speaking for the Arboretum in 1932? It's not clear. Leopold wasn't added to the faculty until June 1933. As late as August 1932, three months after the six land parcel deeds had been transferred to the Regents, there was still no general agreement on the Arboretum's future and, even more disconcerting, there was no one from the university directly in charge of operations.

On April 29, 1932, three days after the Regents had accepted the deeds, Bud Jackson sent a conciliatory letter to Richard H. Marshall, president of the Madison Parks Commission, in which he responded to Marshall's request for updated information on the Arboretum's progress. Jackson also laid out his own predictions on the Arboretum's future. There appears to have been no agreement at that early date even as to what title would officially identify the new facility. Jackson referred to it in the letter as "The Wingra Park Arboretum and Wild Life Refuge." He told Marshall that if he wanted more information on the general plan for the future of the new Arboretum, he should invite Paul Stark to a Parks Commission meeting. Stark, of course, was unable to speak officially for the university since he had no direct connection to it. Jackson, in an attempt to assure Marshall and other skeptical members of the Parks Commission that the city of Madison really did stand to gain from the presence of the Arboretum, suggested that there were "features of the entire program which were not fully presented to the Regents" but which were "extremely important to the City of Madison." Jackson also informed Marshall "of a system of drives and park areas" that he was certain would eventually surround Lake Wingra. He added that the State Conservation Commission and the federal government were also keenly interested in cooperating with the Arboretum's plans for a wildlife refuge and a government-funded migratory bird refuge.[4] Jackson left the impression that government assistance—Jackson's persistent solution to Arboretum problems—was pending. But it was merely wishful thinking.

Confusion about who was in charge and how the Arboretum was to be developed continued through May 1932. Jackson, again speaking boastfully but totally without authority, was even telling his International Rotary correspondents by June 1932 that an "International Forest of Friendship," consisting of trees to be planted yearly by University of Wisconsin foreign students, was destined to be an outstanding future attraction of the new Arboretum.[5] It seems likely and understandable that Jackson, the strongest supporter of the Arboretum, after Olbrich, now had a difficult time coming to terms with the realization that the Arboretum was no longer his progeny. Nor was it to be either the city of Madison's showpiece or a federal game refuge. Actually, Jackson's long-sought-after Arboretum was now the sole responsibility of the Board of Regents acting for the state and the University of Wisconsin. And to Jackson's dismay, the university seemed noncommittal, almost disinterested.

Edward M. Gilbert and the Arboretum Committee

The administrative hiatus continued into August 1932—four months. Finally, in late August, UW president Glenn Frank took the initiative and, as Olbrich had earlier proposed, asked Professor Edward M. Gilbert, chair of the Botany Department, to form and chair a committee that would be called the Arboretum Committee and that would directly address the matter of governance. Gilbert was hesitant to accept the position.

Gilbert was burdened with his own department's demands and had little time for the challenge of putting together a committee that had no direct administrative charge from either the Regents or President Frank. He would be largely on his own, and this would continue to be the case. He was an easygoing, affable man, a diplomatic administrator. Frank trusted that Gilbert's good sense would guide the committee through the difficult formative years. Reluctantly, Gilbert accepted the position. His long friendship with Olbrich influenced his final decision, as did his love for wilderness, which, as the Memorial Resolution at the time of his death recorded, "permeated his outlook on life."[6] Once Gilbert accepted the challenge, he worked quickly. Within two months, by October 1932, two committees charged with the administration of the Arboretum were in place. The first, the Arboretum Committee, or the Technical Committee, as it came to be known, consisted of university faculty and staff members who were acknowledged "experts" in their fields. The first Arboretum Committee consisted of the following members:

Professor Edward M. Gilbert chaired the first Arboretum Committee from 1933 to 1939. He worked closely with Olbrich, Jackson, and Stark during the Arboretum's formative years. (UW Madison Archives, Image #S05287)

Franz Aust, Horticulture
L. J. Cole, Genetics
James G. Dickson, Plant Pathology
Norman C. Fassett, Botany
Albert F. Gallistel, superintendent, UW Buildings and Grounds
Chauncey Juday, Zoology
Maurice E. McCaffrey, secretary, Board of Regents
Fred B. Trenk, university forester, Agricultural Engineering
George Wagner, Zoology

The second committee, an Advisory Committee, consisted of the following:

E. A. Birge, former UW president, Limnology
Joseph W. "Bud" Jackson, Jackson Clinic administrator and executive director of the Madison and Wisconsin Foundation

Paul D. Kelleter, conservation director, Wisconsin Conservation Commission

Aldo Leopold, former associate director of the U.S. Forest Products Laboratory

Harry L. Russell, dean of the College of Agriculture

C. P. Winslow, director of the U.S. Forest Products Laboratory

Raphael Zon, Forest Service, U.S. Department of Agriculture

Noticeably missing from the list appointed to the first Arboretum Technical Committee was the name of G. William Longenecker, who in 1932 was finishing his master's work as a graduate student in landscape architecture. Longenecker was also working for Albert Gallistel in the Department of University Buildings and Grounds. The Gallistel connection would serve Longenecker well. By 1933, Longenecker would be appointed executive director of the Arboretum. Noticeably present on the list appointed to the first Advisory Committee was the name of Aldo Leopold, who accepted the committee appointment on August 31, 1932, almost a year before he joined the faculty.[7] He would be appointed to the newly endowed chair of game management in the Department of Agricultural Economics in June 1933.

Once the two Arboretum governing committees were in place, Gilbert called a meeting of both the Arboretum Technical Committee and the Advisory Committee over lunch at the University Club on Saturday, November 26, 1932. Gilbert set as his one agenda item "a general presentation of the entire situation with reference to the Arboretum and the formulation of plans relating to the project."[8] And so, sometime either before or after that noon luncheon, the future of the UW Arboretum officially began to be shaped by UW faculty and staff. Jackson, particularly, felt relieved.

As the year turned from 1932 to 1933, outside operations at the Arboretum got underway. There was the continuing irritation of the road extension east through Lake Forest property. Efforts to get to work on that troublesome road would consume considerable time and energy for planning but—much to Bernie Chapman's consternation—as explained above in chapter 2, the efforts produced few results in 1933.

Charles E. Brown and the First Arboretum Restoration Effort: The Native American Effigy Mounds

The Arboretum Committee (AC) had better luck with Charles E. Brown's ongoing concern over the future of the effigy mounds in the Arboretum woods.

Brown, curator and director of the Wisconsin State Historical Museum, surely was delighted when he received a message on January 5, 1933, from Jackson, writing for the newly formed Arboretum Committee. Jackson, an old friend, told Brown that he thought Brown might "agree that everyone of these mounds should be carefully located, marked and preserved as a permanent matter of genuine interest to all." He queried if Brown would take the time to locate the mounds and then sketch them on a map so that the AC would know exactly where they were and would take measures not to disturb them. He closed his letter to Brown with the following aside, "I am sure the entire committee would much appreciate it if you would be good enough to do this for us."[9] Jackson, always in good humor and always up to a jest, obviously knew that Brown would need little persuasion complying with the request.

The fact was that Brown, for years, at least since 1915, had been doing his utmost, short of throwing his body on them, to see to it that the mounds around Lake Wingra survived. Now he had his chance to guarantee the preservation of the three mound groups in the Arboretum woods along with the endorsement and support of the AC. Brown replied to Jackson on January 10, 1933. He confirmed what Jackson already knew: that Brown was fully aware of the existence of the mounds and that he had "always hoped that something could finally be done to secure their permanent preservation." When the weather improved, Brown started the restoration work. He worked on the mounds at limited times with a limited crew during 1933 and into 1934. By September 1934 he reported to the AC that the restoration of the twelve mounds in Wingra Woods was near completion. The restoration involved removing invasive weeds and brush, filling in numerous holes caused over the years by animals and relic hunters, and seeding the tops of the mounds with grass. He actually excavated two of the mounds and found traces of fires (ashes and charcoal) and unusually large, out-of-place boulders. He added that he felt certain that in restoring the mounds he and his crew had "accomplished a valuable work for the Arboretum." Nevertheless, he added a complaint about his inability to obtain trucks to transport dirt when he needed them.[10] Unknown to Brown, however, was the fact that no trucks were available.

Brown was an interesting figure, to say the least. He was a talented, dedicated archaeologist and renowned collector of Wisconsin folklore, particularly Wisconsin Native American lore and Wisconsin lumbering and logging tales. He had helped to organize the Wisconsin Archaeological Society and was an authority on the history of Southern Wisconsin Native American tribes. His extensive firsthand collection of Native American tales and folklore, many based on personal interviews, though poorly documented, has never been surpassed. But Brown also had an abrasive side to his personality that tended to grate on people. His association with Arboretum management, consequently, was not always harmonious. He was a dedicated preservationist, convinced of the nobleness of his cause, and had a difficult time understanding why university faculty members—science professors, in particular—were not as committed to the cause of preservation as he was. He was convinced, for instance, that as soon as UW officials claimed ownership of the six parcels on Wingra's south shore, they should have invited Native Americans who once camped there to return to their old campgrounds. He also urged that an old dugout canoe retrieved from Lake Wingra and stored in the Historical Society Museum's basement be returned to the Wingra shoreline to be put on permanent display.[11] He also thought that wigwam frameworks should be constructed in the Wingra Woods for the benefit of visitors.

As time passed, the faculty in charge of the Arboretum, Leopold, in particular, grew impatient with many of Brown's requests, one of which—particularly problematic—was his insistence that the Arboretum mounds be labeled. He approached the AC in May 1936 with the complaint that yet another season was passing (he had been working on the mounds since 1933) "without our getting the . . . mounds . . . properly marked." He added that without markers, "distinguished visitors" he had sent to the Arboretum had been unable to find the mounds. "A plain bronze tablet on a boulder should long ago have been supplied," he complained. Brown insisted, in a manner that surely did not sit well with the Arboretum Executive Committee, particularly Leopold, that he wanted "action in this matter."

It appears that at some point in 1935 the Executive Committee did pass a resolution authorizing Longenecker and Leopold to work out a design for bronze tablets that would satisfy Brown's request. And therein lay the conflict, because it also appears that neither Longenecker nor Leopold were particularly keen on producing the tablets. Their response became an exercise in passive resistance. Leopold, in a handwritten postscript to a letter from Jackson on the issue, noted that he was "of course doubtful of the value" of appropriating money for tablets, and then deferred to Longenecker with the ironic aside, "Bill—I am at your call on this."[12]

Brown sent Leopold a letter on June 5, 1936, taking both him and Longenecker to task for "overlooking" his request and for what he referred to peevishly as their "several well intended admonitions." Brown noted that he and others held "that a proper tablet marker for the Indian mounds" was desirable, and that, to his mind, it would "be far less conspicuous than the . . . wholly unnecessary stone structure now being constructed in the small stone quarry near the mounds." The "unnecessary stone structure" he referred to, however, was a shelter designed by Longenecker that the CCC, under Longenecker's supervision, was building in Wingra Woods. Brown's sarcastic remark about the stone structure near the mounds likely stemmed from an earlier rebuff in 1934 when Jackson, at Brown's urging, tried to get a Ho-Chunk Indian statue or statuary group approved by the Civil Works Administration (CWA) and the AC for construction in Wingra Woods. Jackson had hoped that he could convince Lorado Taft, the famous Chicago sculptor, to do the work, but he also realized that money for the project did not exist. The Ho-Chunk display, however, which appears never to have gotten beyond the planning stage, was to be constructed in the woods "on the crest of the hill just above the big springs, and near a group of 13 Indian mounds"—almost exactly where Longenecker's "unnecessary stone structure" was being built when Brown sent Leopold the antagonizing letter in 1936.[13]

Personal differences aside, the real objection for Leopold and Longenecker, both in the case of the proposed Ho-Chunk statue and the bronze tablet advertising the mounds, was their mutual aversion to signs or statues—anywhere—on Arboretum land, an aversion that continues among Arboretum staff today and to which Brown responded in his letter by stating that though he agreed with Leopold and Longenecker's concern about signs, he didn't see how they could be avoided "if you really desire to contribute to the education of . . . uninformed visitors."

Brown also tended to be fractious and quick to take offense. He told Jackson, for instance, that though he was uncertain of who was or was not chosen for membership on the 1933–34 AC, he was certain that when the appointments were made, he had been "somehow overlooked." He chided Jackson: "you know of my keen interest and of my desire to assist in every possible way in the preservation of the Arboretum."[14] Jackson, good soul that he was, saw to it that Brown, after having initially been passed over, was finally appointed, on Jackson's recommendation, to the Arboretum Advisory Committee for the 1934–35 year. Brown's concern for the Native American historical connection to Lake Wingra and the Arboretum mounds, well intended although often frustrating for him, would be ongoing right up to the time of his death in 1946.

The Arboretum Pines

The planting of pines along the Arboretum's southern boundary, an exercise in forestation but not restoration, was another matter taken up by the new AC early in 1933. The 190-acre Bartlett-Noe property was acquired on April 27, 1933. Even before the deed was officially transferred, members of the AC had decided that they would plant pines in 40 to 80 acres of what had once been open grazing land. The pine planting was the Arboretum's first serious forestation effort.

On April 5, Jackson wrote to Raphael Zon of the U.S. Forest Service (USFS) and a member of the Arboretum Advisory Committee, asking for assistance with the planting. Understandably, the USFS at this point looked like a good federal prospect for financial and professional help. Zon's position on the Advisory Committee was strategic. Jackson urged Zon to send along as soon as possible his suggestions for the planting. He added: "if we are to go forward now, then every day counts. Of course, our thought is to get the cooperation of the Federal Forestry program, and they have rather indicated that Wisconsin is a good place to start their work."[15] Jackson never gave up. The Federal Forestry program may have been interested in Wisconsin, but not in efforts to plant an 80-acre pine forest in the UW Arboretum. Jackson, the consummate optimist, would simply turn to another outside source as a possibility for help. Not one to sit by and wait for opportunities to present themselves, he sought the opportunities himself . . . for better or worse. Federal assistance or not, the pine planting on the 80 acres of former Bartlett-Noe grazing land went forward. By mid-April 1933, Edward Gilbert, AC chair, had contacted Forester F. G. Wilson at the Wisconsin Conservation Department (WCD) and had ordered 10,000 white pines, 12,000 Norway pines, 2,000 white spruce, and 5,000 jack pines. He also asked Wilson if he knew where the Arboretum could acquire black spruce, tamarack, white cedar, and hemlock. He asked Wilson to notify him of the possible shipment date since he had to arrange for the labor. He also suggested that it would be helpful if someone from the WCD could accompany the pine shipment and supervise the planting. That "someone," he added, would be working "in part with Mr. Longenecker," who, Gilbert noted, "will act for the Committee."

An Arboretum Directorship?

Gilbert's reference here to Longenecker is important, as it is the first official reference to him in the Arboretum correspondence that I have reviewed. In 1933, Longenecker, still completing his graduate work, was not yet a member of the AC, but he had obviously caught Gilbert's eye as a potentially valuable addition to the budding Arboretum family. Longenecker had also earned the support of Superintendent Gallistel of UW Building and Grounds, for whom he had been working. By June 1933, Gilbert would recommend Longenecker for the Arboretum directorship. In fact, by mid-year 1933, the AC was intensely debating the issue of who should be in charge. Did the committee want a single director or some other form of executive management? The most obvious choice would have been the appointment of a single director. As early as August 1932, word had somehow gotten out that the University Arboretum was official and in business, that Edward Gilbert would soon be forming a committee to oversee Arboretum matters, and that the university was looking for a director.

On August 16, 1932, weeks before the AC was even formed, Gilbert received a letter of application for the directorship, which must have surprised him. The letter came from Wilfred E. Chase, whose family had a long history of property ownership on the north shore of Lake Wingra.[16] Chase listed as his qualifications: a degree in liberal arts, some writings "along scientific lines," and an independent study of forestry and landscape architecture. He also claimed to be an "all-around naturalist" with a familiarity with Lake Forest land that he traced to his childhood. I think "I am more familiar with its flora and fauna than any one else," he boasted.[17]

Chase's application was premature, but it underscored the issue, early on, of exactly what kind of director would suit the position, if in fact the AC decided that it actually wanted a single director. Chase had no direct connection to the university and he lacked scientific credentials, but he was impressively familiar with the Arboretum's flora and fauna and he had management experience. Six months after receiving the application, in February 1933, the question of how the Arboretum should be managed was still not settled. In a talk to the Wisconsin Nurserymen's Association in mid February, Gilbert discussed Arboretum plans and projected proposals. He also spoke of having no director as yet, but that filling the position was a first order of business.[18] "We have the land [but] not a single cent for furthering any project as yet," Gilbert told an acquaintance.[19] Even if the Arboretum Committee decided that it wanted to hire a single fulltime director, the money was not there. The Arboretum could not afford to pay a fulltime managerial salary to anyone.

Aldo Leopold's Short-Lived Fifty-Day Directorship

The lack of immediate funds for the hiring of an Arboretum director fed directly into a plan that had been engineered by the dean of the College of Agriculture, Harry L. Russell, and professors in the college and promoted by trustees of the Wisconsin Alumni Research Foundation (WARF) with the eager support of Jackson.[20] In 1933, the nation was embroiled in the Great

Depression, and given the problems with soil erosion in the dust bowl and a failed agricultural economy, conservation, especially land management and reforestation, was receiving a lot of government attention from Franklin Roosevelt's New Deal administration. In a May 23 letter to William Kies, a WARF trustee living in Chicago, Jackson explained why WARF was interested in establishing a "Chair of Conservation" and who the foundation had in mind for the position. Jackson also pointed out that the WARF proposal for the conservation chair also included the directive that the occupant of the chair was "to serve as Director of the University Arboretum and Wild Life Refuge."

WARF's choice for the position, not surprisingly, was Leopold. Jackson had confided to an old friend as early as March 1933 that he thought Leopold could be secured for the position.[21] Now, in May 1933, he assured Kies: "It is our good fortune that there is . . . available for the position the one man who stands out above all others as the best qualified for the work." Jackson, eager to restore some semblance of the days when open land and ample game were abundant, added nostalgically, "as a State and Federal Migratory Bird Refuge we plan to bring back the ducks, geese and swans by the thousands just as they were there in our boyhood days."[22] Jackson and others, including the WARF trustees, Dean Russell, and professors in the College of Agriculture, wanted the Arboretum to become a government-funded migratory bird refuge. A migratory bird refuge was also in Leopold's sights. Jackson had been proposing and hoping for a federally funded game refuge at least from April 1932 when the deeds for the six parcels were transferred to the Regents. Actually, according to Jackson's records, the prospect for government funding went back farther, at least to 1927, five years before the Arboretum's founding. In 1927, as he explained, Leopold, Russell, and professors from the College of Agriculture proposed at a meeting that the federal government establish a game refuge and forest experiment station at Madison in conjunction with the University of Wisconsin and the U.S. Forest Products Laboratory. At the time, Leopold was working for the USFS as the associate director of Madison's Forest Products Laboratory. Quite surprisingly, however, it was Michael Olbrich, according to Jackson, who first proposed to the Regents the idea of establishing a working connection between the prospective Arboretum and the Forest Products Laboratory. The possibility of a federal connection was essentially why the Regents decided in 1927 that the $83,000 from the Tripp Estate would be set aside for the creation of a Forest Preserve Arboretum and Wild Life Refuge. Jackson recalled that while going through Olbrich's papers after his death, he had come across an "excellent brief presented by Leopold setting forth the reasons for establishing such a Forest Station." As a result, Jackson added, "we have revived the plan."[23]

Professor Aldo Leopold, Arboretum director of research and professor of
wildlife management from 1933 to 1948. (Wisconsin Historical Society,
WHi-34893)

The revival of the plan for government funding for a game refuge and forest
station now, in May 1933, had a direct bearing on WARF's and Jackson's choice
of Leopold for both the proposed conservation chair and the Arboretum directorship. Also factored into the decision was the reality that the Arboretum
lacked the money to hire someone whose sole job would be fulltime director.
Placing a UW professor occupying an endowed chair in the position of director meant that, at the very least, the endowment would pay most or all of the
salary—and the Arboretum would have its director as well. Leopold's professional credentials were excellent and qualified him both for the conservation
chair and the directorship. By 1928 he had given up his Forest Products Laboratory position to work for the Sporting Arms and Ammunition Manufacturer's Institute. Also in 1928, he added to his academic experience by delivering,
with Dean Russell's blessing, a course of lectures on game management in the
College of Agriculture. At the time, Russell, with the assistance of Fred Trenk,
the UW forester, was spearheading, for the first time in the country's history,
an idea he had brought back to Wisconsin in 1925 after a trip to Australia.

While in Australia, Russell had watched school children planting trees on public land as an educational project. He came home and introduced legislation in the Wisconsin Senate that gave birth in 1928 to the Wisconsin School Forest Movement, the first in the nation. The movement encouraged school districts to purchase clear-cut or burned-over forest land, often acquired for the payment of delinquent taxes. The damaged land would be restored by students and teachers as an exercise in outdoor education. In the spring of 1928, the 80-acre Laona School Forest, much of it blackened stumps near the northeastern Wisconsin town of Laona, was the state's first School Forest restoration project.[24] Leopold was closely associated at the time with Russell and was available to counsel Russell's efforts to promote the School Forest Movement. From 1928 to 1933, Leopold had worked as a consulting forester on issues involving conservation policy, reforestation, restoration, and game management. In May 1933, he published his book *Game Management*. By 1933, he had an established national reputation and was widely known as a forester, a conservationist, and a game management specialist. But more to the point, his UW supporters, including particularly Dean Russell, believed that once Leopold was director, his stature among conservationists and his growing national reputation would enhance their chances for government support of a federal game refuge and an experimental forest program in the Arboretum.

Jackson was particularly hopeful. On June 8, 1933, however, less than three weeks after Jackson's May 23 letter to Kies, Edward Gilbert, the influential chair of the AC, told George W. Mead, a member of the Board of Regents, that if the AC's decision was to put one person in charge, his choice was G. William Longenecker—"a very capable young man." Gilbert added that though Longenecker was young and lacking experience, once he was "given some time for travel and observation," he would be ideally suited, when the funds became available, to carry forward "the planning of the project." Gilbert also appears not to have been as staunch a supporter of potential government funding as was Jackson, WARF, Dean Russell, and other influential members of the AC, including Gallistel and Board of Regents secretary Maurice Mc-Caffrey. Nor was Gilbert sold on the idea that becoming a federal game refuge was in the best interest of the Arboretum.

His choice of Longenecker as a director obviously had something to do with Longenecker's background in horticulture, particularly his work in landscape design. Longenecker had been the first graduate student in the Department of Horticulture to receive an M.A. in landscape architecture, an option that the department had just introduced. Even though the deck was stacked against him, Gilbert believed that landscape design, botanical planting, and biological research, rather than game management and wildlife, were what an arboretum should emphasize. He worried, under the new policies that he saw developing by June 1933, that the Arboretum—"our project," he called it—would "develop into something quite different from the beautifully . . . landscaped areas" distinguishing other arboretums.

Gilbert was a mild-tempered administrator who preferred diplomacy to confrontation. He never once admitted openly that he did not want Leopold as director and that he was dismayed by the direction the Arboretum was heading. He would support the wishes of the AC. He told Mead, in his letter of June 8, 1933, that if he wanted more information on Arboretum policy, he should now turn to McCaffrey or Gallistel, who, Gilbert observed, were "much more influential in determining policies than I am." The implication here is that the key Arboretum policymakers by June 1933 had become McCaffrey and Gallistel. They also had the support of Jackson, and Leopold was their choice for the directorship. But he was not Gilbert's choice, and Gilbert believed by June 1933 that he would "probably not be retained on the permanent committee."[25]

Events proved Gilbert both right and wrong. He was right to think that Leopold would be the choice for the directorship, but he was wrong to think that the decision would determine the Arboretum's future. He was also right in his assessment that control of the Arboretum, toward the end of 1933, was in transition and that one of the newly emerging dominant policymakers was Gallistel. In 1939, Gallistel would be appointed to the chairmanship of the AC, a position he would retain until his retirement in 1959.

In June 1933, the AC voted to support a crucial shift in Arboretum control, one that provided for a shared caretaking responsibility between the AC and the UW Buildings and Grounds department. The formal motion approved by the Board of Regents on June 17, 1933, reclassified Gallistel's Buildings and Grounds as the Committee on Constructional Development. The move, a sensible one in light of the many upcoming construction and groundskeeping issues posed by the Arboretum, also increased the decision-making influence that Gallistel would have and would continue to have until his retirement.

Gallistel, by profession, was an architect who was hired in 1907 as a campus planner and who in 1919 was named superintendent of Buildings and Grounds. His architectural skills, in particular, were put to good use at the Arboretum, especially in 1934 when the camp, which a year later would house the Civilian Conservation Corps, was first constructed. He was also a crafty town-and-gown politician who, like Jackson, was a pragmatic man of action who managed to get things done—one way or the other. He knew the campus, literally, both inside and out. From 1933 until his retirement very little

Albert F. Gallistel, an architect by profession, was superintendent of UW Building and Grounds and chair of the Arboretum Committee from 1939 to 1959. (photo by UW Department of Photography, Arboretum Photo Collection)

was accomplished at the Arboretum that did not somehow reflect Gallistel's influence. He would be particularly productive and effective in acquiring laborers for the Arboretum in the early years. Gilbert was also right to think that Leopold would be appointed to an academic chair and eventually be named Arboretum director. On August 4, 1933, the Regents voted to approve the establishment of a chair, not in conservation but in game management, funded, as expected, by WARF. The Regents also voted to appoint Leopold to the chair as the first professor of game management. The motion also automatically directed that Leopold would "serve in the capacity of Director of the University Arboretum."[26]

Jackson was delighted with the decision. Four days after the appointment, on August 8, 1933, he asked his good friend Aaron M. Brayton, editor and publisher of the *Wisconsin State Journal*, to delay publishing the story of Leopold's appointment even though the Regents, according to Jackson, had "buttoned it all up" four days earlier. He told Brayton that UW officials had delayed releasing the story because they were waiting "for the director to get here." Leopold, at the time, was working with the Forest Service in New Mexico but was expected back in Madison on August 10. So that Brayton would have all the information for what Jackson termed "a corking good story," Jackson, in "strict confidence," outlined for Brayton the university developments leading up to Leopold's appointment. He told Brayton the "story"—that efforts to get Leopold appointed director had begun several years earlier in 1927 in the College of Agriculture, when Dean Russell "conceived the idea of establishing a Chair of Conservation at the Agricultural College in order to sell conservation, especially of upland and aquatic game birds to the farmers of Wisconsin." The hope was that if the college could show farmers how to grow "crops of game on their unused wood lots and marginal lands," they "could add to their . . . income." It was called "game cropping." Dean Russell, Jackson added, "had the complete set-up and the man in sight for the job, but the money was not available." Russell's purpose for promoting the idea of a chair of conservation was intended to help Wisconsin agriculturally but at the same time to provide improved hunting for Wisconsin sportsmen. Leopold, given his recent experience with the Sporting Arms and Ammunition Manufacturer's Institute as well as his extensive work on conservation issues, was a natural for the position. It would be up to him to build a program within the university that would convince farmers of the value of creating habitat on their farms that would enable them to crop game.

Jackson also shared an important document with Brayton that likely had originated in the College of Agriculture and likely, prior to Leopold's appointment, had circulated among faculty evaluating the rationale behind the creation of the new chair of conservation. The three-page document in the University Archives, entitled simply "Proposed Chair of Conservation," is unsigned and includes no reference to any specific point of origin. It is a detailed, well-thought-out explanation of why the university should create a conservation chair and also how the university, the Arboretum, and the entire agricultural community would benefit from the effort.

The document outlined two directions in which the Arboretum could develop:

a) As a natural area, relatively unimproved, where citizens can take walks and research men could find samples of undisturbed biological conditions. This conception would require relatively small development funds. It [would] suffice to have a caretaker, a few thousand dollars for plantings, a few hundred dollars for feeding water fowl, and an initial expenditure of $1,000 for an artificial island in Lake Wingra. Given these resources, the proposed conservation officer could supervise such a development with a small fraction of his time.

b) As a botanical garden and outdoor public museum, highly improved, and visited by many people for many purposes. Such a development would require a resident full-time director, and an annual budget in five figures. It could not be supervised on a part-time basis. Since there are no present funds for this type of development, however, the part-time supervision here proposed will suffice.[27]

The document notes that at the time it was drafted, no immediate decision between those two options needed to be made. But it cautions that if the second scheme of development—the botanic garden and outdoor public museum—is undertaken, full-time personnel will have to be hired. And, of course, funding being almost non-existent, money to hire full-time help was impossible.

Given the two options for development offered to the Arboretum, Edward Gilbert's uneasiness and frustration with the plans for the Arboretum's future, though subdued, become more understandable. He preferred the second option, the Arboretum as a botanical garden and outdoor public museum. That is also why he preferred having Longenecker, a professor of horticulture and landscape architecture, in charge, and why he told Mead in June 1933 that he thought he would not be retained as chair of the AC. But he was wrong on the last point. He would continue on as chair through the 1933–1934 academic year.

Through August and most of September 1933 Leopold's appointment as Arboretum director went unchallenged. The future direction of the Arboretum, it followed, would be geared primarily to the interests of conservation and game management. Option a, the Arboretum "as a natural area, relatively unimproved," with an emphasis on the possibility of government funding for the creation of a game refuge and bird sanctuary seemed where the Arboretum was heading. And Leopold, the new professor of game management, who was very interested in establishing a game refuge, particularly a migratory bird sanctuary, would be in charge. Toward the end of August, Gilbert (always the gentleman), assuming that Leopold's directorship was a certainty, told Jackson that he was "pleased" that Leopold would now have "a real opportunity to try out some of his ideas." He also confided to Jackson that he actually had not expected Leopold to accept the directorship since, as he understood it, Leopold had expressed that he was not interested in the position. But, Gilbert added, now that Leopold was in the position, "I personally feel that I should take no stand on any matter during the few days that remain." There would soon be a new AC. Leopold's feelings, Gilbert told Jackson, were now the major concern, not the feelings of the members of the old Arboretum Committee, including himself. Gilbert still assumed he would not be retained on the committee, let alone be named chair. He had prepared his final report for President Frank, in which, he graciously credited Jackson for his efforts, "seconded only by McCaffrey and Gallistel." And, indeed, the policymaking power had shifted on the AC; it was now mainly in the hands of Jackson, Mc-Caffrey, and Gallistel.[28]

But there were mixed feelings and considerable aggravation among the members of the AC over Leopold's appointment as director, particularly when it was learned that the August 4 recommendation to the Regents to appoint Leopold as director had not originated with the AC, since it had not met during the summer. The August recommendation, surprisingly, had come directly from the College of Agriculture. Specifically, the Regents had acted totally on the recommendation of Dean Chris L. Christensen, who, in 1931, had replaced Russell as dean of the College. Solely on Christensen's recommendation, the Regents had established the chair in game management, had appointed Leopold to the chair, and had also appointed him Arboretum director. The realization of the hasty action of Christensen and the Board of Regents led to serious misgivings and concern on the part of Gallistel and McCaffrey, in particular.

The College of Agriculture had overstepped its bounds. It had jurisdiction over the appointment of a professor of game management, but it was not delegated to make decisions directly affecting the Arboretum, especially the appointment of a director. And to make matters worse, Jackson, in his enthusiasm, had taken it upon himself, without authorization, to spread the news around Madison that Leopold had been appointed as the new Arboretum director. He also, as noted above, had leaked the story to Brayton, who finally published it in late August. For many of the members of the AC, the first time they learned that Leopold had been appointed director was when they read it in the newspaper. It was now left mainly to Gallistel to clear up the confusion. On August 28, Gallistel contacted Gilbert, who was on vacation. He told Gilbert that he had conferred with McCaffrey and that both of them were convinced that there had been a serious misunderstanding. However, "no real harm" had been done, and "the matter" could be worked out. There was particular concern over Longenecker's reaction to the sudden news of Leopold's appointment as director. Gallistel wrote to Gilbert:

If my memory serves me correctly, I believe the committee at one time expressed itself as being favorable to Leopold's appointment, although formal action may not have been taken. Later on the idea was developed of placing Leopold as Research Director and Longenecker as Resident Director. Jackson knew this of course.

Longenecker seems somewhat disturbed . . . but he is plodding along waiting the time when the committee can meet to get every one straightened out. Mr. McCaffrey and I are hoping that he can be given some additional compensation from the Alumni Research Foundation fund. . . .

We must remember that the Regents have gone on record placing the administration of the Arboretum in the hands of the committee. In an informal talk with Leopold I gathered that he understood that the College of Agriculture was not to have any jurisdiction over the project. . . . I think his attitude will be O.K. and he expressed high regard for Longenecker. Of course, the whole situation needs clarification, due largely to a very dynamic personality outside of the University, and as I see it, this clarification must wait for a committee meeting.[29]

The "dynamic personality outside of the University" was, of course, Jackson. The "clarification" to which Gallistel referred came on Saturday, September 23, 1933, when the Arboretum Executive Committee met and voted to ask the Regents to override the August 4 motion with a new one designating Leopold as research director and Longenecker as executive director. Significantly, it was Leopold who made the motion to have Longenecker designated as executive director. The earlier choice of resident director, it was decided, would be held in abeyance until such a time as the Arboretum had proper housing on the grounds.

And so, as it turned out, Leopold *was* the director of the Arboretum—for fifty days! Leopold's salary was covered by the endowment from WARF. He would function both as the professor of game management and as the Arboretum's research director. Longenecker at the time was working with Gallistel in Buildings and Grounds and occasionally teaching in the Department of Horticulture. He was also serving under Gallistel as a member of the committee charged with landscaping the UW grounds. His appointment as executive director provided no additional funds. The AC did go on record as requesting an increase in Longenecker's salary from the Department of Buildings and Grounds, adding the following qualification to the request—"if such be possible at this time."

On October 2, 1933, the Executive Committee of the Board of Regents, upon the recommendation of Gilbert, still chair of the AC, voted to approve the following:

> 1. That the title of Aldo Leopold be changed from Director of the Arboretum to Research Director of the Arboretum.
> 2. That G. William Longenecker of the Department of Horticulture and the Department of Buildings and Grounds be appointed Executive Director of the Arboretum.[30]

Gilbert reported to President Frank that Leopold was heartily in favor of dividing up the administrative responsibilities and was hopeful "that both the landscaping and the research work" could be carried on simultaneously.[31] Longenecker would be executive director, but he would also be in charge of landscape design and planting, as Gilbert had hoped. There would be a lot of tree planting beginning in the fall of 1933, which Longenecker would supervise. Leopold would immerse himself in research, particularly research on migratory game birds and the creation of habitats designed to encourage farmers to experiment with game cropping. Leopold was anxious to get to work. The second order of business at the Executive Committee meeting on September 23 was Leopold's presentation of his plan for attracting migratory geese and establishing feeding stations throughout the Arboretum. He requested funding for a caretaker and for feed. Leopold was wasting no time.[32]

And so, as it turned out, neither option a nor option b dominated plans for the future. And there was to be no bad blood between Leopold and Longenecker. They would work in harmony even though their expertise and their interests often moved in different directions. They also would become Wisconsin's first exemplars of a brand new ecological tradition developing out of the 1920s and 1930s of foresters and landscape architects restoring and rehabilitating natural areas in tandem.[33]

CHAPTER 4

❦

1934–1935

TWO DIRECTORS, DEDICATION,
CAMP MADISON, AND THE CCC

G. William (Bill) Longenecker and Aldo Leopold were different personalities with distinctly different professional styles. Longenecker worked within the limits of the Arboretum; Leopold divided his time almost equally between the Arboretum and areas of the state, such as Coon Valley, where his land management skills were needed and regularly tested. But in spite of their different specialties and professional commitments, they worked well together and kept the Arboretum moving forward. They both were also ecological designers, well suited and in step with international developments in the nascent science of ecology emerging out of the 1920s and 1930s. As Marcus Hall has observed in his study of the transatlantic history of environmental restoration, the developing interest in ecology in the 1930s posed "*new* challenges," especially to students of the *new* science of landscape architecture, who were attempting to design "with nature." And naturalistic gardens and parks had become fashionable internationally as early as the 1920s, particularly in the United States.

American naturalists, in particular, who were caught up in the movement, as Hall notes, "were becoming more attuned to the ways in which natural sciences could help them produce greater naturalness." Consequently, they often paired up with research scientists in other biotic disciplines in order to create more diverse natural areas, with, as Hall observes, "the city or university arboretum being a favorite place" for combining interests in science and aesthetics. To support his argument, he points to "some of the most famous" paired American park and arboretum designers. Included in the list are Charles Sargent and Frederick Law Olmsted Sr., Henry Cowles and Jens Jensen, Edith Roberts and Elsa Rehmann, and Aldo Leopold and John Nolen.

65

"Where the scientists focused on natural processes, the landscape architects were concerned with natural patterns."[1]

Hall is right about the last point. The landscape architects were mainly concerned with regional design and natural patterns, and so was John Nolen. But it was not John Nolen who paired up his talents in landscape design with Aldo Leopold. In fact, Leopold and Nolen never worked together on any project. Nolen died in Massachusetts in 1937. He may have met Leopold once, perhaps, at the Arboretum dedication in June 1934, but there is no extant evidence of any written correspondence between them. By 1937, when Nolen died, Leopold was only four years into his tenure as the UW professor of game management, but he had indeed partnered-up at the Arboretum with a very talented professor of landscape architecture—Bill Longenecker. Hall appears to have had no knowledge of Longenecker's role in the early design of the UW Arboretum. That is not surprising given what is now a long, disillusioning history of environmental historians and journalists who continue to credit Leopold almost solely with the early design of and early commitment to the restoration of Arboretum ecosystems, particularly the restoration of the Curtis Prairie, as well as to the creation of the Arboretum itself.

Leopold was important, of course, but he worked at the Arboretum in tandem with Longenecker. They were partners; they shared the directorship. The bifurcation that Hall observes in the case of Sargent and Olmsted and the others applied equally well to the bifurcation between Leopold, the forester-scientist, concerned mainly with natural processes (e.g., game cropping and creating habitat for migratory game birds) and Longenecker, the landscape architect, concerned mainly with design and natural aesthetic patterns (e.g., the combined aesthetic effect gained by planting magnolia trees and lilac bushes next to crabapple trees or visualizing how a conifer forest might look in the distance at the edge of a tall grass prairie). And Leopold and Longenecker also shared the assumption "that the combination of scientific and aesthetic insights would provide more precise ways to reinstate naturalness."[2] They worked well together, and the record shows that they worked as a team.

Leopold's and Longenecker's Arboretum Commitment

By the early 1930s, Leopold was a nationally recognized environmentalist, with a broad-based cultural commitment to the conservation movement globally and to the promotion of a relationship between people and nature built on long-range ecological principles. Much of his work from the mid 1930s until his death in 1948 was geared to the propagation of a "land ethic" that evolved mainly out of his concern for the disastrous and tragic condition of the nation's farmland in the 1930s and the greed and self-interest that largely accounted for the devastation. During his Arboretum years, from 1933 until 1948, he worked industriously but with modest success to turn the Arboretum into a viable, nationally recognized wildlife refuge. As noted in chapter 3, he was committed particularly during the early Arboretum years to sustaining and increasing the numbers and availability of migratory game birds. The effort was part of a much larger program backed by the federal government and the UW administration directed at selling the benefits of game cropping to Wisconsin farmers, who never did totally embrace the idea. His efforts at increasing the game bird population at the Arboretum and for the future also had limited success. However, he persisted, even as early as 1933, in promoting ecological reform, a vision of what "could be," both within and beyond the Arboretum.

One famous example from 1933, Leopold's first year as a UW professor, was the Coon Valley project, "An Adventure in Cooperative Conservation," undertaken in the Coon Valley watershed in Wisconsin's southwestern driftless area. The driftless area had been for decades highly prone to soil erosion. In 1933, Leopold was asked to work on a solution to the problem. The project, strongly supported by Hugh Hammond Bennett, director of Roosevelt's newly established U.S. Soil Erosion Service, promoted the construction of demonstration areas, the first in the nation, built specifically for the study of erosion control and land restoration. The project was an impressively successful cooperative effort on the part of Leopold, working as the extension adviser along with colleagues from the UW College of Agriculture, and local farmers who learned the value of contour plowing, crop rotation, keeping livestock away from steep slopes, and other preventive measures.[3]

Leopold's efforts in the Coon Valley to encourage farmers to create feeding stations or "food patches," as they came to be known, also met with considerable success. He claimed, with great satisfaction, that as a result of the new feeding system, the Coon Valley population of quail in 1934–35 doubled from 1933–34, and the population of pheasants quadrupled. If he could achieve desirable results, Leopold was ready to expend the effort—both inside or outside of the Arboretum.

Leopold, according to Ted Sperry, the Arboretum's first prairie ecologist, was a hard worker and easy to get along with. But he was "business," Sperry claimed; he "kept to business."[4] And his business, throughout his academic career, was wildlife preservation, conservation, land ethics, wildlife management, and soil restoration. He was a scientist—a research scientist—in the best sense of the term. Sperry credited him ultimately as having been the "brains" behind the vision, particularly for the restoration incentive at the Arboretum. He was "the driving force," Sperry observed, largely because, as a promoter of game management, he "held tight to the link between grasslands and game." Open prairies provided habitat. That fact essentially was the primary motive for Leopold's enthusiastic support of prairie restoration. Yet, as Sperry, who directed the prairie restoration fieldwork, noted, Leopold "never turned over a shovelful of earth; never worked with me. But we discussed the project," and it was done under his watchful eye as the Arboretum's director of research. He was Sperry's "boss."[5]

To Jackson, Leopold was "the *thinker*—the *planner*—and the *doer* . . . the human question mark, eager to discover the *why*—the *where*—the *when*—and the *how* of nature in all of her manifestations." Indeed, his focus was on developing a better understanding of natural processes. He was also, to Jackson, a man of "great character," a moralist who consistently advocated the need for increased harmony between people and the good earth.[6] To achieve that objective, he continued throughout his Arboretum years to insist on the need for more conservation education, both in the schools and on the land. The CCC enrollees who got to know Leopold in the late 1930s thought well of him. James Hendricks, a corporal serving with the CCC at the Arboretum's Camp Madison, recalled Leopold's visits to the barracks area to talk about archery with the enrollees. He would come at night, Hendricks recalled, with bows and arrows in order to teach them how to shoot an arrow. He also taught some of them how to fashion handmade bows. "Everybody liked him," Hendricks noted.[7]

Longenecker, on the other hand, was the "stay-at-home" executive director of the Arboretum. Most of his responsibilities were centered on day-to-day management in the immediate Arboretum area. The Arboretum was his "life's work," he admitted, and tree and shrub planting were his responsibilities. Almost all of the trees planted between the 1930s and the 1950s, including most of the white and red pines, were planted under his direction. He also staked out most of the locations within the Arboretum for the trees, the shrubs, and the bulk of the stone work. When work commenced on the prairie restoration in 1936, Longenecker chose the location, the area east of the Noe Woods with a wide range of trees in the background. Longenecker also served as the public relations connection for the Arboretum. He designed all of the stone shelters. He also designed and supervised the construction of most of the foot trails and ponds that went in during the 1930s, including Ho-nee-um Pond on the west end of Lake Wingra. He had a simple philosophy about trail building

Professor G. William Longenecker was Arboretum executive director from 1933 to 1966. Longenecker Gardens was originally planted under his supervision, implementing his own designs. (Wisconsin Historical Society, WHi-57704)

that matched his low-key personality. As he put it, "trails should take you in a pleasant sort of way to where you want to go, seeing the most there is to see on the way."[8]

It would be left to Longenecker to plant the lilacs and the trees, now world famous, that flower each spring in the horticultural garden that has borne his name since 1967. The collection includes over 200 different lilacs, 90 crabapples, and 40 viburnums. The first lilac went into the ground in March 1935, on Good Friday, according to Professor Ed Hasselkus, one of Longenecker's former graduate students, who, upon Longenecker's retirement in 1966, followed him as curator of Longenecker Gardens. In 1996, Hasselkus received the Arthur Hoyt Scott Garden and Horticulture Award from the Scott Arboretum at Swarthmore College for his contribution to the science of gardening. The planting of the first lilac in 1935 marked the beginning of the now famous display, designed for its aesthetic value but also as a scientific testing area for Wisconsin and, more generally, midwestern climatic conditions. The equally famous and colorful crabapple collection was begun in 1942, according to Hasselkus.[9] The Longenecker Gardens remain a visible living legacy that Longenecker left to the Arboretum, to the world, and to the abiding and loving care of Hasselkus, who retired in 1994 but who continues to work beside a cadre of loyal volunteers to keep the garden maintained and beautiful. The Longenecker Gardens also remind us of the extensive time and effort that Longenecker put into his work as executive director of the Arboretum and, one might say, as senior horticulturalist and landscape designer.

Upon Longenecker's death in February 1969, Grant Cottam, chairman of the Arboretum Committee (AC), who had taken courses from Longenecker, would say of his commitment to the Arboretum: "he was so closely related to the Arboretum . . . that it is not possible to consider any aspect of the Arboretum without considering Bill's contribution. The Horticultural Area . . . named in his honor . . . was probably his favorite area, but his influence was felt throughout all of the Arboretum."[10]

Twelve years younger than Leopold, Longenecker was born in 1899 in Neillsville, Wisconsin, but spent most of his early childhood in Utah and the Dakotas. His father was a Congregationalist minister. His mother, an avid gardener and plant collector, encouraged her young son in the collecting and study of plants. At age fourteen Longenecker returned to Wisconsin in legendary fashion—the family making the journey riding in a mountain wagon and on horseback. In 1924 he earned his bachelor of science degree in horticulture from the University of Wisconsin. After graduation, he went to work for a time in a nursery in Louisiana, Missouri. When he returned to Madison in 1926, he was offered a job in Buildings and Grounds by Albert Gallistel. He also entered graduate school about the same time to begin his master's work on a landscape architecture degree. His was the first M.A. in landscape architecture granted by the UW.

He was a master teacher and meticulous designer advocating the value of concentrated natural or ecological groupings in his landscape creations. The fact that the Arboretum is a collection of different regional plant types, yet not a hodge-podge, is the result of Longenecker's landscape design talents. He was also a skilled artist who first sketched his designs and then painted them in pastel watercolors. He liked to see his plantings in natural or native groupings blending from a distance into simple settings that suggested, more often than not, colorful still life paintings. Ted Sperry would say of Longenecker that, unlike Leopold, he was primarily into woods and forests. The one big reason why he approved of building a prairie, when the Arboretum finally got around to it, was because it was "low vegetation" and enhanced the perspective of being able to see the design of his beloved trees beyond the grasses. Consequently, the location of the first Arboretum prairie, which would eventually be called the Curtis Prairie, was purposely left up to him.[11]

Leopold's "Wildlife Management Plan"

In October 1933, Leopold, as the new Arboretum research director, was anxious to begin working. The immediate need was to establish feeding stations—"food patches"—throughout the Arboretum, and the food patches needed to be protected. To that end Leopold requested funds from the university to hire a caretaker, but funds were unavailable. The AC was obliged, consequently, to undertake a search and eventually managed to find a man named Walter F. Hanson, who agreed to assume the position sans salary and with the specific understanding that his appointment would not obligate the AC in any way as to the permanency of the position. In return for policing the Arboretum, watching for fires, routing poachers, removing wild-ranging dogs and feral cats, operating Leopold's feeding stations, and keeping the public from trespassing especially on Leopold's game bird research areas, Hanson was able to live rent-free in the small house built in 1932 as a temporary caretaker's residence. He was also given permission to cut wood for fuel, providing the location and the amount of wood needed to be cut were first approved by Longenecker.[12]

With a caretaker in place, Leopold began to plan a research procedure for the migratory game bird project that would consume most of his daily Arboretum working time. In October 1933, he circulated among the members of the AC his "University Arboretum Wild Life Management Plan." It was designed mainly to accommodate his needs as both Arboretum research director and professor of game management interested particularly in assessing the feasibility of increasing migratory game bird populations throughout the state. His time at the Arboretum now included planning and stocking food patches and creating protective cover, habitat, for his birds. His "Wild Life Management Plan" was mainly, as he acknowledged on the cover page, a "rough draft . . . offered to the Committee for preliminary criticism and discussion." Contained in the plan, however, is a section that speaks to what Leopold considered the general educational mission of the Arboretum. He suggests, as a start, that mimeographed leaflets called "Conservation Guideposts" be issued regularly from the Arboretum offices. The "Guideposts" would explain items of current ecological interest found in the Arboretum that could "be exhibited to the public without damage to the Arboretum." Leopold was persistently wary of the "public" having unlimited access to the Arboretum. He apparently would have preferred keeping it, or most of it, off limits to anyone but Arboretum personnel, researching faculty and students, and those who came to the Arboretum for instruction on game preservation and habitat development. On this point, not surprisingly, he clashed often during the CCC years with National Park Service (NPS) expectations. When he learned in 1940, during the waning CCC years, of the NPS plan to cut Arboretum research appointments (particularly the appointment of W. S. "Bill" Feeney, who Leopold believed was indispensable to his wildlife bird program), he sent a heated letter of protest to the NPS headquarters in Omaha. The appointments under consideration, "essential to the Arboretum idea," needed to be protected, he insisted. He went on, underscoring one of his consistent complaints about NPS policy generally:

> I think I detect in this proposal a basic conflict between two opposing viewpoints. The NPS has developed, for use in ordinary parks, a set of "canned" procedures which give priority to construction. For ordinary parks there is doubtless reason for this. When forcibly applied to the Arboretum, however, these procedures simply do not fit. They will force us to abandon . . . wildlife work . . . in favor of certain constructions which to my mind are of very minor importance. If the Arboretum becomes an ordinary park, my interest in it will cease. Any City Council can run an ordinary park; a university has no call to duplicate such functions.[13]

In Leopold's eyes, the Arboretum's primary mission was not and never should be recreational. For Leopold, its primary mission was pedagogical—scientific wildlife research. He was adamant that the Arboretum was not a park. His concern with keeping the public, particularly intrusive trespassers and poachers, at a distance was also closely linked to a sentiment that in the 1930s had become a hallmark of his developing land ethic.

Significantly, his benchmark essay titled "The Conservation Ethic," parts of which were later worked into "The Land Ethic" in *A Sand County Almanac*, appeared in the *Journal of Forestry* in October 1933, the same month the Arboretum "Wild Life Management Plan" was distributed. The essay marked a distinct turning point in Leopold's feelings about land-use ills and public priorities. As Marybeth Lorbiecki has observed, "In this address, Leopold made the jump publicly from speaking of society's norms and laws to the values of a culture and the choices of its individuals; from policies for public lands to the moral use of private lands; from the economics of self-interest to land-based, community-centered ethics."[14] In the essay, he registers his impatience particularly with legislation that is "strictly economic, entailing privileges but not obligations." Conservation, he argues, should be as much about educating and improving humanity as it is about restoring and protecting the land. "Only recently," he wrote, "has research made it clear that the implements for restoration lie not in the legislature, but in the farmer's toolshed. Barbed wire and brains are doing what laws alone failed to do."[15]

In his speech titled "Conservation Economics," a companion piece to "Conservation Ethics," delivered to the Taylor-Hibbard Economics Club at the University of Wisconsin in 1934, he leveled some tough talk at those who permitted the land to be exploited and abused—again—in the name of public appeasement. "Conservation Economics" was one of his first public endorsements of New Deal conservation and of the role of individuals and government in efforts to correct the problem of land erosion.[16] The "dust bowl" did not have to happen; it was not an act of God, Leopold notes, but of economic exploitation and civic neglect: "The disease of erosion is a leprosy of the land." He added: "I see no more reason for running a national or state park to please the mob than a public art gallery or a public university. A slum is a slum, whether in the Bowery or in Yellowstone. . . . It is inconceivable to me that the 'leisure for all' revealed to us in Mr. Hoover's dream can be spent mainly, or even in large part, on public recreation grounds."[17] Very tough talk, indeed!

Sample entries for a "Conservation Guideposts" issue that he included in his "Wild Life Management Plan," hence, were never designed with the general public in mind. They were directed at a select audience of farmers and conservationists that he hoped to interest in the possibility of game cropping and the subsequent sale of hunting privileges. "Our biggest task in the arboretum," he challenged, is to build up "a food supply for . . . game birds and mammals." The Arboretum would show people "how game preservation [could] be conducted on a large scale."[18] His sample "Guideposts" leaflet was loaded with game cropping advice about what worked in the Arboretum that might also work on farms, such as the recommendation for a "Pheasant Food Patch" that could be implemented by simply "leaving a few rows of uncut corn along an ungrazed marsh" or the recommendation that "every farm should plant a few mulberry trees in open situations to furnish summer berries for birds."

Leopold's 1933 wildlife plan was ambitious and farsighted and anticipated many experimental developments on Arboretum lands during the 1930s and 1940s including the creation of the Teal Pond and the plant communities that surround it. The plan also foreshadowed some of the remarks Leopold would make in his speech at the Arboretum's official dedication on June 17, 1934.

Arboretum Dedication, June 17, 1934

Aldo Leopold was on the program for the dedication as the fifth of seven speakers scheduled to address an audience of approximately 150 invited guests gathered together at 9:30 that bright Sunday morning in June in the Nelson barn. They were there to dedicate what was now officially labeled as the "University of Wisconsin Arboretum, Wild Life Refuge and Experimental Forest Preserve." An eighth speaker, Chief Albert Yellow Thunder, a member of the Wisconsin Ho-Chunk tribe and also a Native American historian and lecturer, although not on the program, was also invited to speak that morning by Edward Gilbert, who served as master of ceremonies. The addresses would have to be brief, Gilbert advised—kept to a limit of twenty minutes per speech. He had promised John Nolen, the featured speaker who would speak last, that they would be through before noon. Nolen had to catch a train to Chicago at one o'clock.

Nolen, "the godfather of . . . Greater Madison," was finally returning to Madison to address city officials on Madison's future and to attend the dedication of the University Arboretum that he had proposed twenty-five years earlier. His Madison visit was two-tiered. He spoke to the Commerce Association and other city officials at a dinner on Thursday, June 14, at Madison's Loraine Hotel. For the dinner he chose the topic "The City of the Future." Thursday morning Nolen toured the city with Jackson and Mayor James R. Law. That evening, his talk centered on suggestions for making Madison into the city that he had envisioned twenty-five years earlier. He had high praise for Madison's efforts to expand its parks and playgrounds. But he was critical of what he deemed the haphazard fashion with which the city had progressed. He had expected more. The city looked different, but "I cannot say," he added, that it looked "altogether better." He had expected to see State Street widened, and he was critical of the ubiquitous presence of railroad tracks. His address at the Arboretum dedication on Sunday, June 17, was more upbeat.

On that Sunday morning the heavy-beamed old barn was flag draped. The audience sat at tables running the length of the barn eating breakfast while listening to the speakers. The barn was crowded and musty. Outside, in the open air, students from the university prepared breakfast over a charcoal fire.

One young man in attendance, Harold Tarkow, a member of the university's 1934 graduating class, recalled crossing the Mills Street Bridge over Murphy's Creek on the eastern perimeter of the Arboretum lands and walking the old rutted dirt road from Lake Forest west to get to the event.[19] Most of the others in attendance had come by cars that were parked in line on the road next to the barn.

Speakers for the Regents, the university, the state, and the Arboretum, in that order, were to be recognized first. George W. Mead, a member of the Board of Regents, speaking for the Regents, and Ralph M. Immell, directing commissioner of state conservation, speaking for the state, pledged the support of the Regents and the state in the promotion of the new Arboretum. John Callahan, Wisconsin state superintendent of education, representing the university, read a message from President Glenn Frank that envisioned "many thousands of tired Americans" soon to "turn their backs upon the stress and strain of the year's work" to go "to the woods for the healing medicine of rest and recreation." If the trees had tongues, Frank claimed, he was sure they would "turn all these tired vacationists into ardent apostles of reforestation." He urged the need to "administer and advance . . . with vision" the interests of the new Arboretum.[20]

Longenecker spoke fourth and, appropriately, as executive director, he spoke for the Arboretum and of plans for its development. According to the *Wisconsin State Journal* account, he "explained the present extension of the arboretum, and the various growths which had been promoted in it." He also acknowledged, as a harbinger of things to come, that the Arboretum management had left "a prairie . . . in the center as a coordinating unit."[21] And then it was Leopold's turn. He spoke for the Arboretum also, but as the director of research, he spoke specifically for its designation as "a Wildlife Refuge."

"What is the Arboretum?" was the title of his speech. Shortly after the dedication, Leopold sent a typed copy of a late edited draft (with lines crossed out and a few insertions) to Jackson with the following attached note: "Dear Colonel—This is a popularized version of the speech I gave at the dedication. Should you have any use for printed copies, you might want to give it to the papers. It will be some time before I can get the other one out in printable form." Leopold would revise the speech a number of times. Lines cited below are taken from the edited copy he sent Jackson. Leopold began his address with a brief summary of how one might describe an arboretum. He noted that, ordinarily, arboretums were viewed as "collections of trees." Some arboretums also served as outdoor libraries, exhibiting a catalog of "horticultural varieties," everything from apples to lilacs to roses. Of late, more advanced arboretums, he noted, often combined "natural associations"—"a Douglas fir forest of the Northwest," for instance, exhibiting all the trees, ferns, and shrubs that grow in one. We want all these things in our Arboretum, Leopold maintained, but he also felt that the UW Arboretum should aspire to be something "new and different." And perhaps, as a result, he added, when we finish we should not call it an arboretum at all. Whether our objective is worthy or not, he told the audience, "I will . . . leave you to judge." In essence, he went on, our idea "is to re-construct . . . a sample of original Wisconsin—a sample of what Dane County looked like when our ancestors arrived here during the 1840s." It will take time to do this, he stated, at least fifty years, but it will not be done for "amusement." It will be done for "research." He might have added here, in keeping with his developing sentiments at the time on the need for a viable "land ethic," that the work would not be done for either appeasement or the economics of self-interest."[22]

"I want to try and picture today," he continued, why it is important to the welfare of the state to know what the land in the 1840s was like. In the 1840s what one would have seen here were "oak-openings"—"open orchard-like stands of oaks, interspersed with copses of shrubs, and the profusion of prairie grasses and flowers which grew between." Along with native grasses and oak-openings, the land was populated with sharp-tailed grouse, copses containing partridges and ruffed grouse, and an abundance of elk and deer. What is now the Wingra Marsh was, in 1840, a tamarack forest with an undergrowth of sphagnum moss and orchids. Lake Wingra itself was home to an abundance of waterfowl; wild rice beds lined its shore.[23]

And so, Leopold asked, why should we try to discover what Wisconsin of 1840 looked like? Why "dig up these ecological graves?" His answer: because in the process we are made aware of the "unintentional and unnecessary changes which threaten to undermine the future capacity of the soil to support our civilization." The erosion of topsoil, the result of "too much wheat and too many cattle," had carried "the best parts of southwestern Wisconsin to the Gulf of Mexico." The loss must be repaired. Modern tools—meaning, modern engineering techniques—have become so powerful, he added, that "their use . . . has unexpected consequences." We are tearing things down as fast as we build things up. The remark echoed his stark warning in "The Conservation Ethic," a hard lesson for the dust bowl era, that if people continued to "*rebuild* the earth," without a plan, "without understanding . . . the increasingly coarse and powerful tools which science" had placed at their disposal, they would be doomed to continue in the ruinous process of "remodeling the Alhambra with a steam-shovel."[24]

He concluded his presentation by pledging that the new UW Arboretum would attempt to provide "a reconstructed sample of old Wisconsin" that would "serve as a bench mark, a starting point, in the long and laborious job of building a permanent and mutually beneficial relationship between civilized men and a civilized landscape." His message that morning resonated. The next day, the *Wisconsin State Journal* reported that Leopold had addressed the need for "a new science which realizes the interdependence between men, plants, and animals" and that he had underscored the fallacy of believing "that when the destructive forces are withdrawn the healing process begins automatically."[25]

Leopold was followed on the podium by Harry L. Russell, then the director of WARF, who delivered a brief five-minute address on the value of scientific research. Russell was followed by Chief Albert Yellow Thunder, a Ho-Chunk chief, looking stately and dignified in a trailing feather headdress and a red-beaded tunic. Yellow Thunder's grandfather, also named Yellow Thunder (Wakunzagah), was a chief who fought in the Blackhawk wars of 1832. To date there is no record or copy of the exact remarks of Yellow Thunder, who did not speak from notes or from a prepared text. All that remains of his speech are short paragraphs quoted in the June 18 *Wisconsin State Journal* and *Capital Times* accounts of the dedication. In actuality, since Gilbert had urged the chief to be brief because the program was behind schedule, and since Russell had limited his remarks to five minutes, these quoted paragraphs in all likelihood comprise much of what the chief had time to say. Indeed, his speech was brief, but it was also plaintive and powerful. He told the audience: "Like the trees, we are a dying race today. Some of you may see the day when an Indian is considered a relic. The only monument we leave is America, and its natural forests, and the streams which my people traveled. Take care of those natural beauties."[26]

Gilbert wrote to Yellow Thunder on June 18, the day after the ceremony. He thanked him for his "splendid talk," apologized for the limited time he had been given, and promised that the next time he was in the Wisconsin Dells, where Yellow Thunder lived, he would look him up and thank him properly. He concluded his letter with the observation that he hoped that Yellow Thunder's grim prophecy regarding his race "may not come true."[27]

The final speaker that morning was John Nolen, who had been asked to address "the Future." Nolen began by elaborating on his "love of Madison" and its people and his desire to share with his audience his belief in "the high importance" of arboretums "in the solution of grave and unprecedented problems in our national life." He quoted John Muir: "The world needs the woods, and is beginning to come to them." And reminiscent of Olbrich's memorable 1928 address to the Madison Rotary, he quoted Thoreau on "wildness" as "the preservation of the world." He also told the audience that he was there to talk about the unlimited possibilities for the "future growth and development of the city as the capital of the state, and the home of its university." He cautioned, however, that one needed to know the past in order to understand the present, and "enough of the present to forecast the opportunities of the future." He quoted passages from his 1911 report on the significance of the university in future plans for the city. And he held up the example of Harvard's Arnold Arboretum as the model for future direction.[28]

For Nolen, the "achievement" of the Arnold Arboretum was largely the result of the foresight of its talented director, Charles Sprague Sargent, appointed in 1873. "Sargent was not only a maker of farsighted plans," Nolen stated, "but he gave more than fifty years to carrying them out himself, overcoming all obstacles. He was of the opinion that at least a century should be used in research." Sargent believed "that the observation and study of trees, to be of highest value, must often extend over . . . 300 years." He succeeded in establishing a policy that guaranteed the Arboretum's "continuity for a thousand years, then to be renewed for another thousand years, and so on forever." The lesson Nolen urged his listeners to take with them that morning was that "our failure" not "to plan ahead or to make our program for public works no longer than five or ten years seems petty." The message was clear. Planning ahead effectively for the Arboretum meant planning ahead for centuries, not decades. The University of Wisconsin Arboretum had a profound model to follow but ample time with which to develop its forests, prairies, wetlands, and variety of plant communities and ample time also with which to develop a global identity as a research center for land reclamation and ecologically based techniques of land restoration.[29]

Gilbert had promised to keep the program moving so that Nolen would not miss his train to Chicago. Fortunately, the event concluded soon enough and Nolen had ample time to catch his train. That afternoon Nolen bade goodbye to Madison friends, old and new. He never would return to Madison. However, Jackson stayed in touch with him. Nolen asked to be kept up-to-date with Arboretum developments. On July 24, 1934, Jackson told Nolen that he still wanted Nolen's input on long-term Arboretum planning. He spoke of possible federal money that would pay for Nolen's services. In August 1934, Nolen expressed his approval of the "practical public spirit with which" Jackson and the others were proceeding. Through 1935, Jackson wrote regularly, providing Nolen with progress reports. He also kept Nolen apprised of the list of projects for Madison's development that Mayor Law was sending to the government, many of which were in line with Nolen's recommendations. Jackson continued to hope that Nolen might return to Madison to consult and sit in on meetings of the Arboretum General and Advisory committees. He told Nolen on April 30, 1935, "You well know we are still most hopeful of having you come back to Madison and finish up at least for the time being what you began twenty-six years ago." He concluded, "Madison is Nolen-minded today more than it has been since I returned fifteen years ago." Nolen must have been pleased to hear that.[30]

Toward the end of 1935 and into early 1936, Jackson asked Nolen for his advice on a proposed memorial to Michael Olbrich. Jackson knew that Nolen and Olbrich had corresponded and that Nolen would be interested in efforts to honor Olbrich's contributions. He sent Nolen a watercolor sketch of a proposed memorial entrance to the Arboretum and asked for his response. Nolen replied that he liked the "character" of the sketch. Jackson shared Nolen's response with members of the AC. In less than a year, on February 18, 1937, Nolen died at the age of sixty-seven. Although Nolen never met Michael Olbrich, somehow it seems fitting that what appears to have been Nolen's last substantial correspondence with UW Arboretum officials was his supportive endorsement of the design of the memorial to Olbrich's memory. An era was passing.

Leopold's Wildlife Refuge

Efforts to get Leopold his government-funded migratory game bird refuge were ongoing during those early Arboretum years. As early as November 1933, Gilbert contacted Paul Redington, chief of the U.S. Bureau of Biological Survey (BBS). Gilbert was following up on earlier correspondence that Jackson had with Redington about the possibility of the BBS establishing the Arboretum as a federal migratory bird refuge and then subsequently designating the University of Wisconsin as its headquarters for midwestern research. Gilbert told Redington that the Arboretum had sufficient funds for the development of the wildlife refuge and that a goose bar would be installed in the winter (1933–34), that duck ponds were being excavated, and that range improvements for upland game were underway. The AC, he added, desires to "make it financially possible for you to work here. It hardly seems likely that the large expansion of waterfowl research needed to meet the present emergency can all be administered from Washington."[31] On February 19, 1934, Redington responded. It was another disappointment. The BBS turned down the request.

On March 17 Gilbert wrote to Ralph Immell, directing commissioner, Wisconsin Conservation Department (WCD), airing his regrets and proposing that, in lieu of the federal rejection, the WCD, the UW scientists, and the Arboretum staff combine efforts, reinforce each other on a more congenial *state* level, and get on with the business of establishing Madison as a premier conservation research center. With that prospect in mind, Gilbert proposed that the Arboretum Executive Committee "sit in" at one of the WCD's early meetings in order to discuss mutual problems and ways to work together. He even offered to give the commissioners a guided Arboretum tour. They could "look it over," he suggested, and then we will "have lunch out on the grounds."[32]

Jackson, on the other hand, undaunted as usual, decided to test Redington's and the BBS's resolve. In a March letter, he told Redington that turning down the request was a mistake and that if he and the survey were to become personally familiar with Arboretum research, they would surely change their minds. We're on our way, he added, "to becoming one of the real biological research centers of America." He concluded by calling Redington's attention to the fact that since Leopold was now holding the chair of game management at the University of Wisconsin, that knowledge alone should encourage the BBS to change its mind.[33] Little did Jackson know, however, that pinning the conclusion to his argument on Leopold's position at the university likely had the effect of waving a red flag in Redington's face. Jackson, as he acknowledged to Congressman George W. Blanchard of Wisconsin's first district, was aware that the BBS was in trouble, that conservationists around the country were upset with Redington and the survey, owing mainly to its inability to deal with the dwindling numbers of waterfowl. What Jackson didn't know was that, by December 1933, Leopold, among others, was advocating nationally for a radical overhaul of the BBS and a radical change in leadership. In December 1933, Secretary of Agriculture Henry Wallace put together a special three-man committee to look into the availability of new initiatives in wildlife management. The action was an implicit judgment of *no confidence* in the work of Redington and the BBS.

The three-man government committee headed by Thomas Beck, a personal friend of Roosevelt and chairman of the Connecticut State Board of Fisheries and Game, would be known as the Beck Committee. By the end of 1933, Leopold was one of its three members. The other was Jay "Ding" Darling, prominent conservationist cartoonist and newspaper icon. The committee continued its scrutiny and criticism of the BBS. Beck was in favor of completely abolishing the survey. Leopold and Darling were not. Eventually, the pressure wore Redington down, and he resigned. Darling was named the new director.[34] There is no concrete evidence that Redington turned down the Arboretum proposal for federal status as a migratory bird refuge in 1933 because of the attacks on him and the BBS, including, in particular, Leopold's. But it stands to reason that holding Leopold up to Redington, as Jackson in his innocence did, as a reason for supporting the request had little effect in convincing Redington to change his mind and may well have influenced the judgment against the Arboretum in the first place. But Jackson, undeterred, was not discouraged.

In April 1934, Jackson was still at it. This time he went after funding from the U.S. Forest Service (USFS). He told George W. Mead, of the Board of Regents, on April 23, that he had sent a letter asking for government assistance to Chief F. A. "Gus" Silcox, who had been appointed director of the USFS in 1933. Silcox replied and asked for more information. "The next move is up to us," Jackson told Mead. What Jackson wanted was nothing less than a federal appropriation that would enable the AC to purchase all of the land east of the Arboretum then owned by Bernie Chapman and the Lake Forest Company. The money would enable the committee to buy practically all of the remaining acreage blocked out for the UW Arboretum and Wild Life Refuge, Jackson noted.

At the time, the USFS, under Silcox's leadership, was systematically adding millions of acres of forest land to U.S. holdings. Funding from the Forest Service would necessitate designating the acreage as a forest experiment station. Nevertheless, and in spite of what it was called, the Arboretum would have the acreage and could use it as a resource for other forms of conservation research including finally Leopold's migratory bird refuge and his wildlife management program. Jackson told Mead that the establishment of a forest experiment station in the Arboretum might necessitate additional commitments but that Leopold understood the situation and its ramifications and would be glad to confer with him about them. Jackson also told Mead that there was a possibility that the Arboretum might receive some kind of funding assistance in the near future from Wisconsin's federal allotment.[35]

Unfortunately, Jackson's lack of success pinning down federal government assistance and recognition of the Arboretum as a migratory bird refuge continued. The USFS never did allocate any money for the purchase of Lake Forest land. And developments at the Arboretum after April and into June 1934 began to shift the AC's focus away from the hope for federal government recognition and funding toward a brand new direction that few would have predicted prior to June 1934. To begin with, on June 19, 1934, two days after the June 17 dedication, Leopold finally did get his state government-sanctioned game refuge. The AC's proposal sent by Gilbert to the Wisconsin Conservation Department (WCD) had paid off. Perhaps it was "sitting in" on the WCD's early meetings that did it; or perhaps it was Gilbert's invitation to give the commissioners an Arboretum guided tour with lunch on the grounds. Regardless, on June 19 a *Wisconsin State Journal* story made the following announcement: "The marshland area adjoining Lake Wingra comprising the best feeding grounds for ducks was designated today by the state Conservation Commission as the Lake Wingra Game Refuge." The designation took effect on August 1, 1934.

The new game refuge was designated specifically for the protection of waterfowl. The area, beginning at the Wingra Spring and extending northwest and then south along the Lake Wingra shore, consisted of 80 acres.[36] That 80 acres added to the 90-acre plot known as the Island and the 190 acres from the East Marsh (i.e., Gardner Marsh), when acquired in 1936, would give Leopold a total of 360 acres of marshland in which to conduct his migratory game bird research.

On August 25, 1936, Leopold and the rest of the Arboretum staff were taught a lesson about "being careful what you ask for." In July Leopold had applied to the WCD for a permit to shoot pheasants on Arboretum grounds in connection with a study of the palatability of grains they had been eating. On August 25 he and his staff were informed verbally that permission could not be granted owing to a restriction on shooting wildlife on a state game refuge. Leopold appealed. He argued that actually the request for permission to shoot pheasants had a two-fold purpose. One was to study the palatability of grains, but the other and more pressing motive was to reduce the increasing pheasant population that was, for the first time, threatening the survival of quail and other Arboretum game birds. He wished to remove roughly fifty pheasants for the palatability study and for the population reduction. He added that once the winter snow arrived, the birds would be trapped. Prior to the snow, however, the only efficient way to take them was to shoot them. The WCD, however, held to its decision.

In an August 27 letter of appeal to H. W. MacKenzie, director of the WCD, Leopold posed the critical question: as a result of the restriction, could the Arboretum, simultaneously, be both a game refuge and a research area? He wrote: "The whole Arboretum is developed on the idea of its use for research," and if it cannot get permission for the occasional "killing of specimen material," then "we had better write off the Arboretum" efforts to conduct wildlife research.[37]

The predicament touched on a much broader issue that was to raise a question about the role of the Arboretum as a wildlife game refuge.

The question was also directly linked to a professional challenge that Leopold was undergoing personally in the late 1930s. In August 1935, he had traveled to Germany as a Carl Schurz fellow. The Schurz Foundation had invited Leopold and five other foresters to study and evaluate forestry methods then current in central Europe. Leopold, specifically, was to study the relationship between game management and forestry. The foundation grant covered all living and traveling expenses. Leopold couldn't pass it up. He was in Europe from late July until late November 1935. While he was away, as Curt Meine points out, nothing less than a national revolution in wildlife conservation had taken place. New conservation initiatives were drastically altering American thinking about the relationship to land use as the premise upon which all other conservation programs must depend.[38] The shift in thinking—Leopold's thinking also—would require him upon his return in November 1935 and throughout 1936 to readjust his perspective on the mission of the Arboretum, on what Leopold had called "the Arboretum idea," and particularly on the nature of *game management* in that mission.[39]

Throughout the latter half of 1935 and into 1936, Leopold's interest in game management began to shift. Now largely from necessity and the lack of "natural" or "unspoiled areas" for research within the Arboretum, he moved from a preoccupation strictly with the management of game birds to a more encompassing, much broader interest in wildlife management overall, including non-game wildlife studied in a series of sub-stations scattered throughout the state.[40] Not surprisingly, it was also in 1935 that Leopold purchased the "Shack," the family weekend retreat in the Baraboo Hills of south-central Wisconsin, and set out, stimulated by projected plans for the Arboretum, to experiment with the restoration of his own "unspoiled" private piece of land. By 1940, the shift in his ecological focus would account for his official change in title from professor of game management to professor and chair of the new wildlife management department.

There were also other significant changes occurring in Leopold's professional life and at the Arboretum during 1935. When Leopold left Madison in late July 1935, the Arboretum was called Camp Arboretum, and approximately two hundred transient Wisconsin Emergency Relief Administration (WERA) laborers who had been working there since July 1934 were preparing to leave. When he returned from Germany three months later in late November 1935, he returned to a different setting, an Arboretum that was now called Camp

Madison, with a contingent of two hundred uniformed Civilian Conservation Corps (CCC) enrollees who had arrived in mid-August, a contingent
of U.S. Army officers that accompanied them, and a National Park Service
(NPS) crew of foremen and heavy equipment operators who would supervise
their work.

The Arrival of the CCC

The announcement of the new state-sanctioned Lake Wingra Game Refuge
in June 1934 was good news, but the news got even better for the AC that
June. Less than two weeks after the announcement, Gallistel notified Gilbert
that behind-the-scenes efforts had paid off, and the federal government was
proposing to build in the Arboretum a "transient camp," under the auspices of
WERA. The timing was perfect. The Arboretum had lost its labor pool when
the Civil Works Administration (CWA), which had been funding the laborers
on public relief, was disbanded by Roosevelt in spring 1934. But now, only a
few short months later, another and better option presented itself. Unlike the
CWA laborers whose job assignments varied from day to day and who had to
be trucked in, the WERA "transients" would live in the camp under regular
supervision and would work regular shifts.

On June 28, 1934, shortly after receiving news of the transient camp proposal, an excited Albert Gallistel sent the following telegram to Gilbert,
vacationing in his summer home in Hayward, Wisconsin: "To E. M. Gil-
bert/ Government proposes to build transient camp at Arboretum for 200
men/ Administration building . . . / Mess Hall . . . / 8 Barracks . . . / Recreation Hall . . . / Amphitheatre/ Design subject to our approval/ We have use of
men when desired/ Probable duration of camp 2 years/ Buildings to become
our property then."[41]

Gilbert was thrilled; the AC was supportive. Not everyone was elated, however. Longenecker had reservations. Leopold suspended his judgment. His experience with government-sponsored labor, the CCC in the Southwest, had
not been totally satisfying. The effort to employ the enrollees in anything
productive had been full of disconnects; there were crews working at cross
purposes and crews sitting idle waiting for someone to find work for them.
And New Deal conservation programs tended to be narrowly focused on
single track issues, Leopold thought.[42] John Nolen also had cautioned Jackson
that federal transient camps he had visited in the South were disillusioning.
He spoke of a lack of initiative among the transients and a recurrence of
drunkenness.[43]

But federal assistance for the Arboretum, even with reservations, now meant that funding would be available and, even more important, laborers would be available, and work could begin in earnest on reclaiming and restoring the Arboretum land and building "that road." The Arboretum for the time being would be known as Camp Arboretum. The emphasis on the Arboretum as a field research station—as Leopold's wildlife refuge and migratory bird sanctuary—was about to shift to include a broader and more encompassing challenge involving the stewardship and, at times, restoration of pasturelands, prairies, grazed woodlots, forests, ponds, nursery gardens, and fens as well as Leopold's marshes. There were even plans in the offing for serious stone quarrying and the possibility of building ornamental stone gates and walls around the Arboretum, none of which Leopold or anyone else had envisioned as a possibility a year earlier.

WERA officials proposed an impressive work list that involved the completion of the main road through the Arboretum, the excavation of ponds, the stock piling of marl material, the construction of new spillways, the construction of an island in Lake Wingra for feeding birds, the rental of large trucks and other heavy equipment for use in planting trees, and the construction of foot paths. There was even a proposal for hauling dirt excavated from the pond sites to playgrounds east of the Arboretum's entrance. The list was promising and ambitious, but the first major challenge was to clean up the Arboretum. To that end, the "transients"—men who initially worked in a labor pool connected to a downtown shelter, a refuge center—were used, and they needed close supervision.

Enter Harold Madden, who for the next six years would be one of the most valuable members of the Arboretum staff. From August 1935 when the CCC arrived until December 1941, Madden, whose background was in construction, was camp superintendent in charge of all of the CCC work details. The year before, in 1934, he was the work supervisor for the downtown Madison refuge center known as the Transfer Building. The Transfer Building, located at the corner of Wilson and King Streets, was close to the railroad yards that accommodated the freight trains that brought most of the transients to town. Most were "boxcar Willies," hoboes, displaced by the Depression and riding the rails looking for something better or simply "someplace else." The Transfer Building housed anywhere from five to eight hundred transients, who received room and board and a hot meal. But if they wanted to stay longer, they had to join the labor pool. As the work supervisor, Madden transported men from the labor pool to jobs around the southern part of the state.[44]

In late July 1934, slightly more than a month after the Arboretum's dedication, a contingent of transient laborers, under Madden's supervision, set up tents in the Arboretum and started to clear out the old farm and pasture land. By October 1934, ten wooden barracks were ready for occupancy, just in time for winter, and the men moved in. The barracks had been constructed by local union carpenters with the help of the transient laborers. The camp's design was drafted by Gallistel, who was an architect. All construction costs were paid for by WERA. Camp Arboretum was to house up to 350 transients for a two-year period. The transients were paid one to three dollars a week, depending on their work assignments. Madden's salary was covered by WERA. Once again the university got a deal, as Camp Arboretum cost it nothing. But high hopes for the camp did not work out. To begin with, WERA had a problem securing funds from the federal government for any costs beyond what their monthly budget could afford. Hence, for the completion of the controversial road through the Arboretum, their resources were so limited that renting a truck necessary for the work was not possible. Almost half of the work projects planned for the camp required heavy equipment and other material that WERA, which was being phased out federally, could not afford. The result was that the transient laborers were not working on the projects that had been planned.[45]

There were also problems with the laborers, who were, in fact, "transients" and could not be counted on to stay around long enough to see a job completed. And as John Nolen had warned Jackson, they were not very motivated. Their first task at the Arboretum had been a general cleanup. Yet when an advance party from the 2670th Company of the CCC arrived in early August 1935 to get the camp ready for the mid-August arrival of the company's main body, they were first obliged to take care of what the WERA transients were already supposed to have cleaned up. John J. Wawrzaszek, company supply sergeant from 1935 to 1939, recalled cleaning up litter, mostly whiskey and beer bottles. He also recalled holes in some of the barracks walls and mattresses infested with bedbugs and lice.[46]

Within six months, by the end of 1934, plans were underway to replace the transients with a CCC company. The CCC, a work force of unemployed and unmarried young men between the ages of eighteen and twenty-five, had been in existence since March 31, 1933, when Congress authorized Roosevelt's request for the formation of the corps. The CCC was a natural choice for the Arboretum, since the primary mission of the CCC was geared to conservation and reforestation projects. It appears to have been Gallistel who first conceived of the possibility of turning the camp over to the CCC under the supervision of the National Park Service and the U.S. Army.

As the story goes, Gallistel got the idea while talking with an NPS representative and Harry Sauthoff, Wisconsin's 2nd District congressman, during a walk around the Arboretum.[47] The date of the walk is unclear, but it must have occurred during or prior to early January 1935, because on January 14, 1935, Gilbert, speaking for the Arboretum Executive Committee, which included himself, Gallistel, Longenecker, Leopold, and McCaffrey, sent a letter to inform the Board of Regents of a motion passed by the Executive Committee recommending that the board "consider the feasibility of securing the establishment of a Park and Forestry Service Civilian Conservation Camp in the vicinity of the Arboretum." The letter also addressed two major complaints about WERA and the transient camp: the poor showing by the transient laborers and the inability of WERA to adequately fund project needs.

By contrast, the letter continued, men in the CCC were "physically fit" and remained in camp long enough to "work on long-time projects," with "trained men who fully understand what we desire to accomplish." Also, CCC camps, unlike the WERA camps, had resources for materials and access to heavy machinery.[48] In March 1935, Harold Madden was officially hired by the NPS as the new camp superintendent. Madden also saw to it that some of the laborers who had worked in the transient camp, those with essential skills, especially stone cutters, could stay on as NPS employees. The CCC enrollees called these skilled workers LEMs, or Locally Experienced Men. The LEMs worked for an hourly wage. The CCC enrollees worked for a dollar a day—a welcome wage during the Depression. They earned thirty dollars a month, twenty-five of which went home to their families.

By July 18, 1935, the transients began vacating Camp Arboretum.[49] It was also in late July that Leopold left for Germany. Before he left he sent a well-thought-out but skeptical letter of proxy to Gilbert alerting him about potential possible conflicts with the new camp based on his past experience with the CCC in Arizona and New Mexico:

> There appear to be difficulties in the selection of overhead personnel for the Arboretum CCC camp. It is possible that during my absence a situation might arise in which the University might have to accept unsatisfactory personnel to execute technical work on the Arboretum or else forgo the services of the camp. This letter is my proxy in the event such a question should arise.
>
> The camp was established with the understanding that there was technical work to be executed and that the overhead for such work had to meet more than usually exacting specifications. . . . Should it prove to be otherwise, and should there be further evidences of political influence in selections, then I think the

University would be morally obligated to renounce its connection . . . with the camp. As a member of the committee it is my opinion that we should give up the camp rather than to assume responsibility for a lot of work . . . which must be competently executed or not at all.[50]

Leopold's skepticism was based on personal experience, an experience that no one else closely connected with the Arboretum shared. Consequently, there is no evidence that I can find that indicates that anyone else close to the Arboretum felt the way he did. In time, Leopold appreciated the help of the enrollees, especially with his migratory bird projects, but his disaffection for politics generally continued to color his attitude toward the danger of political or bureaucratic cronyism contaminating CCC supervision and ultimately its conservation programs. He was particularly worried that the Arboretum's ecological research mission would be sacrificed to the more immediate NPS engineering mindset accustomed to attaching priority to outdoor construction projects, log picnic shelters, campground facilities, rest rooms, and other park-like recreational buildings.

The main body of the 2670th CCC Company arrived at Camp Madison by truck convoy from Camp Honey Creek in West Allis, Wisconsin, on August 16, 1935. Close to 50 additional enrollees arrived a few days later from Racine and Milwaukee. The full company complement eventually included 180 plus enrollees and approximately 20 Army Reserve officers and noncommissioned officers—mess sergeants, supply sergeants, property sergeants—who were transfers from CCC camps near Milwaukee. Some new enrollees, called "rookies" by the others, came directly from Fort Sheridan, Illinois, where they had just gone through a four-week boot camp, a "conditioning period" under the supervision of regular army personnel.[51]

At Fort Sheridan the recruits learned how to march and how to live and work together as a military unit, although they were never to be, strictly speaking, "military." They never trained with firearms, and they were not required to salute officers or the flag. Also, they only enlisted for six-month periods, during which they had the option of leaving the corps at any time without penalty, an option most of the enrollees never chose. They also had the option, if their records warranted it, of reenlisting two more times for a total of three enlistments or eighteen months of service. At Fort Sheridan, which was also the district headquarters of the sixth CCC district, which included Camp Madison, the recruits were issued a *Handbook for Enrollees* that they were to study and carry with them. They were also issued a full complement of summer and winter clothing. Since a qualification for enlistment required that

A group photo of CCC enrollees from a Camp Madison barracks, 1937–38.
(photo by Norman Schimelfeni, Arboretum Photo Collection)

almost all of the enrollees come from the nation's neediest homes and families, the clothing issue included more good-quality clothes than most of them had ever seen. Most of the clothing was originally World War I issue. They worked in regular army fatigues. Dress uniforms included black ties that they tucked in near the top of their khaki shirts.

When Camp Madison's 2670th Company moved into their barracks—now, with guaranteed clean mattresses—each enrollee was issued a locker box with a lock, a canteen, towels, clean bedding, and wool blankets. They lived like soldiers, a garrison life controlled by U.S. Army regulations and discipline. Reveille was at 6:00 a.m., and the enrollees formed up in ranks for a morning roll call. By 7:30 they reported for work details, at which time the NPS took over and supervised all of the outside work, which ended at 4:00 p.m., when they were once again under the command of the U.S. Army. Lights out was at 10:00 p.m. The enrollees walked fire watch around the barracks and the other camp buildings. They served guard duty and took turns on mess duty, a shift beginning at 6:00 a.m. and ending at 8:00 p.m. They had to have a signed pass to leave the camp. They marched in formation from the barracks area to the camp mess hall—the renovated Nelson barn—three times a day for meals. They also marched in formation to the flagpole opposite the mess hall, where they formed up and stood at attention as the flag was lowered during evening "Retreat." They also washed their own clothes by hand at concrete sinks, although some of the more enterprising enrollees had bought a washing machine and made extra money renting it out.

But life at Camp Madison was not all work. For a nickel they took a bus downtown on weekends to see movies. The university provided musical instruments. Initially there were high hopes among members of the AC that the enrollees would benefit from the close proximity to the university. With that in mind, members of the AC, particularly Jackson, approached the UW in September 1935 with a request that the Extension Division set up "a co-operative educational program."[52] The university, however, made a half-hearted effort to provide courses, likely because there actually was little enthusiasm among the enrollees for taking courses—free or not. And few of the old enrollees returning in 1983 to a Camp Madison reunion at the Arboretum recalled courses ever being offered on a regular basis. Most who were interviewed during the reunion maintained that they had little contact with the university and saw very little of UW professors. Leopold came by occasionally at night to visit and talk about archery. Longenecker was with them in the field as an overseer at times, but not often. Yet, Corporal James Hendricks, who joined the CCC in 1936 and spent three years at Camp Madison, recalled that some of the enrollees managed to finish high school while at Camp Madison.[53]

The most popular and most accessible form of entertainment for the enrollees was sports. Jackson wanted them to be given free tickets to the University of Wisconsin football games. After some resistance, the Athletic Department, owing mainly to Jackson's dogged persistence, decided that the enrollees could work as ushers, hence, enabling them to get in free even if they didn't have seats. Baseball at Camp Madison was particularly popular. According to enrollees who returned for the 1983 reunion, the camp baseball teams ranked with today's major league farm teams. They were good, and they practiced a lot because competition among the CCC camps in the district was intense. The staff and the army officers responding to the heated district sports competition regularly excused players from work so they could practice. The Camp Madison basketball teams also were exceptional, winning the district conference in 1939.

There were ten wooden barracks at the camp when the 2670th Company moved in that August 1935. The enrollees occupied seven of them. One barracks, the only one still standing on the Arboretum service road, housed Army Headquarters on one end and a medical dispensary on the other. Medical doctors—Reserve Army officers—were assigned to the dispensary. Some of the

A CCC issue of blankets and bed sheets at Camp Madison's company supply warehouse, which also housed the camp commissary. The supply warehouse pictured here still stands at the end of the Arboretum service road. (photo by John Wawrzaszek, Arboretum Photo Collection)

enrollees with medical training worked as corpsmen. The camp also housed a traveling dentist. The barracks across the way and to the immediate east of the Army Headquarters barracks was designated as the Camp Education Building and contained a library, tables for study, and an area set aside for lectures and classes. The barracks immediately behind the Army Headquarters barracks housed NPS workers and was designated as the Forestry Quarters. The company bath house, which still stands on the service road, was west of the barracks area on the edge of the bend in the road leading to the mess hall and out of the camp. The Officers' Quarters, originally the little security residence built in 1932, was on the road between the mess hall to the north and the bath house to the south. The mess hall covered the bottom floor of the renovated Nelson barn. The NPS had its offices in the old Nelson farm house east of the Officers' Quarters. There were four NPS garages just slightly north and east of the barracks area. One was used as a Park Service tool and repair shop; one housed heavy equipment; and the other two were used for trucks and cars. The eastern half of the northernmost garage was a warehouse that contained the company supply warehouse and the commissary. The commissary sold soft drinks, candy bars, crackers, cigarettes, and personal hygiene items that the enrollees paid for out of the five dollars a month on which they lived. Two of the original garages are still standing.

Camp Madison was atypical of CCC camps. It was the only camp in the nation connected with a university campus; and though they didn't attend many academic lectures, the enrollees did attend dances with coeds in the "old barn," as they called it. To Robert Herbert, stationed at the camp in 1939, it was a "show camp."[54] To Eugene Adler, assigned to the camp in 1935, Camp Madison was a part of something big and historic. There was a vision of something happening for the future, he recalled.[55] Camp Madison's university connection made it special, but there was something else that set it apart. Most CCC camps, especially the early ones dating from 1933, were involved in some kind of soil work. As Lieutenant Theodore Rathje, the camp's commander through most of 1940, observed, 1934 had been the worst drought year, a sentiment shared by Timothy Egan in *The Worst Hard Time*, his study of the dust bowl years. As Egan points out, in early May 1934 "the temperature reached one hundred degrees in North Dakota. Sections of Nebraska were starting to blow away. Fields that had yielded twenty bushels of wheat per acre were lucky to get a single bushel. On eight million acres, crops were so withered that there was no harvest."[56] Hence, from 1934 on, most of the CCC camps were geared to drought relief and were either soil conservation camps or state and Department of Agriculture National Forest Service camps working on game control and forest management.

Camp Madison was also the only CCC camp in the nation involved in creating an arboretum.[57] From 1935 until 1941, work done by the CCC permanently changed the face of the UW Arboretum. Almost everything that distinguishes the Arboretum today can be traced back to those CCC years. The enrollees finished building McCaffrey Drive, "that road" through the Arboretum. They constructed maintenance buildings that are still used. They built foot trails. They dredged most of the ponds that now dot the Arboretum. They planted tamarack trees around the Teal Pond. They also planted vetch and sweet clover on the south border of the Curtis Prairie—feed for Leopold's migratory game birds. They also helped to build the Kenneth Jensen Wheeler Council Ring and the Stevens Memorial Aquatic Gardens. Working with Joe Elfner, a NPS landscape foreman, they planted trees and shrubs throughout the Arboretum, according to designs that Longenecker produced in consultation with Elfner.[58] They dug the holes and planted the early lilacs for the Arboretum's horticultural area. They quarried and learned to cut stone professionally under the supervision of Swen Swenson, the NPS foreman in charge of stone work. They built stone shelters. They also built stone walls on

Monroe Street and the Olbrich entrance on the Arboretum's western end. And under the watchful eye of Ted Sperry, NPS technical foreman, they cut and transported prairie grass sods, dug the transplant holes, planted the seeds and, around the clock, fought the 1936 drought that menaced the beginning of the Curtis Prairie.

They also worked with Sperry on the first prescribed, experimental prairie patch and line burns in spring of 1938 and 1939.[59] Sperry depended on his crews, especially during the drought periods. He jokingly noted that, over time, they took to referring to his prairie as "Sperry's weed patch."[60] And, at times, the enrollees literally did work under his "watchful eye"; otherwise, as he put it, when the little prairie plants started to come up, "they'd walk all over them."[61] The enrollees, in turn, took to calling Sperry "the little owl."[62]

After six years of measurable productivity, in November 1941, with very little forewarning, Camp Madison was closed. The original arrangement the Regents had with the federal government called for a five-year limit on the camp. However, as late as April 16, 1940, the U.S. Army had approached the Regents with a request that the five-year limit be extended four more years.[63] On May 28, 1940, the Regents entered into a formal contractual "Extension of Agreement" with the federal government, extending the life of the camp to August 1944.[64] It looked like Camp Madison and the CCC would be around for a while. But by mid-October 1941, seventeen months after the extension was authorized, Lawrence Merriam, regional director for the NPS, notified Longenecker that owing to "necessary" reductions "in the CCC camp program, SP-14 University of Wisconsin Arboretum would be terminated, effective October 31."[65]

On November 1, 1941, Camp Madison officially closed. The staff and the AC were distressed, and the enrollees were shocked. Rumors would abound over reasons for the abrupt closing. Some at the 1983 reunion talked about the enrollees en masse marching to the recruiting centers to "join up." The story is colorful folklore, but it never happened. Others insisted that the start of the war had something to do with the camp shutdown, but the reality was that Camp Madison was closed and the enrollees were gone a solid month before the Pearl Harbor attack of December 7 and the subsequent formal declaration of war. Yet, the nation did feel the war coming.

So why was Camp Madison shut down with so little warning in November 1941? Quite possibly there is no single reason—certainly not the war, except indirectly. By 1941 the CCC program had outlasted its attraction. It was getting harder and harder to find young men who wanted to enlist. The economy had improved; more jobs were available. The military draft was drawing thousands of young men into the service. Congress had systematically been cutting back on funding for the CCC since 1939. Finally, in 1942, the 77th Congress stopped all CCC funding, precipitating a total shutdown on June 30, 1942. The end of the funding meant the end of the CCC at Camp Madison and elsewhere in the nation.

CHAPTER 5

1935

The Arboretum prairie, renamed the Curtis Prairie in 1962, in honor of John T. Curtis, University of Wisconsin professor of botany (1937–1961), is the oldest ecologically restored tall grass prairie in the world. Curtis, who was also the director of Arboretum plant research, is given credit for the planting techniques that greatly expanded, during the late 1940s and 1950s, the quantity and stability of the prairie grasses and plants that now cover 60 acres of what in 1932, when first deeded to the university, was little more than worn-out pasture land. The history of the effort to plant a prairie in the Arboretum, however, did not begin with Curtis. The planting history began in the mid-1930s with one of Curtis's mentors, Norman Carter Fassett.

Norman C. Fassett's Early Experiments

Norman C. Fassett, UW professor of botany, was a pioneering plant ecologist, a specialist in worldwide aquatic vegetation, a conservationist, a successful teacher, and a driving force behind efforts to preserve in place the Earth's rich and diverse flora, *wherever* it existed. In 1945, he chaired the first state committee charged with the task of designating Wisconsin Natural Areas for preservation. He was also the curator of the UW-Madison's herbarium. A native New Englander with graduate degrees from Harvard, he was the author of *Spring Flora of Wisconsin* (1931), *Grasses of Wisconsin: The Taxonomy, Ecology, and Distribution of the Gramineae Growing in the State without Cultivation* (1951), and *A Manual of Aquatic Plants* (1940), all of which are still in print. His *Manual of Aquatic Plants*, revised in two volumes and edited by Garret E. Crow and C. Barre Hellquist in 2006, remains one of the most useful identification keys for aquatic plants in North America.

Professor Norman C. Fassett first experimented in the fall of 1935 with the transplanting of prairie grass sods on Arboretum land. (UW Archives, Image #S04292)

Fassett has also been credited with being the first person to have come up with the idea for the planting of a tall grass prairie in the Arboretum. The claim, however, is disputable. Where and with whom the idea originated, owing to a lack of any solid documentation, is unclear. From the outset, however, Fassett was deeply involved in Arboretum developments. He was one of the original members appointed in 1933 to the first Arboretum Executive Committee. After serving one year on the committee, he declined a second appointment for 1934–35. He agreed, however, to serve in 1934–35 on the less active and hence less time-consuming Arboretum Advisory Committee. Fassett was not a campus politician. Curt Meine notes that, like Leopold, he was "a sensitive man, sometimes withdrawn."[1] And though he was committed to the Arboretum's development, he was not particularly interested in Arboretum administration and had limited patience with committees and their time-consuming committee meetings. He was a scientist, first, who guarded his time for his research, his courses, and his students. Nevertheless, he accepted, albeit with reservations, Gilbert's request, made on December 2, 1933, that he also serve on a Special Committee on Arboretum Planning that had just been mandated by the Arboretum Committee (AC). The Special Committee, chaired by landscape architect Franz A. Aust (Horticulture), was limited to five of the most influential Arboretum figures: Fassett, Aust, Leopold, Longenecker, and Fred B. Trenk, at the time the campus forester.

The formation of the Special Committee in December 1933 coincided, by chance, with the formation of the Civil Works Administration (CWA) in Washington. As pointed out in chapter 4, the CWA was the Roosevelt administration's short-term economic fix designed to get the country through the critical winter of 1933–34 while other long-term programs were being developed. The CWA funded money for labor for which the Arboretum qualified. Gilbert, AC chair, seized the opportunity and asked the Special Committee to compile a list of potential research projects for CWA funding.

On January 10, 1934, Leopold sent a letter to Gilbert with an attached copy of his and Longenecker's recommended projects and recommended project supervisors.[2] The projects included "Wild Rice Study" to be supervised by James G. Dickson (UW Plant Pathology); "Status of Carp in Lake Wingra," supervised by Chauncey Juday (UW Zoology); and "Pioneer History of the Lake Wingra Region," supervised by Charles E. Brown (Wisconsin Historical Society curator). The list also included a project described as "Prairie Grass Dissemination by planting sods in land now occupied by agricultural weeds and exotic grasses." The supervisor, not surprisingly, was Fassett, who likely had proposed the project.[3] Leopold forwarded this list to Gilbert on January 10, 1934, yet, it wasn't until fall 1935, over a year and a half later, that John W. Thomson and Roger Reeve, under Fassett's supervision, began to collect prairie planting materials to be transported and transplanted in the Arboretum to see if they would grow in a selection of experimental plots.[4] An explanation for the lengthy delay is discussed in some detail below.

The idea, however, for establishing a prairie, as the archival record shows, predates Fassett's efforts in fall 1935 by almost two years. Actually, the desire for a prairie was in the air in late 1933 and early 1934. Bud Jackson, for instance, prior to 1934 had designs on the creation of a tall grass prairie somewhere within the Arboretum. He alluded regularly and nostalgically to the majestic beauty of the open prairies that he recalled from his years on his North Dakota ranch. Also, early in March 1934, Charles E. Brown told Jackson that he hoped among the plans for the future of the Arboretum would be "the setting aside of a tract to be developed as a typical Wisconsin prairie area with its characteristic wild flowers and grasses." Our prairies are gone, Brown added.

"Almost the only places where we still find examples" of the wild flowers and grasses "are along the railroad rights-of-way."[5] Hope for a tall grass prairie in the Arboretum was also shared by Leopold and Longenecker. At the June 1934 dedication of the Arboretum, Leopold, as discussed in chapter 4, echoed Brown's sentiments as he spoke of reconstructing "a sample of original Wisconsin" with its oak-openings and "a profusion of prairie grasses and flowers." Longenecker, also speaking at the June 1934 dedication, acknowledged, as a harbinger of things to come, that the Arboretum management, collectively, had agreed to leave "a prairie . . . in the center" of the Arboretum "as a coordinating unit."[6]

We will probably never know for certain who first proposed the plan for the planting of a tall grass prairie. What we do know, however, is that Fassett's scientific experiments with prairie grasses commencing in the fall of 1935 most assuredly marked the beginning of the *research* phase of the effort, though not the actual attempt to turn that worn-out 60-acre pasture into a prairie.[7] That came a year later, in 1936, with Ted Sperry working with the CCC. Another question related to the early years of the prairie experiment, alluded to above, is why Fassett's CWA sod transplanting project, proposed originally in January 1934, was delayed for two years. For a possible answer, we need to review some history. In January 1934 when the CWA research proposals were sent to Gilbert, the AC had no idea that within four months the CWA would be disbanded. Four months hardly left time for conducting planting experiments, especially ones that required transplanting sod. Also, from April 1934, the month the CWA was disbanded, until July 1934, the Arboretum had no labor pool available.

In July 1934, transient laborers, as described in chapter 4, under the employ of the Wisconsin Emergency Relief Administration (WERA), moved into the Arboretum. Camp Arboretum was officially activated, and the transient laborers were put to work. No accommodation, however, was made by WERA for Fassett's research on the transplanting of sods. The WERA officials approved road work, construction projects, tree planting, and pond dredging, but they did not include in the list of projects the transplanting of prairie grasses. WERA, as a state extension of FERA (Federal Emergency Relief Administration), was unable to secure funds from the federal government for any costs beyond what their monthly budget could afford. WERA was in the process of being phased out. Their resources, consequently, were so limited that renting a truck was impossible. Almost half of the work projects planned for Camp Arboretum required equipment and material that WERA could not afford. Since Camp Arboretum was short on both labor and equipment, particularly trucks, and since prairie sod removal and transplanting would require both labor and trucks, particularly flatbed stake trucks that would need to travel extensive round-trip distances, approval of a prairie restoration project was unrealistic in 1934. The shortage of equipment, as discussed in chapter 4, was one of the main reasons why Arboretum officials, spurred on by Gallistel, decided early in 1935 to abandon WERA and to turn the camp over to the CCC.

When the CCC arrived in August 1935, they brought along eager young enrollees as workers and army officers to command the camp. When the NPS arrived, they brought along work supervisors and heavy equipment operators, but they also brought along something that must have warmed Fassett's heart—a fleet of trucks, particularly flatbed stake trucks! Having proper trucks for the work of hauling sods must also have pleased John Thomson, who, as an eager first-year graduate student in 1935, had the opportunity to work with Fassett on a field project that never before had been attempted—the potential restoration of dominant prairie grasses and flowers in a worn-out field. After almost two years of waiting, Fassett's study of "prairie grass dissemination by planting sod," first proposed as a CWA project in January 1934, was to be realized.

You can hear the satisfaction in Thomson's voice as he penned his "Notes on Collecting." The first sentence sums it up: "Collecting was done with the aid of the CCC who furnished men and trucks." He went on enthusiastically to provide a litany of what he and the CCC had loaded onto those flatbed trucks and where the material was found: "Collecting of hay and sod . . . done in a low prairie near Mazomanie." Sod collected "on a dry hill slope and also a wet meadow between Sauk City and Mazomanie [and] . . . on a dry hill slope 4 ½ miles west of Middleton." Hay and sod collected on "sand plains and beside the road at Arena, and . . . in the sand plains northwest of Spring Green." He also described how he and Fassett had collected "shrubs and seeds on the bluffs of the Mississippi River at Hager, seeds on the Mississippi River terraces near Lake Pepin . . . and north of Portage."[8] Thomson left the project in 1936 to research another prairie project in central Wisconsin, but he carried with him fond memories of working through the fall of 1935 and into 1936 with the CCC, including sharing a beer or two with the enrollees after gathering the sods.[9]

Fassett's contribution to the Arboretum's prairie experiment in fall 1935 was a promising start, but it was just that—a start. The research produced results that would make the business of prairie restoration, when it finally happened, just that much easier. As Thomas J. Blewett and Grant Cottam have observed, when Fassett started "little was known about planting procedures" for complements of prairie plants in old fields and pastures.[10] Fassett and Thomson (Reeve had transferred to another project) demonstrated, after the experimental plots were planted and the data recorded, that the best survival rate for the dissemination of prairie plants was based on sod transplanting accompanied by the placement of prairie hay. And according to Thomson, the best examples of transplantable prairie plants in the final analysis in fall 1935 were found in those large chunks of prairie sod taken from the sand prairie near Spring Green, Wisconsin.[11] When Ted Sperry initiated the effort to plant what is now the Curtis Prairie, he consulted Fassett's and Thomson's records. And Sperry began his restoration work by transplanting sods.

Ted Sperry Hired as Prairie Ecologist

On June 20, 1935, two months before the CCC moved into Camp Madison, Leopold forwarded an FYI memo across the UW campus to Fassett, Longenecker, and Aust, all members of the Special Committee on Arboretum Planning. The memo contained a simple bibliographical reference. It read:

Root Systems in Illinois Prairie. By Theodore M. Sperry, U.S. Forest Service, Harrisburg, Illinois. Ecology, Vol. XVI, No. 2, April, 1935, pp. 178–202.

The reference was to a condensed version of Ted Sperry's doctoral dissertation on the root systems of prairie plants that had been published in the journal Ecology in April 1935, and that Leopold had happened across while randomly paging through scientific journals. He was eager to share the discovery with the Special Committee because in June 1935, two months before Fassett's field experiments began, the Arboretum was already in the market for a prairie ecologist—and prairie ecologists were rare. Sperry looked promising.

He had the credentials. He was born in Toronto, Ontario, on February 20, 1907, but he grew up on the outskirts of Indianapolis as a country boy close to nature.[12] He did his undergraduate work at Butler University in Indianapolis. His graduate degrees were from the University of Illinois, a master's degree in botany in 1931 and a PhD in botany in 1933. His dissertation, "Root Systems of Prairie Plants," a new research field at the time, was directed by Arthur Gibson Vestal, an early authority on Illinois sand prairies. Vestal encouraged Sperry to emulate research on prairie soils and root development that was being conducted in Nebraska in the 1930s by John E. Weaver. Weaver was a strong influence, but Sperry also was in touch with and received advice from Frederic E. Clements, the famous plant ecologist.

Sperry's research resulted in the discovery that Illinois prairies were unlike prairies farther west—the grasses were taller, yet the roots were shorter because they stopped where the water table came up.[13] Most of the prairie plants in his study came from "little places," railroad rights of way and old Illinois cemeteries. He observed, in a 1982 interview, that Illinois in the 1930s had very few prairies left. "Not a piece as big as this room," he claimed. And, he added, he was happy with the prospect of moving to Wisconsin because the state had more relic prairies than Illinois.[14] After finishing his graduate work in 1933, Sperry worked for the U.S. Forest Service as a foreman in a CCC camp in the Shawnee National Forest in southern Illinois. The work involved mainly forest surveys and timber management. On July 18, 1935, a full month before the CCC arrived at Camp Madison, Leopold sent a letter to Sperry asking if he would be interested in taking on the responsibility of constructing a prairie in the Arboretum. He added: "The area is for research rather than for recreational purposes. We now have a CCC camp. . . . On the area is an open flat formerly in farm fields which may have been originally partially prairie or near prairie. . . . We want to construct on this a flat Wisconsin prairie, together with its accompanying 'oak openings'. . . . You can appreciate the difficulty and variety of the technical problems involved. The National Park Service, which operates the CCC camp, has assured us of a foremanship on the camp overhead to take charge of this prairie restoration work."

In the final paragraph, Leopold asked Sperry if he would be interested in the position. He also told him that he and Fassett would act as advisers on the technical aspects of the job, and that before any decision could be reached, they wanted a personal interview. Consequently, if Sperry was interested, he should arrange an interview with Fassett. Leopold was going to Europe at the end of July and would be gone until November. Leopold added that after the interview the decision to hire would be Fassett's, and that he supported whatever decision Fassett made. He concluded with a postscript based on a recommendation from Longenecker that cautioned that "the incumbent of this position" would need to have the ability to manage work crews as well as to guide the scientific work.[15] Leopold ended the letter with the question, "do you think you have this?"[16]

When Leopold left for Germany on July 30, 1935, he left the matter of hiring a prairie ecologist in the hands of Fassett, Longenecker, and Aust. The hiring should have been a simple matter, but it turned out not to be. After receiving Leopold's letter, Sperry responded quickly that he was indeed interested in the prairie ecologist position. On August 17, Fassett, while traveling through Illinois, met with Sperry and Vestal on the University of Illinois campus. Prior to the meeting, Fassett had received the ominous news from Longenecker that the AC was having a difficult time filling NPS positions at Camp Madison. The jobs necessitated official congressional appointments. Longenecker told Fassett that he had been in touch with Congressman Harry Sauthoff and was hoping that Sauthoff alone could make the appointment. But "we cannot . . . tell [Sperry] he is hired," Longenecker added, "until we hear definitely from Washington that Sauthoff will be given permission to select him."[17] Fassett came away from his meeting with Sperry convinced that he was the best choice. Academically, he was fully prepared; he was interested in the challenges posed by the job; and having been trained in Illinois, he "knew more about Wisconsin prairies" than the other out-of-state applicants. By late August 1935, the AC applied to the NPS for Sperry to be appointed to the position of prairie ecologist, a senior NPS foremanship, at Arboretum Camp 14 (Camp Madison) at a salary of $2,000. Fassett, Longenecker, and Aust were hoping for prompt action so that Sperry and the CCC would have the precious fall months to begin transporting and transplanting prairie sods. But it was not to be.

Three months later, in late November 1935, when Leopold returned from Europe, winter had set in, and the appointment still had not been made. Actually, the whole matter was far more complex in November than it had been in August when the decision was made to hire Sperry. First, the Arboretum's "prairie project" had to be investigated by the NPS before it could be approved for funding. As usual, with government inspections and such, the action forced a delay. Finally, in mid-October 1935 an inspector arrived who then filed a report with Paul Brown, the NPS regional director, approving the project. On November 12, a letter arrived from the NPS turning down the application for Sperry's appointment for two reasons: he was a nonresident of Wisconsin and he was "already employed." The NPS letter was awaiting Leopold upon his arrival home. He was not pleased.

Given what Leopold had learned about other bureaucratic obstacles in the way of the appointment, he was out of patience with the NPS and downright angry. On November 23, 1935, he penned a heated letter to Paul Brown. As to Sperry being turned down because he was a not a resident of Wisconsin, Leopold responded, "If there were any Wisconsin ecologist qualified to do this work, we here would certainly not have gone to the trouble of sending Fassett to other states to interview prospects." As to the second reason—Sperry being "already employed"—Leopold noted that Sperry's Forest Service superintendent had acknowledged his willingness to release him. We hope, "no doubt," Leopold added, that he does this "on the perfectly sensible grounds that our position is in line with [Sperry's] training, whereas the Forest Service position is much less so."

But there was another, more subtle reason underlying the refusal to appoint Sperry that Leopold also set out to expose—and he did so masterfully. There had been for years a regulation prohibiting any employee of the National Forest Service (Department of Agriculture) from accepting a position with the National Park Service (Department of the Interior). But as everyone knew, there had also been many exceptions to that rule over the years. Leopold's dismissal of that questionable inter-bureau regulation merited this pointed query: "Why should an inter-bureau taboo on taking men apply to mutual consent cases. . . ? This seems to me the blindest sort of bureaucracy." Then Leopold went on to question another suspected development related to Sperry's application. When the rule prohibiting a move from one forest agency to another was combined with an argument that someone *inside* the designated state should be chosen for the position over someone from *outside* the designated state, then one might begin to wonder if the real worm in the bud had a state-centered source. And that was exactly what Leopold suspected when he wrote to Brown. The NPS letter of November 12, 1935, had advised the members of the Special Committee on Arboretum Planning to select somebody from inside Wisconsin for the NPS position at the Arboretum, a recommendation that, as Leopold observed in the letter to Brown, "seems to tie in with the recent addition to the Wisconsin eligible list of a Mr. O. M. Osborne." Now what Leopold also had learned after he returned from Germany was that the eligibility of Osborne, a Wisconsin citizen, had been endorsed by Wisconsin's U.S. senator Robert M. LaFollette Jr.. LaFollette's rationale was simple—a Wisconsin state position should go to a Wisconsin state resident. Sperry was from Illinois. But Leopold was having none of it. Here is his reply, and it is classic:

> Four of our faculty have interviewed Mr. Osborne and are unanimous in our appraisal as follows: He is a kindly elderly gentleman of rather wide experience in horticulture and soils, with a good botanical background, but no ecology, since there was no such science in his day. . . . As a test of his command of ecological science: he had never heard of a quadrat.
>
> My personal guess is that as a crew boss and as a practical man to move dirt, Mr. Osborne might, for the first month or two, be better than a young scientist like Sperry. But as a planner of ecological experiments . . . he would help us not at all.
>
> The issue, then, is this: if we take Osborne, we will move a respectable lot of

dirt but may get nowhere (Route No. 1). If we take Sperry, we have at least the chance of doing something new and valuable, of great future import to conservation (Route No. 2).

We are interested only in the second alternative. Which are you interested in? We are at a crossroads . . . and . . . this "prairie ecologist" venture is not the usual piece of high-sounding scientific terminology, but rather a serious and difficult venture entered into by mutual consent of yourself and the University, and now in danger of being sidetracked.

Leopold added, as a postscript, that the Arboretum faculty had "not been twiddling their thumbs," that Fassett, with the help of his graduate students, had "installed a preliminary series of test plantings which should yield some indications next spring and give us valuable light on further tests." Nevertheless, Leopold added, Fassett's work in no way lessened the need for a prairie ecologist who knows where and how to establish an actual prairie with those plants.[18]

The conflict with the NPS over Sperry's appointment remained unsettled. Weeks passed. When the AC learned that both the Department of Agriculture and the Department of the Interior had to approve a permit transferring Sperry from the Forest Service to the Park Service, Longenecker volunteered to travel to Washington to confer with Congressman Sauthoff and to secure the signatures of the secretaries of Agriculture and the Interior on the transfer permit. His journey produced results. Once the permit was signed, Sperry was appointed senior NPS foreman at Camp Madison He was to be the first of the Arboretum's distinguished ecologists.

Sperry learned of his appointment in February 1936. He was "as surprised as a chipmunk," he recalled.[19] Wasting no time, he left for Madison. He arrived at Camp Madison in the dead of night after a winter snowstorm. The temperature was 20 below zero, with two and a half feet of snow on the ground. No one at the camp, including Harold Madden and the army guard on duty, had any idea that Sperry was hired and on his way to Wisconsin. The army guard found a bunk for him. Sperry moved into the camp the next day and would live there in the NPS staff barracks most of the time from 1936 until he left for military service in the spring of 1941. Sperry liked life at Camp Madison and later claimed that he considered his Arboretum prairie work his greatest professional achievement.

Winter 1936 was bitterly cold. Summer 1936 would prove to be one of the hottest and driest in the history of southern Wisconsin. When the winter weather started to break in April, Sperry and his CCC crew went out searching for patches of prairie to transplant. Leopold and Fassett knew where the

Ted Sperry (*far right*), the Arboretum's first prairie ecologist, directing CCC fieldwork in the initial planting of the Curtis Prairie. (photo from Arboretum Photo Collection)

prairie remnants were, and Sperry followed their direction. As he liked to describe it, he had been given a small crew of ten to fifteen CCC workers, some long-handled shovels, and a truck and had been told by Leopold, his "boss," to "go make a prairie." Leopold was the director of research, and the prairie project, from the time of Fassett's experiments, was considered "research."

Sperry's efforts at planting the prairie, like Fassett's earlier experimental efforts, were mainly trial and error. He admitted that "he didn't know much about prairie projects," but he was willing to learn.[20] Fassett knew more about prairie plants than anyone else, Sperry recalled, but not even Fassett knew exactly how to turn a plowed field with "corn stubble, quack grass and ragweed" into a native Wisconsin tall grass prairie.[21] Leopold had no idea how to "make a prairie" either and, consequently, let Sperry develop his own techniques, which provided a learning experience for Leopold and the rest of the Arboretum staff. And so early in the spring of 1936, Sperry and his crew found a patch of native prairie on a glacial terminal moraine on the east side of the Wisconsin River, opposite Prairie du Sac. He also located other spotty remnants in Sauk County. Because the ground at the moraine prairie was so gravelly, Sperry recalled, large pieces of sod fell apart, and so the best they could do was to cut small pieces around four inches in diameter and one to three inches in thickness. They loaded the pieces four to six layers deep on the truck bed during the morning and returned to the Arboretum by noon.

Once at the Arboretum, the pieces were transplanted on the drier western slopes below the Noe Woods in spots approximately three feet apart. By July 1936, he estimated, they had transplanted nineteen loads or approximately twenty-five tons of sod that were planted in over ten thousand spots. Most of the transplanted Wisconsin River prairie sods contained a variety of grasses and flowers: little bluestem, big bluestem, Indian grass, two grama grasses, some goldenrods, asters, and violets.[22]

His methods for transplanting were primitive. When asked what techniques he used, Sperry responded coyly: "We took spades and mattocks. We dug a hole. We got rid of as much quack grass as possible and we put the sods right in there."[23] When asked how he seeded, since they also brought seeds back with them, Sperry explained, "We got out the quack grass, loosened the soil and put the seeds a quarter of an inch below the soil." He added that during the entire planting season, from warm-up in May until November, he was guessing what to do as he went along.[24] When asked where he started putting in the plants and how he arranged them, he responded that he had "no specific arrangement." They worked eastward. "We had to decide how we were going to put the plants in," he observed. "Were we going to make a carefully organized garden or going to just try to cover an area?" He decided to spread the prairie thin over as much area as possible; "it would look like we were getting the job done," he added.

So they put grasses in. "Blue stem here, Indian grass there, etc. It was hit and miss with no plan about what came next. We did what we could with what we had in the time available." The object was to "cover the territory."[25] When asked how he decided what species to plant, Sperry replied that he tried to mix the plants. He didn't want large areas of a single species, so he "put in patches—patches of blue stem, etc." When asked if he thought he was re-creating a natural looking prairie, he responded: "I hoped that eventually it would be. We were just trying to get the plants established at that time and assumed that once established they'd spread themselves and would eventually form a natural mix." When asked about "his goals for abundance and distribution," he replied, "We didn't have any goals." They brought in tons of sod and bushels of seeds and worked at creating a mix. He never knew specifically what he was supposed to be doing.[26]

Sperry started the planting in April 1936. Most of the plants that went in during that first spring and summer were lost. The temperature in southern Wisconsin during the 1936 summer reached an all-time high of 107 degrees. For twenty-seven days during the drought, the temperatures reached daily highs above 90 degrees.[27] Water was carried up from Lake Wingra in barrels on flatbed trucks in an effort to try to salvage new plantings of both prairie grasses and trees. The trees took precedence. The prairie plants, since they came free of charge, were low on the watering list. The CCC crews carried water that summer from the lake around the clock for three weeks. Individual shifts, including night shifts, covered a full twenty-four-hour period. Since there was no road providing truck access directly to the lakeshore, the enrollees cut a temporary road through the woods. Water was pumped into barrels and trucked, much of it spilling, up to the planting areas. Sperry estimated that only 3 percent of the transplanted prairie plants survived that wretched summer.[28]

Nevertheless, Sperry and his CCC crews managed between 1936 and 1941 to put in a prairie that actually began to look like a tall grass prairie. He eventually developed a planting system that involved both seeding and sodding. He gathered most of the seeds himself. They were planted early in spring and summer and put in storage in the fall. He estimated that during any given year, he had thirty to forty packets of seeds tucked away. He also set up a prairie nursery just off the edge of the Nakoma golf course on the west end of the Arboretum. "They grew well," he said. He concluded that "seeds grew better when just plain planted in the prairie area," and that "transplants from the nursery did not do well."[29] For plant spacing, he planted patches of prairie plants four to five feet apart. There were some blue stem grasses and brown-eyed Susans, along with some roses—Carolina roses—scattered around along with some narrow-leaf milkweed. Sperry planted prairie dock, compass plants, and rattlesnake master. By the summer of 1938 purple coneflowers were actually beginning to bloom.[30]

It was also in 1938 that Sperry got permission from the AC to experiment with controlled burns, the first in the history of the Arboretum. On November 14, 1938, in an advisory Arboretum staff meeting chaired by Leopold, Sperry was directed by the Arboretum Committee to begin the experimental burns in spring 1939 on limited prairie plot areas.[31]

Sperry's Legacy

Progress on the Arboretum prairie moved slowly during the years between 1936 and 1941. But by 1941, before he left for military service, Sperry managed to plant all 60 acres.[32] Yet, there were problems, little annoyances that kept getting in the way. One, the problem of the oaks nearly got him fired. The story speaks to Sperry's forbearance and to his unflagging sense of humor. As he explained it, there were several old oaks scattered around the prairie area. Since shade trees keep prairie plants from growing, he decided to cut the oaks down. After he did so, he was "called on the carpet" and asked for an explanation. As it happened, one of the AC members who adored the old oaks was upset by their removal. He accused Sperry of rampant "vandalism." Sperry explained his motive, weathered the conflict, and kept his job, although he knew he had soured some people on the prairie project. As an amusing follow-up to the story, Sperry remarked that he took to referring to the incensed AC member as "a Joyce Kilmer votary." Why? "Because," as Sperry quipped, "he never dreamed that he would see a prairie as beautiful as a tree."[33]

In later years, in 1982, when he returned to the Arboretum, Sperry, still uncertain about the long-term success of his experiment, was pleased to see the plants thriving. He had never expected the prairie to turn out as well as it did. He was particularly surprised by the large number of surviving flowers. It was "truly a restored prairie," he claimed, and likely the oldest. But he also cautioned that it was not "a natural prairie," because "we are not at all sure just exactly what a natural prairie ever looked like." "I don't know of anybody who has ever described one," he added. And anyway, he continued, flashing a sly smile, "it has no bison."[34]

Sperry's work on the prairie project began to show lasting results by 1939, results that begged to be shared with people outside the Arboretum. What was once a worn-out field now actually looked like a prairie. Jackson, in September 1939, was so pleased with the results that he proudly advised Thomas J. Pattison, secretary of the Wisconsin State Highway Commission, to consider the possibility of seeding or reseeding highway shoulders with prairie grasses. It was a novel idea in 1939. Jackson told Pattison the idea had first been suggested by Leopold owing to the success of the Arboretum's prairie project. He added that the federal government has "attached considerable value to the work" since "it may possibly help in the solution of the dust bowl problem" by demonstrating a practical technique for restoring original prairie flora.[35]

Indeed, by 1939, word of the success of the UW Arboretum prairie project was spreading. In spring 1940, commissioners of the Forest Preserve District of Cook County, Illinois, decided that a prairie restoration in the Cook County Forest Preserve might be desirable. Roberts Mann, the superintendent of maintenance for Cook County, turned for assistance to the one place in the country where he knew an actual prairie restoration was successfully in progress—the UW Arboretum. Mann corresponded with Leopold about the possibility of "borrowing" Sperry for thirty days. "It is imperative," Mann told Leopold, "that we get the services of Dr. Sperry some time during the month of May and early June if we are to go ahead with this project within the coming year." However, in order for Sperry to leave Camp Madison, he had to have NPS permission. Mann wrote to the NPS district headquarters requesting the loan of Sperry, but the request was denied. On April 25, 1940, Mann received a memo from Donald B. Alexander, assistant regional director of region 2 of the NPS, offering the following rationale for the denial; "We regret to advise that the park authority of Wisconsin SP-14" would not be able to spare the services of senior foreman (ecologist) T. M. Sperry "at this time, as it will seriously handicap the program at the University Arboretum now in operation." Sperry was now truly "one of a kind." And his services were in demand. Unfortunately, 1940 was his last year of work on the Arboretum prairie.

In 1941 Sperry was drafted into the Army Air Corps. He was commissioned and trained in meteorology. In 1943, he was transferred to England as part of a weather section providing pre-flight briefings for bomber pilots, including pilots who were part of the D-Day invasion of Normandy. After the war, in spring 1946, he returned to the Arboretum. He surveyed his plots and found that the prairie plants left under the care of Leopold, Robert McCabe, and a few wildlife management graduate students were thriving. In need of a job, in June 1946 Sperry accepted a professorship in botany at Kansas State Teachers College, later renamed Pittsburgh State University, in Pittsburgh, Kansas.[36] From 1982 to 1992, Sperry returned to the Arboretum for regular visits during which he graciously consented to a series of historically informative taped interviews. He died in March 1995.

CHAPTER 6

❧

Late 1930s

FRUSTRATIONS, "GOLDEN YEARS," EFFORTS TO INCREASE FUNDING

In many respects, the late 1930s were golden years for the Arboretum, filled with successful first-time efforts. Work on the 60-acre prairie continued at an encouraging pace. But the late 1930s also was a time of growing concern over burdensome problems, many of them old issues that had gone unattended for years.

Trout Fishing in Lake Wingra?

One old issue that the Arboretum Committee (AC) felt obliged to embrace was the ongoing problem of Lake Wingra's excessive carp population. Carp, regrettably, had been introduced into Wisconsin lakes during the waning decades of the nineteenth century by the U.S. Fish Commission. Lakes and rivers throughout the nation, but predominantly in the Midwest, were stocked with over two million carp during the 1880s and 1890s. Anglers and economists praised the effort as a boon to both the fishing industry and the U.S. food industry. Yet, carp, once established, were—and are—almost impossible to get rid of without endangering other game fish. And by the 1930s, carp domination in Lake Wingra had seriously reduced the game fish population. Leopold, as professor of game management and the Arboretum's director of research, acquired research grants designed to help with the problem. Tagged as "Aquatic Surveys of Lake Wingra," the grants called for studies that would provide scientific guidance for ridding the lake of carp. For years, Leopold complained, supporters and staff had continued to stock Wingra with fish without any scientific grasp of what they were doing.[1] In December 1934, the AC, hoping to remedy the problem, came up with a possible solution. The plan, devised by Professor George Wagner of the UW's Zoology Department, called for driving the carp and other fish from west to east across the lake ahead of a large seine (a vertically hanging fish net) stretched end to end. Once the fish were across the lake, the seine would be staked a few hundred yards from the east end. Cross seining would then trap most of the large carp that subsequently would be removed. The hope was that once the carp were removed, the lake could be overstocked with pickerel and bass that would eliminate the carp minnows. Jackson noted that Lake Wingra already contained a sizable population of rainbow and German brown trout, some as large as four pounds, and that four hundred brown and rainbow trout from six to ten years old had been released into the lake in 1934. More would be released in 1935. Hopes for trout survival were high, he added, because the lake was entirely spring fed with reasonable flowage. But, surprisingly, what Jackson and other angling enthusiasts appear not to have realized was that without the presence of tributary streams, how would the trout reproduce?

The carp eradication plan, once implemented, would also provide an opportunity—the first in the lake's history—to conduct a survey of the number and variety of fish in the lake. Leopold, never one to pass up a wildlife research opportunity, saw in the plan a chance for conducting a scientific carp study.[2] Leopold and the AC believed that the experiment, as the first "scientific" attempt to eradicate carp from a freshwater lake, would have singular importance statewide and nationally. With that distinction in mind, in December 1934, Jackson asked Frank T. Bell, commissioner of the U.S. Department of Commerce, for professional advice and, in typical Jackson fashion, for federal assistance. The plan, however, received no federal support. It would take two years before the Arboretum had adequate materials and money to implement the plan. Finally, in 1936, with the assistance of the Wisconsin Conservation Department (WCD), the plan went into effect. On November 15 and November 18, 1936, a five-thousand-foot net seined Lake Wingra in two hauls. A total of 71,550 pounds of fish were trapped. Of the total, not surprisingly, 58 percent were carp. Also netted but released back into the lake was a selection of garfish, pike, black bass, sunfish, crappies, white bass, and trout.[3]

The number of trout, in spite of the 1934 and 1935 stocking, was small in comparison to other game fish netted. Yet the WCD, always optimistic, decided, since some rainbow and brown trout had survived, that it would continue to stock the lake with trout just as rapidly as the carp were removed. A preliminary report on the seining effort, circulated in October 1936, noted for the civic minded that in regard to stocking the lake with trout, "the value of that demonstration to the people of Madison can readily be appreciated." In 1937, the Bureau of Fisheries, Department of Commerce, decided to aid the effort and provided Wingra with a shipment of rainbow trout and black bass. Jackson, looking into an opportunity for also securing from the Bureau of Fisheries breeding stock from the Mississippi River, queried the Bureau on the availability of the service.[4] Longenecker, in his annual Arboretum report for 1937, was so optimistic about the carp eradication that he predicted, in a futuristic wish list that in twenty-five years all carp would be eliminated. The lake, totally restored, would once again have wild rice growing along its shores and would be "a veritable paradise for marsh and water birds."[5] "Hope springs eternal."

In 1937, hopes were still high for the trout experiment as it entered its fourth year. Jackson, clinging to the small but noticeable success, observed that the project was "one more example of what cooperation can accomplish."[6] But despite the carp eradication and the efforts of the WCD to stock the lake throughout the late 1930s, the trout in Wingra were doomed. They could not reproduce. In terms of eradicating carp, however, that netting in November 1936 was a success, and plans were made to continue the operation.[7]

The carp netting continued on an irregular basis through the remainder of the century. In 1947, the WCD decided to stop seining Wingra because the operation was expensive and unprofitable. Gallistel appealed to the WCD to continue, noting that Professor Arthur D. Hasler, a limnologist in the Department of Zoology, had observed that new beds of aquatic plants were appearing in Wingra and that the transparency of the water had increased markedly. Hasler and Gallistel urged continued harvesting of carp, an effort that was continued and that continues to this day. The carp, however, still remain a serious problem. To date, there appears to be no feasible way to eradicate them from Lake Wingra, and they still take a noticeable toll on the pan and game fish. However, there are muskies that have survived impressively and that now haunt the bottom of Lake Wingra, and when caught are often of record size.

Arboretum Fire Worries

By the late 1930s the threat of Arboretum fires was another major concern. The young pines planted south of the prairie earlier in the decade were now particularly susceptible to fire, especially during excessively dry periods. The fire potential was also evident in the CCC camp, which consisted of a renovated old wooden barn as a mess hall, ten wooden barracks with heating stoves in the center, a stove-heated old wooden farm house used as the NPS headquarters, a wooden bath house with hot water boilers, and a series of wooden garages with heating stoves and with trucks and heavy equipment filled with tanks of gasoline. Returning CCC members interviewed in 1983 recalled a destructive and very costly fire in 1938 in one of the equipment garages on the road east of the barracks. The fire destroyed the garage and all of the trucks in it. The trucks had full tanks of gasoline.

In August 1937, severe drought conditions led to three consecutive Arboretum fires. One occurred on a portion of the Lake Wingra marshland east of Monroe Street. It was discovered around ten a.m., giving Madison firefighters an early start on putting it out. Damage was minimal. Another fire occurred at a point along Seminole Highway. It too was discovered early and checked before it got out of control. The third, on August 15, caused more damage. It was blamed on a careless motorist who likely threw a lighted cigarette out of a car window while driving in the evening through the Arboretum. During the night, a strong southwest wind fanned the fire into a sizable blaze that swept northward through two and a half acres of white and red pine plantings. The fire, discovered about 4:30 a.m., took Madison firefighters and over a hundred CCC enrollees well over an hour to extinguish. Most of the young trees, if not burned totally, were scorched badly enough to be lost. Damage was estimated at more than $4,000. A total of 834 young pines were destroyed in that August blaze.[8]

But by far, the biggest conflagration at the Arboretum in the late 1930s occurred on March 17, 1937, when the largest building in camp, the renovated Nelson barn, which was used as the camp mess hall and recreation center, burned to the ground. The fire started in the second-floor recreation center and was discovered at 7:30 a.m. shortly after breakfast. Within an hour the old wooden structure was totally consumed. Madison firefighters, aided by CCC enrollees, managed to keep the fire away from nearby buildings. The officers' quarters located on the south side of the mess hall did catch fire momentarily, but there was no major damage. The camp kitchen, a one-story building behind the mess hall to the west, was more seriously threatened. The kitchen was attached to the mess hall's ground floor by a wooden partition. During the fire, chains were attached to the walls of the partition and pulled away by trucks. The kitchen was scorched but left intact and usable.

The absence of any nearby water source hindered efforts to quell the blaze. In response, the CCC enrollees formed a series of bucket brigades that moved buckets of water from the camp bathhouse, several hundred feet south, to the fire department's water pumper. Other enrollees climbed to the roof of the kitchen and were handed up buckets of water to pour over the kitchen's roof and walls, threatened by sparks leaping from the mess hall. By March 18, the day after the fire, two large canvas army tents were in place next to the kitchen

CCC enrollees throw buckets of water on the roof of the Camp Madison kitchen in a successful effort to save it during the March 17, 1937, mess hall fire. (photo by John Wawrzaszek, Arboretum Photo Collection)

and served as a temporary mess hall for the camp until a new addition housing a dining area could be attached to the kitchen.[9]

The cause of the fire initially was unknown. Sometime in late March, Captain Ralph H. Rusk, camp commander, asked McCaffrey, in his capacity as secretary of the Board of Regents, for a letter relieving the War Department of all liability in connection with the fire. The request led to a further investigation, which concluded that the fire had been caused by sparks from the kitchen chimneys directly behind the mess hall. The army was in charge of the kitchen. McCaffrey told Rusk on April 1 that under the circumstances he could not write the letter. Instead, McCaffrey made him an offer that, under the circumstances, would have been hard for the War Department to refuse. The destroyed building was insured in the Wisconsin State Insurance Fund for $6,000, and payment under the policy amounted to 90 percent of the loss, or $5,400. "We . . . would like to have the building restored," McCaffrey told Rusk, "and there is stone enough on the site . . . to build a building." Hence, McCaffrey added, should the War Department see fit to provide the additional funds needed for reconstruction, the department would be relieved of responsibility, and the army would have a new mess hall. Within months, a new mess hall was constructed next to the kitchen; once again, the cost to the university was minimal.[10]

Bird Banding and Record Keeping

Keeping the trapping and bird banding program going, with adequate record keeping, was another late 1930s Arboretum and Wild Life Refuge issue that particularly concerned Leopold. The nature of the issue, with consequences that extended beyond the Arboretum, was whether or not bird banding and subsequent record keeping were at all useful and productive. What did they accomplish? As an article in the newsletter of the Inland Bird Banding Association put it, "Do we have a program and an objective?" On record, the long-range effort of the Banding Association was nothing less than "systematic banding in all parts of the Western hemisphere." The difficulties developed owing to generally sparse coverage in too many regions, in spite of over two thousand stations scattered over North America.[11] Regardless of the hemispheric shortcomings, bird banding, trapping, and record keeping by Leopold, Sperry, Bill Feeney, and a few chosen CCC enrollees continued faithfully throughout the late 1930s.

Sperry, when interviewed, acknowledged how much he enjoyed helping Leopold with the pheasant banding. When he wasn't working on the prairie, Sperry spent his time with Leopold and his wildlife management staff capturing and banding birds, including song birds. As a self-appointed songbird banding representative and record keeper, Sperry in 1939 was pleased to be able to represent the entire state of Wisconsin in a report to the Inland Bird Banding Association. Feeney, Leopold's assistant and NPS biologist, also kept accurate trapping and banding records. In 1938, Feeney reported trapping and banding 20 species of shorebirds plus blue and snow geese, Caspian terns, and other rare species. "An exceptional list of 204 different species of birds has been recorded on the Arboretum so far this year," Feeney reported. The 1938 list included 20 hawks, over 600 songbirds, and 30 mating pairs of woodcocks, "an unusual number for this area," he added. The presence of these mating pairs of woodcocks particularly appealed to Feeney, who was increasingly drawn to woodcocks. Feeney eventually would make the study of the woodcock, conceived while at the Arboretum, a lifetime commitment.[12]

Leopold and the Need for "Many Hands"

Feeney worked closely with Leopold. Together with the CCC, they trapped and banded most of the Arboretum's pheasants. For Leopold, trapping, banding, and accurate record keeping were absolutely necessary to scientific wildlife studies. To achieve accuracy and to maintain essential credible records, Leopold argued that "many hands" were needed. And "many hands" were what he saw as sorely lacking in the university's commitment, or lack of it, to Arboretum research in the late 1930s. In response to this need, on April 12, 1937, Leopold addressed an open appeal for help to the Wisconsin Alumni Research Foundation (WARF) in which he asked for money to fund a gradual increase in Arboretum research personnel in order to help specifically with record keeping. In deference to NPS and CCC administration arguments that the CCC labor supply and their NPS supervisors were enough to get the job done, Leopold responded that the CCC and their NPS supervisors, though dependable, were not professionally trained in accurate scientific research. The result, he contended, was a serious shortage in the scientific direction that projects at the Arboretum required, particularly in the area of accurate record keeping. Specifically, he listed the need for graduate assistantships and instructorships for 1937 and 1938 in soils, botany, zoology, and fish and game. Filling those positions, he argued, would finally set in motion a system of credible scientific record keeping directed at calculating plant succession, the spread of plantings, the changes in plant life, and soil conditions "essential to future experimental work."[13]

On the same day (April 12, 1937) that Leopold sent the appeal to WARF, he also addressed the AC on his assessment of future needs for the Arboretum's basic research program. His remarks spoke directly to a question that would continue to be at the center of debates over the Arboretum's role as a university research facility long after Leopold was gone. "Is the University Arboretum to be just another park," he asked, "or is the area destined to be a biological laboratory? The answer to this question," he continued, "depends largely upon the initiation of a basic research program for the area within the present year." He added:

> The soil, the plants, and the animals of the Arboretum have been heavily modified by a century of use. Now that a reverse process of recuperation has begun, records should be obtained immediately covering . . . soil type, soil fertility and soil reaction . . . plant types and species on the area. These records constitute the basic information upon which a research program must be placed. Changes are occurring rapidly since the withdrawal of farming and the cessation of fire. More changes will follow when the present heavy plantings become established. Hence the foundational records for this experiment must be obtained during the coming summer if the data are to be complete.

He went on to elaborate on five Arboretum research projects in particular that required a close professional association between the departments involved and an Arboretum research staff that could assist the departments, specifically with record keeping.[14] He made this appeal for funds in April 1937. Three months later, in July, the soil study, upon which he seemed to have banked all the other research projects, still was not funded. On July 24, 1937, he told Professor A. R. Whitson of the Soils Department that the lack of a soil survey was reaching the point of no return, and the situation was serious. "This year we are spending labor in five or six figures on lagoons, plantings, seedings, fencings, and soil treatments, all of which are made in the dark because we do not know what lies underneath. We probably cannot hope to retain the CCC camp more than two years longer and we, therefore, can hardly retrace our steps and rectify our mistakes. Future gifts of funds for land are a probability, but future gifts of funds for labor are highly improbable." He noted that Jackson's attempts to raise money for a soil survey were unsuccessful. The prospective donors consistently posed the same question—"cannot the University make its own soil survey?" Leopold told Whitson that he needed at least enough money for a graduate assistant to work on the soil survey.[15]

Thanks to Leopold's efforts between April and July 1937, Whitson eventually got his soils graduate assistant, and in June 1937, the botany department got money for a new instructorship with a half-time Arboretum appointment that would be filled by John T. Curtis. A confident Leopold told the AC, in support for hiring Curtis, that once his doctorate was conferred, Curtis "would have a flying start" in the study of plant germination.[16] Leopold was fully aware, even as early as 1937, of Curtis's potential in plant research. Leopold had made strides in 1937, but they weren't enough. He continued to beat the drum for additional research money for Arboretum hiring. And he continued to be disappointed.

On July 10, 1938, a year after Whitson got his graduate assistant and Curtis got his instructorship, Leopold, still impatient with the lack of research support, sent a long memo to UW president Clarence Dykstra in which he bluntly posed the question, "Research Now or Later?" After four years of "intense effort to start a research program" at the Arboretum, he noted, with only limited facilities, it was time to give serious consideration to a special research budget. The academic departments had done as much as they could to support Arboretum research projects, but their resources did not even cover basic measurements. In its fullest sense, he reminded Dykstra, the research opportunity "exists only at the outset. The effects of changes, to have significance, must be measured from their beginnings . . . Delay thwarts the whole trend of events . . . and causes the whole enterprise to gravitate steadily toward becoming a park rather than a laboratory."[17] As if to underscore the urgency of his plea for increased assistance with record keeping, in March 1939, less than eight months after his memo to Dykstra on the importance of accurate records, Leopold had occasion to provide an impressive example in support of his claim.

In 1939, he was asked by Tom Wallace, the editor of the Louisville, Kentucky, *Courier-Journal*, for help with a talk on predator control that Wallace was scheduled to give at a national sports gathering. In an accounting, anticipating his classic defense of the wolf as a necessary predator in "Thinking like a Mountain," written in 1944 and later published in *Sand County Almanac*, Leopold related the following interesting information to Wallace.

"I want to give you the history of pheasants on the University Arboretum," he told Wallace, suggesting that it might help with his talk on predators. Since 1934, he told Wallace, Arboretum pheasants have been fed yearly. During that time, "we have purposely omitted all predator control. We wanted to see what the birds would do in an area badly infested with 'vermin' but where food and water conditions were good." Records were carefully kept; a census of the area was conducted at least once a year, sometimes twice. In 1934 the record showed 75 pheasants; by 1935, 175. In 1936, the population was 179; in 1937, it went to 260. In 1938, owing to a wet summer, we "ended up with 182 birds." We can claim conservatively, he added, that the Arboretum averages 1 pheasant per 4 acres. "There are not many pheasant coverts in southern Wisconsin which show . . . better." And, he added, we have increased this population despite a closely measured annual removal as great, or greater, than if the area had been open for hunting. And how was the population increase and control accomplished? His point (which he obviously hoped Wallace would pass on to the sportsmen at the show)—"All of this has happened despite an uncontrolled and abundant population of horned owls, Cooper's hawks, minks, weasels, coons, opossums, crows, stray cats and self-hunting dogs." Summing it up, he noted: "this experiment can be conservatively interpreted as refuting the common assumption that a clean-up of predators must precede any successful effort to build up a good stand of game."

He also conveyed for Wallace's use another reason for not eliminating predators. If predators had been eliminated, he noted, there would have been a serious increase in rodents with all the attendant problems. The record showed that the rodent troubles would have gotten worse because this year the Arboretum logged eleven long-eared owls in one flock and research measurements revealed, from a study of their pellets, that each owl was eating about four meadow mice per day throughout winter. That amounts to "6,000 mice per winter," he noted. And it was a good thing they were eaten, he added, because even without them the mice continued to raid feeding stations and damage cover plantings.[18] For Leopold, and hopefully for Wallace, the Arboretum statistics—scientifically measured—had shown incontrovertibly that bird banding and careful record keeping did make a difference and, ultimately, did have positive results.

Leopold would go into the 1940s still battling with the UW administration for additional funding for Arboretum research. But he was the research director, and keeping university administrators on their toes and his eyes on additional funding opportunities, though unofficial, are implicitly understood as central to any research director's job description.

The "Golden Years" and the Arboretum "Beauty Spots"

The acquisition of the Gay-Way tract in late 1936 marked the start of what one might call the "Golden Years" of 1930s Arboretum growth. After the purchase, the opportunities for development seemed to fall into place in a way they never would again. The Gay-Way Tract was a 28-acre parcel running parallel with Monroe Street along the northwest shoreline of Lake Wingra. Sidney L. Gay, vice president of the Wingra Land Company, and his four brothers (sons of Leonard W. Gay, real estate magnate and old friend of the Arboretum) owned 25 acres of the parcel, and Lulu A. Way owned 3 acres bordering the eastern edge of the Gay property. The deal for the property, shepherded by Regents secretary McCaffrey, was finally sealed on December 21, 1936, at a meeting of the Executive Committee of the Regents.

Money for the purchase, $20,000, was provided by an anonymous donor acting through Jackson's Madison and Wisconsin Foundation. The donor was quoted anonymously as saying that he had utmost confidence "in the public-spirited men . . . guiding this civic organization." And well he should. In fact, the donor turned out to be, not surprisingly, Louis Gardner. The purchase was gilt-edged, a "sweet deal." Once again, Jackson had engineered a masterful acquisition for the Arboretum and the city of Madison WCTU ("without cost to the university"), as Jackson liked to brag. The purchase placed into public ownership all but a few hundred feet of the entire Lake Wingra shoreline. Vilas and Conklin Parks, owned by the city, covered approximately three-quarters of a mile along the northwest shore; the remainder of the shoreline, running south and slightly west, approximately a mile, now belonged to the University of Wisconsin, abutting with the Spring Trail Pond parcel acquired in 1932, giving the Arboretum control of approximately two-thirds of the Wingra shoreline.[19]

Efforts to acquire the Gay-Way tract actually went back to 1933, when Jackson began to look into the possibility of adding the Monroe Street land, then owned by Leonard W. Gay, to the Arboretum holdings. Jackson was making offers, but Gay resisted selling because he had plans to build a row of houses on property facing the lake. Jackson, along with John Icke, a Madison-based civil engineer, and George N. Lamb, an adviser from the NPS, started bargaining with Gay, at one point offering to build a dirt fill roadway for him 350 feet back from the lake extending westerly from Knickerbocker Street and then curving northwesterly to Monroe Street. Gay's housing development would then be on the road. All of the land east of the roadway and from Monroe Street to the lake would be acquired by the university. Between the roadway and the lake, lagoons would be constructed. There were already two large springs on the property that fed into the lake. Jackson and his advisers assured Gay that the entire area would be developed to fit in with the Arboretum's architectural landscape design. It would be beautiful, professional, an ecological gem. Longenecker would see to that. Furthermore, once and for all university ownership would eliminate the constantly threatening danger of a public drive being constructed along the lake shore, as the Madison Parks and Pleasure Drive Association had proposed in 1904 and for which it had been granted a right-of-way option that, in 1936, was still legal.

Jackson's new proposal, with the added assurance of a roadway, was submitted to Gay who, once again, turned it down. Eventually the roadway from Knickerbocker Street curving westerly to Monroe Street would be built; but Gay, in 1933, was not interested. He urged Jackson and his advisers to give up, to drop the whole matter. He wanted his housing development on the lake. Undeterred, Jackson forged ahead anyway and had blue prints drawn up that he circulated displaying the new roadway, the lagoons, and the Monroe Street borderline with its landscaping potential. Gay remained adamant. In spite of local civic and professional pressure, Jackson's proposal was stalled, dead in the water. And then, the unexpected happened. In the fall of 1934, Leonard Gay died.

For the next year and a half, Jackson and Icke attempted to convince Gay's five sons to sell, but the brothers were as unimpressed with the possibility as their father had been. The stalemate continued. Finally, in December 1936, Jackson received Gardner's gift of $20,000. He renewed the effort. After a few weeks of arm twisting, the brothers relented and gave their written consent to accept $17,500 with an assurance that the Knickerbocker to Monroe roadway, with sewers, water, curb, and gutters, would be constructed. They also asked for a sidewalk, but there was no guarantee of one in the bargain. The additional $2,500 was offered to Lulu A. Way for the purchase of three acres that bordered the eastern edge of the Gay property.[20] She signed the option to sell on December 14, 1936. The Way tract was never in contention, but without the Gay property it would have been irrelevant to Arboretum interests.

On December 17, 1936, James R. Law, the mayor of Madison, advised the Board of Regents that the city would supply material and labor and complete the laying of water and sewer mains on the new roadway—WCTU—again "without cost to the university." Putting in gutters and curbs would be the responsibility of the Camp Madison NPS supervisors and the CCC. As a token of thanks, the NPS also told the brothers that they would "throw in" the sidewalk.

The 1936 acquisition of the Gay-Way tract signaled an intense period of development on the Arboretum's western end that emphasized landscape designs—in fact, settings for a "park" that included memorials, aquatic gardens, and ponds designed intentionally as horticultural attractions. These Arboretum "beauty spots" were intended to appeal to the public, but they also provided a marginal opportunity for UW science departments and the Arboretum staff to study and measure the growth and survival rate of a variety of aquatic plants and animals. Nevertheless, Leopold, expectedly, had reservations about the gilded Monroe Street projects and the money and labor that they would cost. His concern was shared by other faculty members and graduate students. In Leopold's eyes, the development would ultimately create more of a park than a laboratory, in spite of claims for aquatic studies connected with ponds. In July 1937, within seven months after the Gay-Way purchase, Leopold, as noted above, shared with Professor Whitson of the Soils Department his consternation over insufficient funding for Arboretum research and hiring. And yet, he would tell Whitson, this year—1937—"we are spending labor in five or six figures on lagoons, plantings, seedings, fencings, and soil treatments," and "we do not know what lies underneath."[21]

In his July 10, 1938, appeal to President Dykstra for additional research funds, as described above, Leopold reminded Dykstra that "the purpose of the Arboretum" was to be "an outdoor research laboratory or 'experimental farm' for the biological sciences bearing on conservation." The object was "*not* to build a museum, but to learn the hidden mechanisms which underlie conservative land use." He went on to make the startling observation that "after four years of intense effort to start a research program with existing facilities," the AC had concluded that the effort to date had been a "failure." If conservation

research were to continue, Leopold warned, the Regents needed finally to provide the Arboretum with an adequate research budget. Not surprisingly, given
the heated tone of that July memo to President Dykstra, Leopold added as an
attachment to the memo his Arboretum budget proposal, which requested
additional research funds for 1939.[22]

A year and a half later, in July 1940, Conrad L. Wirth, NPS supervisor of
the CCC program and a strong advocate of the recreational use of government
lands, asked Leopold to put together something on "Wilderness Values" for
publication in an upcoming NPS yearbook. Wirth probably had no idea that
he had given Leopold a forum for a defense of wilderness preservation over
the expansion of recreational park systems. In response, Leopold formulated a
document, entitled "Wilderness Values," that addressed what Leopold termed
"the wilderness problem." His emphasis on the wilderness problem was an
allusion and response to Wirth's 1938 publication, entitled *The Park & Recreation Problem*. On July 17, 1940, Leopold sent Wirth a copy of a six-page manuscript. He apologized for not sending a longer paper which, perhaps, Wirth
had anticipated, but as Leopold explained, the draft he was sending "includes
all the new things I have in mind at this time." And the essay, though short,
speaks for itself. It gave Leopold the opportunity to vent about his growing
concern over the continuing sacrifice of wilderness to government-supported
recreational objectives: "In measuring the value of recreation," he wrote, "we
are so obsessed with the numbers who now participate that we have forgotten
all about the intensity or quality of their experience." The obsession, he noted,
is particularly evident in parks and forests and other land attractions where,
too often, their value is assessed on the basis of expanding public patronage.
Leopold took issue with the practice, arguing that the value of wilderness, on
the contrary, should not be based on patronage totals but on "the *quality* of
what the public gets." Consequently, the wilderness problem, unlike the park
or recreation problem, was a *development* problem. "As motorized recreation
expands," Leopold wrote, the "wilderness shrinks." When the park builder
"dynamites a road to the river . . . in the name of 'recreational planning,'"
so that the tourists have easier access to a falcon's aerie on a bluff above the
river, by the time the road is put through the falcons will be gone. "Just so,"
he asserted, "does the quality of wilderness fade before the juggernaut of mass
recreation." In what one might take as a couched reference to either the Faville
Grove Prairie or to the Arboretum itself, he wrote that a symptom of "imma-
turity in our concept of recreational values is the assumption . . . that a small
park or forest has no place for wilderness. No tract of land," he concluded, "is
too small for the wilderness idea."[23]

Jackson's "Arboretum Story"

Leopold's "Wilderness Values" essay was published in 1941 in the NPS Yearbook. In March 1938, three years earlier, Jackson, obviously delighted with the
increase in the Arboretum's public patronage, drafted an unsigned promotional
three-page document, headed simply "Arboretum Story," that was clearly a
public relations effort celebrating the 1930s accomplishments that promoted
a perception of the Arboretum as a park, a recreational area, a vision that ran
contrary to Leopold's "wilderness idea." Jackson's document was intended for
public consumption possibly as a press release. It reads like a public relations
flier, a tour guide, summing up a list of Arboretum public attractions, mostly
along the Monroe Street corridor, that also underscore the fact that the years
from 1936 to 1939 had indeed been special—"Golden Years"—truly bountiful
in the evolving history of the Arboretum and a boon for those at the time who
conceived of the Arboretum as primarily a nature-based tourist attraction.

First on Jackson's list of potential sightseeing attractions in 1938 was the
Olbrich Memorial at the entrance on Seminole Highway, which had been
under construction since summer 1937 and was scheduled for completion in
1940. The entrance, Jackson heralded, when completed, will "be one of the
foremost structures of its kind in this part of the country," consisting of over
six hundred feet of low stone wall, all Madison sandstone and all quarried by
the CCC. The entire stone memorial, he went on, "will be of special public
interest" and will exist through time as a fitting tribute to the Arboretum's
founder. Jackson also included in the "Arboretum Story" other attractions that
were open or underway in 1938, particularly in the corridor between Monroe
Street and the Wingra lakeshore. He listed Spring Trail Pond, known also
as the Duck Pond, which by March 1938, to the delight of neighborhood
children, was home to a variety of ducks and ducklings. What Jackson did
not include in the "Arboretum Story," however, was the fact that the pond,
owing to its popularity, was soon to be bereft of any real scientific value. The
mallards, to the dismay of Leopold, stayed all year because of the open water
and daily feeding. Even worse, the mallards were mating with domesticated
ducklings left over from Easter baskets. Incensed, Leopold wrote to the state
superintendent of game management requesting an end to the defilement. He
was told that there was little that could be done. State officials did not want to
risk setting the public against the Arboretum.[24]

Jackson also touted the recreational appeal of the E. Ray Stevens Memorial
Aquatic Gardens that border Manitou Way at the Monroe Street–Nakoma
Road junction. The Gardens' "mallard ducks and Canada geese," Jackson

noted, attract thousands of visitors, and with the coming of summer, "the bog plants and marsh flowers . . . on the lagoon islands will be in full bloom." The public was also encouraged to watch for new Arboretum exhibits soon to be showcased along the corridor. Owing to the acquisition of the Gay-Way tract, the newly dredged Ho-nee-um Pond, a work in progress, would be completed in 1939, and the proposed Wheeler Memorial Council Ring, located a few hundred feet south on Monroe Street just above one of the Gay-Way tract's two natural springs, would be open to the public by summer 1938.

The sad but well-publicized story behind the Wheeler Council Ring added to its popular appeal. The landscape plan for the proposed memorial, as Jackson explained, was designed by Jens Jensen, the famous landscape architect, as a testament to the memory of his grandson, Kenneth Jensen Wheeler, who died unexpectedly three years before, just as he was completing requirements at the UW for a degree in landscape architecture. Jensen designed the landscape plan for the council ring. But the stone placing was the work of Edison Wheeler, Kenneth's father, also a landscape architect. The design was well proportioned. The limestone circle, twenty-seven feet in diameter, featured a continuous stone seat around the inner side. The floor of the circle was flagstone with an open council fire hearth in the center. Wide, natural stone steps led from the circle down to a natural spring that historically, Jackson observed, had been a water source for travelers moving north and south on the old Monroe Road. The final plan for the memorial also included a small parking lot and the positioning of picnic tables and benches for the visiting public. From the tables and benches, Jackson noted, sightseers would have "a beautiful view out over Lake Wingra and the lily ponds and lagoons . . . now under construction—all of which the public, poor and rich alike will be welcome to enjoy." Jackson concluded the Wheeler Memorial section in his "Arboretum Story" with the observation that "as the only existing work in southern Wisconsin of Jens Jensen and his able son-in-law," the Council Ring "will soon become one of the foremost features of the entire Arboretum." Jackson's observation was on the mark. The Wheeler Memorial Council Ring in the years following 1938 was and still is one of the most frequently visited spots in the Arboretum.

Jackson's "Arboretum Story" also listed other noteworthy but less ornamental Arboretum accomplishments for the years 1936 to 1938. They included the planting of over 75,000 trees and shrubs in 1937 by the CCC, including 19,000 red and hard maples, "massed together for the fall coloring effect," 11,000 jack pines, hickory, walnuts, junipers, tamaracks, hemlocks, white pines, red pines, white spruce, black spruce, hawthorns, and elms. At the "east end migratory bird refuge," the list continued, 14 acres of ponds were completed and being used by the 1937 spring flight. Projections for 1938 and 1939 called for additional plantings, including 75 varieties of lilacs "expected to make a rich showing" in the coming spring. Also, as a final item, perhaps with deference to the Arboretum's now beleaguered director of research, Jackson noted that the Arboretum was also widely recognized as a source for serious scientific research and that seven hundred students of botany, zoology, entomology, soils, limnology, landscape design, and game management had used it in 1937 and 1938 for research and outdoor laboratory purposes.[25]

Jackson's "Arboretum Story" could also have included in the list of accomplishments for those bountiful years the addition of 55 acres, located along the southern border of the Lake Forest area, acquired in late July and early August 1937, bringing the total Arboretum acreage in 1938 to over 900 acres.[26] He also could have included the news that the Arboretum in 1938 had announced plans for the development of an east entrance at the corner of Carver Street and Fish Hatchery Road in connection with a proposed memorial to Burr W. Jones, a Wisconsin Supreme Court justice and early Arboretum supporter. The Burr-Jones Memorial, designated unreservedly as a "park" by the AC, would be constructed on 20 acres of land on Fish Hatchery Road directly across from Carver Street. The tract was offered to the Arboretum in February 1938 for $3,500. The corporation that owned the land was liquidating and wanted to wind up its affairs. It was another deal that Jackson could not pass up. He urged the purchase. He also wanted the park as an open and recognized public recreation facility. On March 22, 1938, the Regents accepted the proposal.[27] The Burr Jones Memorial Park, however, was never realized. There was opposition to it, and the developing costs would have been excessive. Eventually the Arboretum sold the property.

But most interesting about the Burr Jones affair were the plans for the tract that developed shortly after the purchase. The AC had high hopes at the end of 1938 and into 1939 for a Memorial Park, a *real* public park with a shelter, fireplaces, toilet facilities, and other picnic and recreational necessities—all the standard equipment that the NPS and the CCC had a long history of installing. The move was seen by some as the solution to the ongoing Arboretum controversy between those, like Leopold and McCabe and Sperry, who clung to the vision of the Arboretum as a wildlife refuge, and those, like Bud Jackson, who were actively promoting the Arboretum's future potential as a public recreational area.

CHAPTER 7

❦

The 1940s

CURTIS YEARS, POSTWAR CHALLENGES,
LEOPOLD'S DEATH

Professor John T. Curtis had a distinguished twenty-seven-year career at the Arboretum extending from 1934 until 1961, the year of his death. He grew up in Waukesha, Wisconsin, where he was born on September 20, 1913. In his youth, he established a close friendship with Albert M. Fuller, the curator of botany at the Milwaukee Public Museum. Fuller inspired young Curtis and encouraged him to study wildlife, especially orchids and birds. As a result, Curtis developed an interest in orchids that led to the publication, while he was still in high school, of a scientific paper on orchids. He attended Carroll College, earning a B.A. in biology in 1934. During his years at Carroll, he began a correspondence with Professor Norman C. Fassett at the UW–Madison. Curtis turned to Fassett often during those early years, particularly for advice on experiments with hybrid orchids. In fall 1934, he enrolled in the botany program at UW–Madison. After earning his doctorate in botany in 1937, at age twenty-four, he worked at the UW as an instructor teaching botany courses, on occasion with Fassett, from 1937 to 1939.

Grant Cottam, one of Curtis's first UW graduate students and eventually a good friend and colleague, worked closely with him in the Botany Department from 1949 until Curtis's death and recalled him as being a very private person who kept his personal life to himself. "He was my best friend," Cottam noted, and although "we spent thousands of hours together, I never knew what his father had done for a living or what originally aroused his interest in natural history." They had much in common, but there were subjects that went untouched. One was politics. "I was a liberal Democrat," Cottam observed, "and he was a McCarthy Republican." Cottam also remembered Curtis as a teacher, "meticulous in his preparation," who "knew and referred to the literature," but who also provided bibliographies that included publications

Professor John T. Curtis chaired the Arboretum Committee from 1959 until his death in 1961 at age forty-seven. Here he examines prairie grass growth after an experimental burn. (photo by Max Partch, Arboretum Photo Collection)

from figures like Henry A. Gleason, who advocated positions on ecology that ran contrary to ideas then commonly accepted by the ecological community.[1] As a classroom lecturer, Curtis was informing and entertaining. His lectures were narratives—continuing stories. He had regular informal brown-bag gatherings with students. He also set aside an hour a week for conferences with each of his graduate students. Although Cottam recalled occasional rifts with some graduate students and some junior professors, all in all, Cottam concluded, most of Curtis's students and colleagues admired him.[2]

Curtis's Dilemma

During the 1940 spring semester, while still an instructor at the University of Wisconsin, Curtis acquired an appointment to teach botany at the University of Pennsylvania. He taught a course in plant physiology and he decided that plant physiology would be his specialty. While at Penn, he also inventoried all the orchid specimens in the university's greenhouses and took on a research project that involved stabilizing a stream bank running through the university's Morris Arboretum. As a result of his experiences at Penn, Curtis faced a crucial professional decision during summer 1940. J.R. Schramm, chairman of Penn's Botany Department, had offered him a tenured professorship. Curtis gave the offer serious thought but ultimately decided to remain at the UW. What made the difference was his connection to the UW Arboretum.

As early as April 1937, Leopold had his eye on young Curtis as a possible addition to the Arboretum staff. He was particularly impressed with Curtis's ability to germinate native Wisconsin orchids. Leopold told Jackson in May 1937 that Curtis's technique held out the possibility in time for an Arboretum lady's slipper nursery that would be a source for restocking the entire state. Leopold deemed Curtis "made-to-order for our purpose." In Curtis, Leopold also sensed a kindred spirit. Curtis was research oriented, the kind of field scientist that Leopold admired. And over the years, Curtis did not disappoint him. He implemented Leopold's vision of a collection of native plant communities in the Arboretum. To accomplish the feat, Curtis spent twelve years (1947–1958), including eleven years after Leopold's death, acquiring extensive data on Wisconsin plant communities and then fostering their restoration at the Arboretum. Cottam observed, in a 1981 interview, that the Arboretum staff originally had the idea that they "could establish in . . . two square miles all the plant communities that existed in Wisconsin. But Curtis knew that before you could establish plants, you had to know what existed."[3] Curtis, consequently, devoted a lifetime to acquiring a comprehensive knowledge of native Wisconsin plant communities, an accomplishment that eventually resulted in 1959 in the publication of *The Vegetation of Wisconsin*, a milestone in the study of plant ecology.

In 1937, however, Leopold and other members of the Arboretum Committee (AC) feared that Curtis was moving on. He'll "get away if we can't make him an offer," Leopold warned. They had tried several ways to convince Curtis to take an offer, "but so far have failed."[4] By spring 1940, the AC and Leopold, in particular, felt certain they were losing Curtis. At the May 3, 1940, meeting of the committee, Gilbert detailed the Pennsylvania offer. Among its perks were better facilities and a higher salary than Wisconsin was offering. In response, the AC passed a resolution that called for Gilbert, as chair, to confer with President Dykstra about the desire to keep Curtis. As part of a package that might convince him to stay, Leopold proposed "that the plant end of . . . research at the Arboretum" be turned over to him. Leopold would continue with the animal research, which he preferred. In offering the planting research position to Curtis, Leopold essentially was proposing a new Arboretum governance arrangement. Longenecker would continue as executive director. Leopold would continue as research director, but the research end of Arboretum business would now be split into two divisions. The AC moved, subsequently, that Gilbert meet with the president to suggest that Curtis be appointed to the AC and given the title director of plant research. The committee also recommended that Curtis's appointment be a cooperative arrangement between the Arboretum and the Botany Department, and that his time not be taken up totally with teaching. A fair portion of his time would be for research and for his Arboretum duties. The Arboretum directorship made the difference. Curtis agreed to take the position. On October 4, 1940, his appointment was formally approved.

Leopold, however, as early as spring 1940, insisted that Curtis's directorship start "informally" and "at once," and that he take over—from Leopold—all immediate Arboretum plant research responsibilities. What Curtis could not handle, owing to his teaching load, would be delegated to Ted Sperry. Sperry, as Arboretum prairie ecologist, would now work under Curtis. Leopold would no longer oversee Sperry's prairie work. After the governance shift at the Arboretum, Sperry and Curtis finally were able to begin working together on the Arboretum prairie in summer 1940. It was a long overdue partnership. On June 14, after a survey of the Arboretum grounds, they put together a "round-up" report on the plant work. The report included the recommendation that along with continuing Sperry's prairie nursery and the woods flower nursery, they install a lady's slipper nursery. Leopold and Longenecker were delighted. After inspecting the prairie planting efforts, they agreed that in order to spread the prairie species now growing in competition with exotic sods, they should begin experimenting with eradication methods that also included controlled burning. The agenda set by Curtis and Sperry, as recorded in their round-up report, involved more extensive experiments with burning both in the fall and the spring in order to see which season's burn worked better for the eradication of "bluegrass, Canada bluegrass, quack, etc." They also experimented with mowing variables: "Very early (as in May) . . . Early hay time (June 20) . . . Marsh hay time (August 10)." They also reported on their experiments with chemicals, such as acidifying, to kill sweet clover in order to determine the "acid tolerance of prairie species." They also tried fertilizing.[5] The 1940 round-up report eventually was given to Leopold, who, duly impressed, forwarded it across campus to Gallistel and to Bill Feeney, who was working with Leopold on Arboretum wildlife management and contemporary Wisconsin deer issues.

Curtis, the Faville Prairie, and Leopold's "Exit Orchis"

Curtis, aside from being a welcome addition to stretched Arboretum management, was a godsend personally for Leopold who, as a consequence of Curtis taking on the plant directorship, now had more time to write.

In spring 1940, as Meine has observed, Leopold decided that he "was going public."[6] Not coincidentally, it was the same time that Curtis took on the duties of plant research director. Finally, with adequate free time, Leopold could put his literary talents to work writing personal essays. He had already written "The Thick-Billed Parrot in Chihuahua" and "Marshland Elegy" (1937), a different kind of writing from the scientific writing he had been used to. This was personal essay writing that provided an early foreshadowing of selections in *A Sand County Almanac*. So now, thanks to Curtis, Leopold had the luxury of time to dwell on his experiences as a naturalist and wrote "When the Geese Return," "Song of the Gavilan," and "Escudilla." It was also in 1940 that Leopold's deep attachment to a 40-acre remnant prairie threatened with destruction produced a brief six-paragraph essay that poetically and eloquently addressed the issue of the disappearance and destruction of the remnants of what the *Milwaukee Journal* referred to as "Wisconsin's Last Remaining Virgin Prairie." The title of the six-paragraph essay was "Exit Orchis," and the prairie in question was a hundred-year-old native prairie of approximately 100 acres, part of the Faville Grove Preserve near Lake Mills, Wisconsin, in Jefferson County.

Leopold's involvement with the Faville Grove Preserve dated from the mid-1930s when, in his capacity as university professor of wildlife management, he was asked by a group of farmers living near Lake Mills to help them get started in the business of "game cropping." The farmers, led by a patriarch in his mid-eighties named Stoughton Willis Faville, were feeling the economic pressures of the Depression and thought that Leopold could show them how to improve their income. Leopold was happy to comply. The Faville Grove Preserve consisted of pastures, swamps, forest, part of a silted-up river bottom, and—its treasure—a 100-acre tract of virgin prairie described as a "tiny island of . . . old Wisconsin, hemmed in by millions of acres of plowed, logged, burned and cattle grazed lands." The prairie had never been grazed; it had been mowed for hay on occasion. "Its profusion of native flowers come and go with the seasons," the news report noted, "in their own peculiar cycles."[7]

Leopold, Fassett, and Curtis saw in the Faville Grove Preserve, particularly in the remnant prairie, a rare opportunity to involve students in original management research in a pristine, untouched research site. Fassett's students, in 1936, would produce a class report on "The Collection of Spring Flora

with Special Reference to Prairie, Woods and Tamarack Bog" that survived
as a reference guide right into the late 1940s.[8] For Curtis, the Faville prairie
provided an opportunity for extensive research into the control and preservation of native wildflowers. The large number of native plants, Curtis observed,
that "greeted the eyes of Wisconsin's early settlers" were still there in the prairie, providing modern spectators with "some idea of the beauty of primeval
Wisconsin." For Leopold, the preserve was a perfect training ground for his
wildlife management graduate students.[9]

One of the graduate students who worked with Leopold, Curtis, and Fassett on the Faville Grove Preserve was Arthur Hawkins, whose graduate project
was a scientific natural history of the preserve from the time of the Aztalan Indians (circa 1200 A.D.) to the early twentieth century. Robert McCabe, a first-
year wildlife management graduate student, also worked with Leopold on the
Faville project. On May 15, 1940, as the story goes, McCabe was in Leopold's
office while Vivian Horn, Leopold's secretary, was typing a manuscript from a
rough draft that the "boss" had given her that morning. McCabe and Hawkins
had been working on the Faville prairie transplanting some of the rare prairie
plants that were doomed by a recent development that had Leopold chewing
on leather.

Horn asked them if they "'would like to see what the boss wrote about their
field trip.'" McCabe read it, and as he explained to Curt Meine in a personal
interview, he was surprised by the "touch for literature" in the manuscript.
"This was not the ordinary scribbling that a scientist would do." This piece
came from a mind quite different, McCabe observed. He retrieved the rough
draft from the wastebasket and asked Horn, in the future, to set aside Leopold's original drafts for him.[10] The essay that McCabe read that morning
was "Exit Orchis," a six-paragraph essay responding to the news that 40 acres
of the 100 acre virgin prairie at Faville Grove were to be destroyed. The AC
had attempted to buy the entire 100 acres but was unable to raise the money.
Unfortunately, at some point during the early months of 1940, the 40-acres of
native prairie did change owners.

"Exit Orchis," written on May 15, 1940, related the tragedy that followed.
Leopold wrote: "Wisconsin conservation will suffer a defeat when, at the end
of this week, 75 cattle will be turned to pasture on the Faville Grove prairie,
long known to botanists as one of the largest and best remnants of unplowed,
ungrazed prairie sod left in the state." Yet, he added, only 30 miles away a
CCC camp at the UW Arboretum had been busy "for four years artificially
replanting a prairie in order that botany classes and the public . . . may know
what a prairie looked like. . . . This synthetic prairie is costing the taxpayer

twenty times as much as what it would have cost to buy the natural remnant at Faville Grove." Yet no one responded to the appeal for funds to buy the prairie "together with other remnants of rare native flora, and set them aside as historical and educational reservations." No one cared enough even to make an offer, Leopold chided. "Our educational system is such that white-fringed orchis means as little to the modern citizen of Wisconsin as it means to a cow. Indeed, it means less, for the cow at least sees something to eat, whereas the citizen sees only three meaningless words." For certain, he went on, all the rarer species of wildflowers will succumb to pasturing as soon as the cows begin to graze. "That is why they are rare." Leopold added that he attached no personal blame to the new owner for converting the prairie to pasture. "No public institution ever told him, or any other farmer, that natural resources not convertible into cash have any value to it or to him. The white-fringed orchis is as irrelevant to the cultural and economic system into which he was born as the Taj Mahal or the Mona Lisa."

Toward the end of the essay, Leopold related a story about John Muir, who had foreseen the impending disappearance of rare wildflowers from the Wisconsin landscape. In response, in 1865, Muir offered to buy part of the family's homestead meadow from his brother, to be preserved and fenced off as a floral sanctuary. His brother, however, refused to sell him the meadow. Leopold added, "I imagine that his brother feared not so much the loss of a few square rods of pasture as he feared the ridicule of his neighbors." By 1965, he concluded, "when the rarer prairie flowers are gone, the cultural descendants of Muir's brother may look at a picture of the legendary white-fringed orchis and wish they could see one."[11] By the end of the week of May 15, 1940, the cattle were indeed turned to pasture on the 40-acre remnant with the expected results. They destroyed the wildflowers and grasses. By the end of summer 1940, the grazed prairie meadow looked like any other cow pasture. As Curtis observed sadly, the white-fringed orchis, with its tall spike of white lacinate flowers, was very rare and was "probably more abundant" in the Faville Grove prairie than in any other place in the state.[12] But the loss of the 40-acre prairie did receive enough attention, thanks to the efforts of Leopold, Curtis, Fassett, the students working with them, and other members of the AC, to raise the concern of environmentalists.

Consequently, the story has a satisfying ending. In March 1941, the adjacent 60 acres of remnant prairie, rich with native wildflowers, was purchased by Mr. and Mrs. Philip E. Miles of Madison in order to preserve the flora and to continue its availability for UW instruction and research in methods of wild-flower preservation and management.[13] In 1945, Mr. and Mrs. Miles deeded the land to the university. Professor Gilbert, who retired from the university in 1946, was instrumental in arranging the transfer.[14] The Stoughton Faville Grove Prairie Preserve was also the first of the UW's outlying properties placed under the Arboretum's stewardship.[15]

For eight full years following the writing of "Exit Orchis," Leopold, owing to Curtis's agreement to direct plant research, had added free time to write. "Exit Orchis" was one of the first of many new titles that he produced during the 1940s. As the decade progressed, he clung to the thought that maybe he could put a book of personal essays together that would appeal to readers and promote his thoughts on the ethics and aesthetics of land use. He did write the book, and the book did appear and has had wide global appeal. Unfortunately, Leopold did not live long enough to see it published. He died in 1948; *A Sand County Almanac* was published in 1949. The hiring of Curtis in 1940 produced many positive results for the Arboretum, not the least of which was the gift of precious time enabling Leopold to get on with his writing.

Acquiring the Grady Tract

1940 turned out to be a busy year for the Arboretum. The Grady Tract, 200 acres of farm land and hard woods adjoining the Arboretum on the south, was acquired in late 1940. The acquisition was another milestone in late 1930s and early 1940s Arboretum history, but the story of the acquisition actually went back to the 1920s and to Michael Olbrich. As noted earlier, Olbrich, acting on the advice of Dean Harry Russell, College of Agriculture, and Leopold, who was the assistant director of Madison's Forest Products Laboratory, in 1927 urged the Board of Regents, while deciding on an official title for the Arboretum, to include in the title the designations "Wild Life Refuge" and "Forest Preserve."[16] Olbrich, Dean Russell, and Leopold were envisioning as early as 1927 the possibility of creating an experimental forest for the university; Olbrich, in particular, had the Grady farm specifically in mind as the primary location. To that end, Olbrich, during the 1920s, often met and talked with Nettie Grady, the owner, about deeding the land to the University of Wisconsin. Nettie, unmarried, was the sole operator for many years of a student rooming house and felt close to the university world. She also liked Olbrich, who had taken a genuine interest in her, and they shared a love of gardens and woods. At some point in the 1920s, Nettie confided to Olbrich that she wanted the old family farm to become a university forest.

The Grady farm dated from 1865, when Frank Matthew Grady, Nettie's father, purchased the property. His eight children had no interest in farming the land and eventually moved away. What little farming there was, since the land was poor, was done by tenant farmers. But Nettie held the title to the land. Fourteen Grady heirs, including a brother, Edward, who lived in Prairie du Chien, Wisconsin, were in line to inherit the property upon Nettie's death.[17] In February 1934, almost four years after Olbrich's death and four months before the Arboretum dedication, the Regents, at the urging of the AC, sent a letter to the U.S. Forest Service (USFS) asking, as Jackson explained it, for an expression of whether or not the Forest Service might be interested in establishing a Forest Experiment Station in the Arboretum. The station would work in cooperation with various UW science departments and with Madison's Forest Products Laboratory. The arrangement would be contingent upon the purchase of land designated for experimental forestry research. The USFS replied favorably to the proposal. They added that it had been their experience that the most successful Forest Experiment Stations were operated in cooperation with state universities. The obvious advantage was the availability of university scientists.[18]

At some point early in 1934, Jackson, following up on Olbrich's efforts, approached Nettie and asked if she was still interested in selling the property to the university. She told Jackson that she had confided that wish to Olbrich before he died and that it still applied.[19] Shortly after that favorable 1934 response from the USFS to the prospect of an Arboretum experimental station, Forest Service officials came to Madison to meet with Arboretum representatives and to survey the 200-acre tract. Impressed with the tract, they confirmed the willingness of the USFS to establish a station for forestry experts, UW scientists, the Forest Products Laboratory, and the Wisconsin State Conservation Department. They advised that if the university acquired the title, they would provide forestry experts and would carry on forestry work, but they were unable to assist in purchasing the land. On the matter of financing, the AC was on its own.

In late autumn 1935, after a series of negotiations with Nettie and her estate lawyers, the Arboretum Advisory Committee and Nettie's lawyers settled on a sale price of $25,000. The general consensus among interested UW administrators and the Board of Regents was that $25,000 was too high. The Regents refused to commit a penny to the project. Jackson, however, optimistic as usual, set out to secure the money. By June 1936, the money still had not been raised. After further negotiations, Nettie, her lawyers, and the AC agreed to a six-month extension, expiring on November 1, 1936. Six months passed. By November 1, in spite of over fifty appeals sent to possible donors, the money had not been raised.[20] Nettie's lawyers agreed to extend the option through

1937. Jackson opined, at the time, that since Nettie was aware of the failed efforts, she might consider lowering the price, at least to $20,000. But it did not happen. The new option, set to expire on January 1, 1938, provided Jackson and the AC with a full year. Nettie continued to hope that the property would go to the university.

By February 1938, Jackson still had not raised the money; the option to buy had expired again. Nettie's health was failing. Jackson met with Nettie's brother, Edward, in Paul Stark's office in early February 1938. He persuaded Edward to ask Nettie to extend the option for yet another year to January 1, 1939. On February 19, 1938, Jackson received a letter from Edward informing him that Nettie had suffered a stroke and that, given "her present condition," was very much in favor of the option to sell. He also told Jackson that given the seriousness of her condition, he thought either Jackson or Stark should see her. He also told Jackson that before their February meeting in Stark's office, Nettie had mentioned Jackson's name to him several times, but he had never realized, as he put it, "that she was referring to Bud Jackson who had pitched the fourteen inning game at Westport on Memorial Day."[21] Indeed, Jackson did "get around" Wisconsin. He had a visibility that made him perfect for his role as the Arboretum fund-raiser. Yet, in spite of his funding talents, he was getting nowhere with the purchase of the Grady Tract.

And so Jackson visited Nettie in March 1938. Her condition had worsened. He talked her into reducing the price to $18,000. The AC informed the Regents of the new $18,000 figure, but they still thought the price too high. "Good souled Nettie Grady," as Jackson referred to her, would not live to see the property deeded to the university she loved.[22] In late autumn 1938 she died. Money for the purchase still had not been raised, and time was running out. In December 1938, Jackson told Edward Grady that he wanted him to know that he had "put forth a sincere and strenuous effort to raise the funds." And to that end, he had appealed "to some 55 people, most of them in Wisconsin, but many of them in other states, in the hope that one or several of them would make the outright gift." And Jackson sincerely did put forth an effort—at times, perhaps too much of an effort.

There's an interesting collection of letters in the University Archives containing correspondence between Jackson and some donors. Some he knew, most he didn't, and the correspondence includes requests for money sent to the Milwaukee beer barons, Fred Pabst (Pabst Brewing Company), Joseph Uihlein (Schlitz Beer Brewery), and Val Blatz (Blatz Brewery). All three turned down his request. In a hopeful, follow-up response, Jackson told Uihlein that he had a "little secret" that he wanted to pass on to him. And the secret was that "my young Madison baker friend Louis Gardner, who gave us the $15,000 to purchase the 190 acre migratory bird refuge at Christmas time last year, has just run a large-size advertisement in the Madison papers stating that in the last twelve months the sale of his various products has increased 255%." Jackson added, that while Gardner had no conception of "the commercial value of his $15,000 gift" at the time of his donation, "the cold facts" have revealed that the donation was "largely responsible for [the] tremendous increase in his sales." Jackson's motive for passing the "little secret" on to Uihlein couldn't have been more transparent, and it didn't succeed. Uihlein could neither be persuaded—nor cajoled—into providing the money.

Jackson also sent a follow-up response to Val Blatz that had a similar lead balloon quality to it. He told Blatz that he had gotten to know Blatz's young brother-in-law, Paul Kayser, quite well, and that he had grown to look upon Kayser "as one of our finest . . . young businessmen." And "better still," he told Blatz, "he has a keen appreciation of civic spirit and a ready willingness to do his full part." And then Jackson unloaded the zinger: "It is truly good to know that youngsters of that sort are coming along to take over the job that some of our older chaps have not done so well."[23] Jackson was never one for subtlety. There's no record of Blatz's response.

Nevertheless, in spite of Jackson's efforts, no donor gifted the Arboretum with $18,000, and the option was about to expire on January 1, 1939. In December 1938, Jackson, the glass always half full, told Edward Grady that although he had not succeeded, he "did arouse a certain amount of interest." Along with appealing to well over fifty-five possible donors, he had "worked through every possible angle of government," contacting the Forest Service, the Park Service and the U.S. Resettlement Administration, but with no success. As a last ditch effort, he told Grady, he had appealed to John D. Rockefeller Jr. and the Rockefeller Foundation. He had also suggested to the Carnegie Foundation and Mrs. Carnegie that they might purchase the 200 acres as a memorial to the great Scots-American John Muir. The tract would be called the "John Muir Experimental Forest Preserve." But it was not to be.

Jackson also told Grady that he was not discouraged; he would find the money. He asked for another year of grace. He added that if Grady, now "the most interested heir," agreed to the extension, he should confer with the estate lawyers and the other heirs.[24] Jackson got his year's extension with a new expiration date, January 1, 1940. He went back to work; 1939 turned into 1940, and the property was still not secured. Luckily, efforts to clear the title after Nettie's death had taken the estate lawyers over eighteen months. The delay gave Jackson and the AC more time. Toward the end of 1940, Jackson met with the

Regents. This time they were interested. Nettie's heirs had agreed to sell the property for $13,500. The price included an existing mortgage of $6,000 on the Grady land north of the railroad tracks, the land the Arboretum wanted. Deducting the $6,000 mortgage from the $13,500 provided the fourteen Grady heirs with $7,500. "They are scattered over the country," Jackson added. "They want their money, and we are reliably informed that they have two standing offers for the property. But they do want the University to get it."[25]

The Regents agreed to appropriate $6,000 to cover the mortgage. The AC was obliged to come up with the additional $7,500. Encouraged by this, Jackson found new strength and began a nonstop campaign. He collected small donations from local citizens and clubs and a few generous UW professors. But it wasn't enough. After the donations were tallied, he was short $3,000, and the deadline was drawing near. Jackson again tried all the angles. He even approached the Wisconsin Alumni Research Foundation (WARF) with the proposition that if they provided the money, the land would be designated the "Russell Forest Preserve," in honor of Dean Harry Russell of the College of Agriculture, who at the time was the director of WARF.[26] Prospects looked bleak. Suddenly and unexpectedly, Mrs. James Bergstrom, Regent Frank J. Sensenbrenner's daughter, donated $3,000. By the end of 1940, the Arboretum could boast of having expanded in size to over 1,125 acres.

Leopold, in particular, was delighted. The purchase provided another unique research opportunity. The Grady Tract would be ideal, he thought, as an area for experimental forestry research, since there was nowhere else, in his estimation, within the Arboretum where serious forestry research could be conducted. Woodlots needed to be restored throughout the state, Leopold argued. Yet restoration techniques were "virtually unknown, because almost all forestry research programs had been directed toward pine, spruce, and northern hardwoods, rather than oak-hickory woodlots." Potential Grady Tract forestry research projects could include comparative studies of direct seeding with nursery transfer planting. Species that adapted best to particular soils, he added, were unknown. Unknown also were the causes of oak disease. The importance of the availability of Arboretum land for forestry research was underscored by the fact that it was unlikely anyone *outside* the university would undertake ecological research on forestry problems. "Only the university had the personnel requisite for these . . . problems," he noted.[27]

But efforts to restore an oak-hickory community in the Grady Tract took years. In 1945, five years after the original purchase, Leopold in a memo to the Arboretum staff reminded them of the importance of the experiment as an ongoing exercise. For years after 1945 and after Leopold's death in 1948, the Grady Tract oaks remained the main object of the Arboretum's forestry research. By 1953, Curtis would refer to the Grady oak woods as the nucleus of a 500-acre UW forest available for research, demonstration, and teaching.[28] Unfortunately, in March 1954, three hundred Grady Tract oaks were destroyed in a fire, discussed in more detail below. The Arboretum's University Forest Research Project, as a result, suffered a setback from which it never fully recovered.

The 1940 acquisition of the 200-acre Grady property, however, also provided the opportunity to experiment with the planting of a brand new Arboretum prairie. The Arboretum would now have two prairies. The new one, along the southernmost boundary, would be planted almost single-handedly by Henry C. Greene, a botanist whose specialty was the study of parasitic fungi, but whose passion from 1940 until the late 1950s was the creation of a prairie in the Grady Tract.

Curtis, Greene, and the Grady Tract Prairie

In October 1940, Leopold proposed the formation of a Technical Research Committee. It included six professors plus Gallistel, who, in 1939, owing to Gilbert's health problems, took over the AC chairmanship. The Technical Research Committee was formed in the shadow of the older Special Committee on Arboretum Planning that dated from 1933. Like the older committee, it existed within the structure of the larger AC.[29] By 1940 Curtis was an active member of both the AC and the Technical Research Committee. Into 1941, he continued to juggle his efforts between teaching and Arboretum plant work. In 1942, he earned a breather. Granted a Guggenheim Fellowship, he undertook the study of lake forests and nature refuges as major plant formations in Michigan, Wisconsin, Minnesota, and Ontario. He worked on the project until October 1942, when he left for Haiti to work on a rubber plantation as the director of a research project studying the possibility of commercially processing latex. The war effort encouraged research into alternative sources of rubber owing to the Japanese occupation of Malaysia, the world's primary source of rubber. While Curtis was in Haiti, he also got interested in vegetation studies.[30]

When he returned to the UW in 1946, Curtis began an extended study of the environmental dynamics of Wisconsin plant communities that culminated in the 1959 publication of *The Vegetation of Wisconsin*. He also went back to the work that Robert L. Burgess tells us he loved the most, the restoration of the Arboretum prairie. By 1946, however, when Curtis returned to his

position as director of plant research, he had two Arboretum prairies to oversee. One was the 60-acre prairie on the western edge of the Arboretum. The
other was the 40-acre Grady Tract prairie that his longtime friend, Henry C.
Greene, was now developing.

Henry Campbell Greene, born on December 13, 1904, was an Indiana native, raised in Fort Wayne. Greene came from a wealthy Hoosier family. He
attended Wabash College in his home state for two years before transferring
to the University of Washington in Seattle, where he earned a B.A. in 1928
and an M.A. in 1929. In fall 1929, he entered the graduate program in botany
at the UW–Madison, completing requirements for the Ph.D. in 1933. As a
Wisconsin graduate student he worked primarily on molds and mold fermen-
tations. In 1937, Greene was hired as an instructor in the Botany Department,
although it was understood when hired that he would not teach since he admitted to having no desire for a professorship or for time-consuming administrative work. The instructorship, however, gave him staff privileges, useful
for conducting his research and for assisting graduate students, which he did
voluntarily and with enthusiasm. Prospective young mycologists, pathologists,
and ecologists regularly sought his counsel.[31] In 1941, he was appointed curator
of the Botany Department's Cryptogamic Herbarium.

Greene's family left him financially secure enough to be able to commit
full-time to his fieldwork and his research. He would publish nearly forty
papers during his lifetime, most of which were included in a continuing series
under the title "Notes on Wisconsin Parasitic Fungi."[32] He also published
three monographs: "Host Index of Parasitic Fungi Collected on Plants in Wisconsin, 1880–1950" in 1951; "Fungi Parasitic on Plants in Wisconsin" in 1957;
and in 1965, an extensive revision of the 1957 monograph, which has been
described as "probably . . . the most complete . . . up-to-date catalogue of the
parasitic fungal flora of any state in this country."[33] Greene and Curtis also
collaborated on a number of scientific publications, including four articles on
Wisconsin relic prairies published between 1947 and 1953 and two articles on
Wisconsin vegetation published in 1953 and 1955. The two were colleagues, but
they were also best friends, a relationship that commenced shortly after Curtis
began his graduate work in 1934.

From 1934 to 1937, the year when Curtis earned his doctorate and when
both he and Greene were hired as instructors, Greene and Curtis frequently
roamed the Waukesha County countryside around Eagle, Wisconsin, together,
looking for pre-settlement prairie remnants.[34] Greene's increasing interest
in low, wet prairies and the variety of plants that survived likely stemmed, in
large measure, from those early excursions with Curtis. When Curtis left for

Haiti in 1942, Greene kept up a regular correspondence. In 1943, in response to a question about the success of his research into the production of latex, Curtis candidly told Greene that the future for latex in Haiti was dismal. With time on his hands, he subsequently shifted much of his energy into the study of plant community groupings in western Haiti.

In 1940, the year the Grady Tract was acquired and Curtis was named director of plant research, Greene began to contemplate the possibility of turning the 40-acre southeastern part of the Grady Tract into a prairie. He knew that there were pockets of native prairie plants along the nearby railroad right-of-way. There were also pockets scattered at random throughout the tract itself. The higher parts of the tract, once sandy hills deposited as ice age moraine, also contained some remnant plants.[35] In 1942, Greene surveyed the 40 acres and came away convinced that he could build a prairie. In 1943, the AC approved his request, and he began work, but on a limited scale. In 1944, anxious to expand the effort, he asked for and received a written precondition, signed by Leopold, Longenecker, and Gallistel, that only he and a few trusted associates would do the work. For the entire duration of the project, no one else was "permitted to plant, experiment or manipulate the small plot of land." One reason he insisted on the terms was because he did not want unskilled laborers on the project. He had in mind specifically problems with the CCC working on the Curtis Prairie in the 1930s.

The odd arrangement, guaranteeing Greene his privacy and solitude, was honored by the AC and Arboretum staff. Nearly twenty years passed before the public had access to Greene's prairie. He staked his grids by himself and hauled water for newly planted seedlings by himself. By 1951, he had planted, *by hand*, over 12,000 seedlings and mature plants representing at least 133 species. He kept records, accurate to the smallest detail, from the first survey and soil maps to a catalog of plants that grew on the acreage before he began his transformation. Over the course of time, he put together eight annual reports, one six-year summary report, one biennial report, and a collection of grid maps showing precisely where plants were grouped. The records alone stand out as unique among prairie restoration efforts.[36]

For Curtis and others, Greene was the ultimate taxonomist. His talents were legendary among the Arboretum staff and graduate students who worked with him. He was said to have had "an innate ability to understand plants," a talent that he possessed "over and above those of most everyone else." His prairie was a textbook marvel, "as precise a reconstruction of a lost landscape as exists anywhere."[37] For Virginia Kline, Arboretum prairie ecologist from 1975 to 1996, Greene's prairie was one of the most successful transformations anywhere, a fine example of "the best we can do." Much of its success, she claimed, in a glowing, lyrical tribute to his accomplishment, was, "measured in esthetic terms—the beauty of the prairie vistas, the colorful flowers in a background of grass, the remote location with its near-relief from highway noise. The songs of sedge wrens mingle with those of yellowthroats and goldfinches along the brushy edges, and an occasional red-tailed hawk calls from above. To complete the sensory impact, mountain mint yields its pungent aroma in response to passing feet, and on a warm day late in summer the air is filled with the tantalizing fragrance of prairie dropseed."[38]

Greene was a hard-working scientist, an inspiring student adviser, a trusted colleague. But he was also a good campus citizen who developed an uncompromising loyalty to the Arboretum. In spite of his request in the early years to be spared the bother of committee work, he willingly served as secretary of the AC for over twenty years. And for ten years, from spring 1952 until January 1962, he single-handedly edited the *Arboretum News*, a follow-up to the informal *Arboretum News Letter* that McCabe, beginning in 1943, had intended as an informational guide to the Arboretum and Arboretum projects.[39] The contract guaranteeing Greene's privacy while working on the Grady prairie was eccentric. It reflected his personality, which was also eccentric. He was, as one account of his life recalled, "solitary . . . comfortable only in his own knowledge of plants."[40] Shy and retiring to the point of timidity, he was happiest with his plants and his prairie. He also appears to have been quite alone. For a period, he was married, but the marriage ended in divorce. In the 1960s, his work on the prairie slowed, and he seemed to slow with it. His reports came less frequently.

On June 7, 1961, his longtime friend, John Curtis, died. Greene wrote the memorial piece for the *Arboretum News*. He remarked that Curtis, despite being exceedingly busy, "was a person upon whom this writer . . . could always depend." But Curtis was much more to Greene than a dependable contributor to the journal. At the end of the tribute, Greene wrote this: "We who have been associated with him in conducting the affairs of the Arboretum feel his loss keenly. . . . had he lived, he would . . . have guided the Arboretum to ever-increasing prestige."[41] In January 1962, Greene gave up the editorship of the *Arboretum News*. In 1966, he retired from the Arboretum and the university. In April 1967, he drove to Tuscon, Arizona, where he planned to retire, to look for a place to live.[42] A dedication ceremony in his honor was scheduled for May 8, 1967, at the Arboretum. The Grady Tract prairie was going to be dedicated to him and renamed the Henry Greene Prairie. But he would never see the Henry Greene Prairie or the Arboretum again. For unknown reasons, on April 27, in Tucson, Henry Greene committed suicide.

Friends and colleagues gathered at the Grady Tract prairie on May 8 to dedicate the prairie to Greene and to honor his memory. They heard of his contributions. "They viewed the simple redwood marker for the Henry Greene Prairie," but as the reporter covering the story noted, "perhaps the crowning tribute came from the prairie itself, spread before them and now approaching spring bloom." Grant Cottam, Greene's friend and colleague, eulogized him as "an eminent mycologist, an excellent ecologist, a teacher of both student and staff." And in his final tribute to Greene's memory, Cottam offered this memorable reflection: "Henry's life was lived among the flowers of the prairie, the trees of the forest, and the microscopic fungi that lived on all these other plants. . . . It is appropriate that we use this prairie as the chapel most suitable in which to remember him. This is where he worshipped. This is what he loved."[43]

The Greene Prairie continues, as Gina Kline observed, to be one of the most successful prairie restorations anywhere. Over the years since Greene completed his planting, the prairie has had limited weed problems. However, recently, owing mainly to a large and controversial housing development built on the hills south of the tract, much runoff from heavy rains now regularly floods the low-lying southeastern end of the prairie. The runoff carries seeds of reed canary grass, a coarse, tough grass, widely distributed throughout the northern midwestern states, that thrives in wetlands and overruns native vegetation. Efforts to thwart the advance of the grass continue, but so far without any guarantee that the grass will finally be eliminated from Henry Greene's carefully hand-planted prairie.

Robert McCabe's Arboretum Years

Ted Sperry left for military service in 1941. Shortly after he left, the AC, in response to dwindling Arboretum staff positions, appealed to the Regents to replace Sperry with a biologist who would also have faculty status. The position was filled by John Catenhusen. When, in 1942, Curtis left for Haiti, Catenhusen, in addition to his other responsibilities, was charged with overseeing the 60-acre Arboretum prairie. With both Curtis and Sperry gone, the future of the prairie seemed uncertain. In 1943, Catenhusen was unexpectedly drafted into the service. That same year, at the behest of Leopold, he was replaced by Robert A. McCabe, from the Department of Wildlife Management, who, as a graduate assistant, had worked with Leopold since 1939. Looking after the plants and the 60-acre prairie was now left largely to McCabe.

McCabe had grown up in Milwaukee. Born in 1914, he lived out the Depression years in poverty in South Milwaukee's German/Polish ghetto. After high school, he wanted to go to college. In order to raise tuition money, he got involved with a circle of Wisconsin rabbit hunters who, impressed with his contagious enthusiasm for nature, subsequently, hired him to take care of their dogs and to skin and clean the rabbits they shot. John T. Emlen Jr. and Thomas R. McCabe (National Biological Service), in a testimonial to Robert McCabe, after his death in 1995, maintained that the early experience with the dogs, the rabbits, and the hunters planted "the seed that blossomed into a life-long pursuit of knowledge regarding natural resource ecology."[44] The money young McCabe earned with the hunters, along with a lifeguarding job and an athletic scholarship, enabled him to cover tuition and books at Curtis's alma mater, Carroll College, in Waukesha. After graduation in 1939, he took a casual pre-registration trip to the UW–Madison, and in "'a master stroke of fate,'" as he later put it, during his visit he met and talked with Leopold. McCabe asked Leopold if he could study with him; Leopold agreed. The meeting initiated a relationship that benefited Leopold, who regularly sought out McCabe for help carrying out research projects. But the meeting had far greater meaning for McCabe, who was profoundly influenced for the rest of his professional life by his early association with the "boss."[45]

In 1943, McCabe earned his master's degree in wildlife management. In 1946, he accepted a teaching position in the Wildlife Management Department and simultaneously began his doctoral work researching ring-necked pheasants with Leopold as director. Throughout the 1940s, he worked in the field with Leopold, establishing feeding stations for game birds, clearing landing areas, rounding up and banding game birds, conducting an annual pheasant and woodcock census, often with the CCC crew members as beaters, and, at times, trapping mink and other small mammals. After Leopold's death in 1948, McCabe, in 1949, was appointed to an assistant professorship in the Department of Wildlife Management. In 1952, he became chairman of the department, a position he maintained until 1979. In 1984, he retired and subsequently died at age eighty-one on May 29, 1995.[46]

In spite of twenty-seven years as chair of the Wildlife Management Department and the constant demands of his research, which often took him far afield from Madison, frequently to Africa and to Ireland, where he served as a wildlife management adviser to the Irish national parks system, McCabe always maintained close but often strained ties with the Arboretum. Nevertheless, he had a commitment to the Arboretum that extended long after the Leopold years. He served regularly on the AC. He planted tamaracks at Leopold's request around the Teal Pond and, for years, kept records of their growth. He conducted a ten-year study of catbirds. He also conducted a ten-year study of

Professor Robert McCabe, Arboretum biologist, and Jim Hale, his assistant, in front of a pheasant trap, studying a band. They used research techniques developed by Aldo Leopold. (photo by Arthur Vinje, Wisconsin Historical Society, WHi-41882)

the alder flycatcher in the wetlands and, among other projects, counted the trilliums in Noe Woods. His woodcock studies, widely recognized, extended over forty years, with much of the census data collected at the Arboretum.[47] For nine years, from 1939 until 1948, he and Leopold worked together. Consequently, in 1943, when the Arboretum's plant communities needed a biologist-caretaker for the war years, Leopold turned for help to McCabe.[48] During the war years, with the Arboretum short on labor, McCabe worked largely by himself. When they could, the staff enlisted the help of UW students and interested faculty. One of McCabe's assignments was to conduct the experimental prairie burns. For that he needed crews. Often, he was assisted by Leopold, Longenecker, Gallistel, Greene, and other volunteers from the AC and the Department of Wildlife Management.

In continuing the practice of controlled burns begun by Sperry and expanded by Curtis, McCabe explained, the objective was "to ascertain which season of the year produced the best results, determined by bluegrass control and prairie plant response."[49] On May 12, 1943, shortly after McCabe's appointment, the AC minutes recorded the first prairie burn of the year.[50] McCabe noted in the November 1943 issue of the *Arboretum News Letter* (the first issue published, which he edited) that prairie quadrat readings and experimental line burns, "as in other years," had taken place for all of the 1943 burns. Areas on the prairie, he added, "which were planted to baptisia and rattlesnake master were staked off for yearly censuses and natural reproduction counts." In 1944, McCabe reported on another prairie burn—in the spring, "made on schedule and without mishap."[51]

"Grass Mowing Is Unnatural," More Problems for Leopold

There is an implicit note of unhappy frustration in McCabe's accounting of his years as the Arboretum biologist that suggests that both he and Leopold were having difficulty with the AC during the mid-1940s. The friction was a harbinger of things to come. For some unaccounted reason, the perception of the Arboretum as a "wildlife refuge" was being questioned and challenged by members of the AC. McCabe observed, in his 1987 profile of Leopold, that gradually, during the course of the 1940s, the "Wildlife Refuge" listing was dropped from Arboretum letterhead. McCabe claimed that the title also gradually ceased being used even in verbal references to the Arboretum. At one point, the Wildlife Management Department employed the figure of a flying woodcock on reprint covers of Leopold's report to the faculty on turnovers in the Arboretum's pheasant population. The flying woodcock design, "incorporating the words 'University of Wisconsin Arboretum,'" was not received with enthusiasm by all members of the . . . Committee," McCabe recalled. It was never used again. "Apparently," he added, "the animal motif was regarded as inadequate for an area intended largely for plant programs." And how did Leopold respond? Leopold "did not argue," McCabe noted.

The conflict within the AC over the "wildlife refuge" designation had been developing, McCabe suggested, at least since the early 1940s when Leopold, Fassett, and Curtis attempted to curtail "the amount of mowed grass and park-like conditions at the Arboretum." They believed that "manicuring" the Arboretum landscape, after the fashion of a public park or the "zoo across Lake Wingra," was not in keeping with an area officially designated as a wildlife refuge for native plant and wild animal research. As late as 1977, McCabe would still maintain that the Arboretum had many more birds "before they took out the shrubs and started mowing the lawns." Mowing grass, he added, was "not natural, and it gobbled up time and money."[52] Yet in spite of efforts to curtail what Leopold, Fassett, and Curtis believed was a cosmetic activity contrary to the Arboretum's original mission, the three outnumbered scientists lost "the battle of the lawnmower."

McCabe also claimed that his efforts in 1943 to promote an *Arboretum News Letter* were "silently rebuffed" by some members of the AC. "After a cool and unresponsive reaction" to his first issue, he tried another, "only to find the climate had not changed." He noted that he and Leopold discussed the problematical reaction to the newsletter and decided to forego any more issues. McCabe observed of that frustrating time, "I never did find out [the cause] of the indifferent reaction among committee members." And as for Leopold's reaction to this indifference: "If he knew, he did not tell me," Mc-Cabe concluded.[53]

McCabe was the Arboretum biologist for only two and a half years before giving up the position in 1946 to return full-time to the Department of Wildlife Management. However, as the record shows, from 1943, the year after Curtis left for Haiti, to 1946, under McCabe's direction a strict schedule was followed for experimental prairie burns in March, May, and October, burns that were done on both planted and unplanted prairie plots.[54] But it would be left to Curtis, upon his return in 1946, to expand the process. Once back, he began to experiment with a variety of burning techniques. Unlike Sperry in the early years, who depended on unskilled CCC workers for labor, Curtis had the luxury of professorial research time and the assistance of skilled graduate students. Consequently, he was also able to collect volumes of scientific data, to carefully record that data, and to make it available to the scientific world. As a result of Curtis's field successes, Sperry, who actually was the first to work on the Arboretum prairie, was quoted, in an interview in 1990, observing that Curtis, both for his accurate record keeping and his published research, deserved to have that 60-acre tall grass prairie named after him. Curtis "gets credit for being first. . . . He did the work," Sperry noted.[55] And eventually Curtis did get that credit. In 1962, the prairie was named the Curtis Prairie.

The Troubled War Years

As expected, the war effort resulted in a steady drain of talent and labor from the Arboretum. Not only did Sperry and Curtis leave, but so did Norman Fassett, who in 1944 went to Colombia with other botanists searching the high Andes for Cinchona, a tree yielding quinine for the treatment of malaria, which was on the increase among troops fighting in tropical zones.[56] To compound Arboretum difficulties, a series of irritating annoyances began to consume much of the beleaguered remaining staff's time. One of the persistent aggravations was the perennial "rabbit problem." At one point, even ferrets were introduced in an effort to control the exploding rabbit population. In April 1941, Longenecker appealed to Leopold, in his capacity as "game manager." It was "imperative," he told Leopold, "that something . . . be done." The rabbits were eating his oaks; over 2,000 seedlings had been destroyed. In winter 1941 alone, rabbits ate or damaged beyond repair 605 bur oak seedlings, 333 red oak seedlings, and 43 white oak seedlings. The entire oak woodland that Longenecker and Curtis had planned to restore was a disaster.[57] McCabe reported in January 1943 that 682 rabbits had been removed and that the removal would continue, but the rabbit problem persisted.

And then there were the neighborhood dogs and cats roaming freely in the Arboretum. As a result, Leopold and McCabe had to stop the pheasant trapping and banding for a time. How does one manage a game program with neighborhood dogs and cats on the prowl? And to compound the difficulties, the city of Madison was rapidly encroaching on the Arboretum. The dog and cat problem was a direct result of the city's growth. By 1946 Madison's population would top one hundred thousand people. Nevertheless, in January 1942, as usual, the annual Arboretum pheasant drive took place. Joe Hickey, who was completing his master's thesis with Leopold, recalled a sizably diverse group taking part that January. The "crew" consisted of Leopold, McCabe, and Hickey from Wildlife; Irv Buss, Fred Zimmerman, and Elton Bussewitz from the Wisconsin Conservation Department (WCD); Longenecker, Catenhusen, and Tom Butzen from the Arboretum staff; two wildlife undergraduate students, and a young man named Frank. Working from 8:30 a.m. to 4:30 p.m., they managed to flush 317 pheasants, as well as 4 to 6 short-eared owls, 4 long-eared owls, a great horned owl, and 9 gray foxes. Lunch, served at the Arboretum, consisted of coffee and hot dogs. "I remember wolfing down five of the latter," Hickey noted.[58]

Into 1943, McCabe still envisioned the Arboretum as a pheasant refuge, with the breeding population holding at 215. McCabe and Leopold were also trying to raise wood ducks from eggs. Fourteen penned Hungarian partridge chicks had been loaned for a research project by the Poynette State Game Farm. Eleven chicks had survived and would be returned. In the meantime, work progressed on a goose and duck refuge in the Gardner Marsh. The Island had been burned, and what didn't burn was cut. Another area, approximately 6 or 7 acres to the south of the Island, which was designated as a landing area, was also burned. A fence was built on the Island for 7 geese and a number of ducks to be used as decoys. The project had the support and cooperation

of the WCD, which agreed to house the geese for the winter at the Poynette
Game Farm. Leopold, pleased with the effort and with the possibility of having a functioning goose and duck refuge in Gardner Marsh, urged the Arboretum Committee to send a note of thanks to the WCD.[59]

In 1942, as a contribution to the war effort, Franz Aust, a landscape architect in the Department of Horticulture, conducted a cooperative project in
camouflage study in conjunction with the Military Science Department.
In 1943, extensive research on fish populations began in the Stevens and
Ho-nee-um Ponds, under the direction of Arthur D. Hasler. Hasler would
report that sunfish grew faster in Stevens than in the Ho-nee-um Pond. A
thousand bluegills were placed in Ho-nee-um. Hasler and H. P. Thomson
also were studying minnow propagation in the Gardner lagoons in the East
Marsh. In 1944, to protect the minnows, they removed a hundred pounds of
carp from the Gardner Marsh.

For Longenecker, the endless planting continued—five thousand two-
year-old tamarack seedlings in spring 1943; one thousand wildflower plants
in spring 1943 in Wingra and Noe Woods. On May 8, 1943, more than three
hundred students organized by the University Work Day Committee planted
about a thousand red and white pines in the northeast corner of the Grady
Tract.[60] By 1943–44, regular workday projects for UW students involved planting trees. To attract students to the campus program, the organizers offered
entertainment and free refreshments. They also initiated a contest to crown a
"blue jean queen," a co-ed who reigned over the festivities with dirt on her face
and a shovel as her scepter.[61] In February 1944, Leopold and McCabe, studying
the Arboretum's expanding mink population, set up a mink-trapping project
and captured seventeen mink. In spring 1944, more wildflowers were planted
in Wingra and Noe Woods. In mid-summer 1944, ten thousand tamarack
seedlings were planted east of the Gardner Marsh.

On April 13, 1944, several boys built a fire in the Grady Tract that, owing
to high winds, went out of control. About 100 acres, including part of the
cherished oak opening, were burned. Concern about Arboretum trespassers
increased. But there was not enough security available to keep trespassers away.
Also, by 1944, there were rumors that the state highway commission was looking into the possibility of putting in a belt highway around Madison. Plans
at the time called for a portion of the highway to extend through the center
of the Grady Tract. The Regents and the Wisconsin Conservation Department passed resolutions opposing the plan. The Arboretum management was
assured, however, that the Wisconsin State Highway Commission (WSHC)
would cooperate, if necessary, in relocating the road.[62]

Between 1944 and 1945, Jackson made an effort to procure the Winslow Tract, a 9- acre parcel of unimproved land abutting the west side of Noe Woods. As late as May 1945, Jackson, still haggling with the Winslow family lawyers, finally gave up. Eventually the tract was divided into residential lots.[63]

Picnic Point and Grazing Cows

In June 1944, the AC, with reservations, agreed to supervise the University's Picnic Point–University Bay property, originally acquired in 1941. Shortly after the acquisition, the University Bay Committee (UBC) was formed, which set an agenda for the whole Picnic Point area. Not surprisingly, owing to the fact that the Arboretum and Picnic Point were both nature preserves with much in common, members of the AC, including Leopold, Longenecker, Fassett, James Dickson, and Arthur Hasler, were obliged to serve on the UBC.[64] Leopold, in particular, played a key role in the committee's recommendations. The UBC produced two early educational publications, one by Hasler, the other by Leopold. The publications suggested possible educational uses for the area. Hasler's "Teaching Exhibits Which Should Be Installed in the University Bay Area" recommended exhibits on plant succession, rodent pressure, shade tolerance, erosion, vegetation understory, and a red cedar plantation.[65] Leopold's "Wildlife in the Picnic Point Program" suggested using the available wildlife for education. The area had an ample selection of owls, pheasants, and shorebirds, and a diverse selection of mammals, including fox, rabbits, muskrats, and minks.[66] Hence, the UBC, at Leopold's urging, recommended developing the area "as an outdoor laboratory for teaching, demonstration and research, and as a museum of natural history and early agriculture of the state."[67] The UBC also recommended discouraging automobile access, removing exotic trees and shrubs, and restoring natural plant associations. In spring 1944 the UBC, at the urging of Leopold, made an effort to have Lake Mendota declared a wildlife sanctuary.[68]

Even though Picnic Point was miles north of the Arboretum and the staff was already overtaxed, the AC agreed, with some hesitation, to take on the added responsibility. Picnic Point would be the first of a series of campus "natural areas" that the Arboretum was asked to manage. After having reluctantly assumed the stewardship, imagine the surprise when two years later in October 1946, the AC learned that the College of Agriculture, with no prior consultation, planned to release a herd of grazing cattle on Picnic Point. Gallistel, as chair of the AC, told A. W. Peterson, the campus director of business and finance, that he was in disbelief. He also told Peterson that Professors

Henry L. Ahlgren and Otto Zeasman, agronomy specialists, had surveyed the targeted area and concluded that the land was unsuitable for grazing. Gallistel hoped that the land would be spared.[69] In response, Peterson explained that the grazing incentive was a postwar matter that had developed after WARF chose to locate its new campus housing project on the university's Eagle Heights Farm. As a result, the Regents agreed to make Picnic Point available for the College of Agriculture. And if the college decided to use the land for grazing, then the supervision of the land would be transferred from the AC to the College of Agriculture. The controversy was temporarily settled when R. A. Brink, from the Agriculture College, proposed a compromise—instead of grazing cattle, the college would settle for three one-quarter-acre plots in which to grow experimental wilt-resistant alfalfa. In fact, the college had already plowed and limed the plots. The AC, frustrated and now with little choice, approved the request. Two months later, in late December 1946, Gallistel was obliged to report to the AC that the College of Agriculture had Picnic Point in its sights once again and now wanted all of the land that was not wooded.[70]

Eventually the AC would be relieved of managing Picnic Point, a decision that the committee members, especially Leopold, Longenecker, Hasler, Fassett, and Dickson, supported enthusiastically.

Nakoma Country Club: An Arboretum Golf Course?

In August 1945, Emerson Ela, a Madison attorney representing a bondholders' committee that owned the Nakoma Country Club, contacted the Board of Regents about the possibility of purchasing the country club. The club consisted of 106 acres of land, with a beautiful golf course, a first-class watering system, and a large stylish clubhouse. This was not the first time that the subject of purchasing the country club had come before the Regents. Since the onset of the Depression, the club had a history of unpaid debts and threatened foreclosures. As early as February 1936, the bondholders asked Paul Stark to suggest a way to raise approximately $50,000 to cover the cost of a bond issue. Once the money was raised, the land, with the support of the country club membership, would be given to the university free and clear of encumbrances. In turn, the country club would be granted a lease on the property for twenty years at a rental fee of $2,500 a year. The Arboretum would have access to the land for the purpose of directing the CCC to plant trees or shrubs. The golf course would become the official University of Wisconsin Golf Course, but the country club membership would retain the right to use it.

Two years later, in April 1938, the issue was still unresolved, and a proposal for the purchase was pending. Neither Stark nor the university, however, could raise the funds. and the deal soured. The Regents, weary of hopeless land ventures, went on record urging the AC "not to entangle the University in any way" in negotiations for additional Arboretum properties.[71] When asked by a *Capital Times* reporter if the university had funds available for adding the Nakoma Country Club to the Arboretum, Gallistel, representing the AC, replied emphatically, "'Not a penny.'"[72]

In summer 1945, the Nakoma Country Club was once again in financial trouble and once again on the block. And, once again, the AC was interested in purchasing it. But this time, so was the city of Madison. Madison mayor F. Halsey Kraege was particularly interested in floating a mutual arrangement between the city and the university whereby they both shared the cost and the privilege of using the golf course.[73] Longenecker encouraged the purchase, warning that if the country club land was subdivided and homes were built, the Arboretum would suffer. He also questioned the benefits of an Arboretum ownership of a golf course. Aware that the Athletic Department and the Division of Physical Education also wanted the property, Longenecker proposed a compromise whereby the golf course area could be used by both the Athletic Department and the Arboretum. Greens, tees, and fairways were expensive to build, and those at the Nakoma Country Club were first rate. The watering system would also help "in establishing and maintaining certain types of planting now almost impossible to grow." He envisioned the clubhouse as a natural history museum or a reading library with additional office space. He urged McCaffrey to convey to the Regents his desire to buy the property. Now was the time to go after it, he advised, adding, that "pressure for building sites in the Nakoma area will keep getting worse."[74] Longenecker was right about the fading prospects for buying property in the fashionable Nakoma area. Unfortunately, the Nakoma Country Club bondholders in 1945 sold the property to a financial group before the university, hindered by the usual bureaucratic delays, could muster up enough money and clout to make an offer. A. L. Masley, acting director of the UW Division of Physical Education, had warned in August 1945 that soon the country would be in a postwar period with a resultant increase in the need for property. He urged the University to act as soon as possible or lose the opportunity once and for all.[75] Masley's prediction was on the mark. The last chance the Arboretum had to acquire the Nakoma Country Club came and went in that summer 1945. It has never come again. But the memory of what had once been a possibility lingers around the Arboretum. Arboretum staff members, on occasion, still often look lustily west to the fertile grass, the many trees, and the handy watering system of the Nakoma Country Club golf course. And they often stop to ponder and sometimes to comment on "what could have been." A golf course? Why not?

An ironic postscript to the Nakoma Country Club story occurred in 1948, when, at the May 6 meeting of the AC, Gallistel reported that the Nakoma Country Club had sent him a proposal offering to buy a portion of adjoining Arboretum land to be used as a parking lot. The proposal was unanimously rejected.[76]

The Postwar Years: Developing a Global Identity, Impending Challenges

By 1947, interest in the Arboretum, both nationally and internationally, had increased considerably. The Arboretum was beginning to develop a global identity. The staff reported receiving inquiries relative to Arboretum matters from as far away as Yugoslavia. An August Technical Committee Report noted that the widening range of subjects being studied at the Arboretum and its increasing use by UW scientists involved research of international significance on a variety of botanical subjects, including the progress of 20 acres of white and red pines planted in the Grady Tract in the spring, studies on the inter-relation of soil moisture and plant associations, the ecological composition of forest stands, and the effect of fire on the maintenance of prairie plants. Greene's continuing work on his Grady Tract prairie was augmented in 1946 by the discovery of 30 Indian paintbrush plants in a remnant prairie at Devil's Lake, from which he and James H. Zimmerman, a botany student who eventually would become the Arboretum's first "ranger-naturalist," collected seeds. A five-year study of the use of fire in preparing prairie seed beds was also underway in 1947 and gaining international interest. Ongoing were studies of the pheasant population, now in its tenth year, and studies of the population and life history of muskrats and frog populations on the Lake Wingra shore-line. Leopold was working on a study of the population mechanics of a flock of semi-wild mallards. He was also guiding a study of the aging character of cottontail rabbits. There was also growing interest in the possibility of publishing a scientific journal.[77]

The pressing need for more laboratory and classroom space came before the AC in 1947 and 1948. There was even talk of building a boathouse on the Wingra shore. A decision to renovate some of the old CCC buildings provided a temporary solution to the space problem. The old CCC bathhouse, for instance, was destined to be turned into the Arboretum laboratory. John Emlen (Zoology) would eventually use one of the barracks for his mice research. That barracks appropriately was tagged "the mouse house." Some members of the AC, thinking ahead, were beginning to advise the possibility of an "outreach" program to local schools. Trenk (Forestry) argued that educating students, especially high school students, on the Arboretum's aims and purposes might help to alleviate continuing difficulties with the public. Leopold suggested hiring a naturalist to conduct public tours. Robert J. Dicke (Entomology), in a notable early effort to urge the AC to take a position on a matter of environmental urgency, alerted the committee to the seriousness with which he viewed the proposed blanket spraying of the city of Madison with DDT. The committee recognized the dangers associated with the spraying and suggested alternatives, but no definite action was taken. Within ten years, the AC and the Arboretum staff would be at the forefront of environmental activism.[78]

In 1949 construction on the Beltline Highway ominously began. The Arboretum was forced to sacrifice 15 acres of land. Before the highway was completed, the Arboretum would sacrifice much more. In 1949 David Archbald was appointed interim Arboretum botanist to succeed Max Partch, who had resigned in July. Partch would be remembered particularly for his extensive work with Curtis on the effects of fire on the propagation of prairie grasses. Archbald, who would also work with Curtis until finishing his Ph.D. in 1954, would be named managing director of the Arboretum in 1962.[79]

Leopold's Untimely Death and the End of the Arboretum as a Wildlife Refuge

In a 1980 interview, Max Partch recalled that in 1946, when he returned to the campus, Leopold was devoting a noticeably larger portion of his time to the Faville Grove area and to ongoing efforts to promote farm practices intended to create habitat for wildlife. He often worked right along with the farmers, Partch observed. There had been growing sentiment in the mid- to late 1940s among Arboretum staff, including Leopold and McCabe, that the Arboretum plant communities were not large enough to support animals. Hence, Leopold began to distance himself from the Arboretum as a wildlife refuge. The 92-acre Faville Grove area offered more possibilities for wildlife field research.[80]

But by August 1947 into the spring of 1948, Leopold's work at the Arboretum, at Faville Grove, and elsewhere in the state was frequently interrupted by illness. He suffered from recurring facial spasms. Leopold kept working in spite of weakness and ill health, which was compounded in fall 1947 by a painful eye problem. After extensive revisions, his book of essays entitled *Great Possessions*, although still in need of editing, had a new foreword and seemed finished enough to chance sending it to publishers. On December 19, 1947, he sent copies to William Sloane Associates and to Oxford University Press. On Christmas day Leopold felt rested enough to take a walk through the Arboretum with his daughter, Nina, and her husband, Bill Elder. There had been a dusting of snow, but they weren't going far, only to the pine forest that had first been planted as seedlings in 1933, the first year of Leopold's tenure as research director. He noticed seeds on the snow, an indication that cones from the red pines were opening.[81] Within five years, the pine forest would bear his name.

On April 14, 1948, Leopold received some good news. Oxford University Press was interested in publishing *Great Possessions*, the essay collection that in 1949 would be published as *A Sand County Almanac*. He looked forward to editing the manuscript for publication. In spite of his weakened condition, he knew he was strong enough to see the book through the press. A week passed. He and his family were staying at the family weekend retreat, called the "Shack," near Baraboo on the Wisconsin River. About 10:30 in the morning on April 21, Leopold spotted smoke filling the air over the farm of his neighbor Jim Ragan. A trash fire on Ragan's property had burned out of control and had turned into a raging grass fire. The family left the Shack and went to help. When they got to the fire, it was heading downhill toward their property. According to a variety of accounts, Leopold disappeared alone into a marsh with a heavy water pump on his back, intending to wet down one of the burning edges. At some point, he suffered a massive heart attack and died. He was sixty-one years old.[82]

His friends and colleagues were shocked by the suddenness of Leopold's death. Irven Buss, a former graduate student, would write, "The announcement of Aldo Leopold's death came suddenly, so suddenly . . . that even now many of us . . . in the halls of the conservation department and on the campus of the university have not yet attuned our hearts and minds to the new world in which he left us."[83] When McCabe first heard the news, he didn't believe it. He thought it was a bad joke.[84] Leopold died at a time when, though his health was diminished, he still had much to offer and much to gain. He was also in the midst of establishing an international reputation as a conservationist. Shortly before his death he had been asked to be a discussion chairman at the Inter-American Conference on Conservation of Renewable Natural Resources and to serve on the Advisory Committee on American Participation at the United Nations Scientific Conference on Conservation and Utilization of Resources tentatively scheduled for either 1949 or 1950.[85]

On May 6, 1948, within two weeks of his death, a grieving AC took up the matter of a memorial. At the meeting, Professor Charles Bunn (Law School) distributed a draft, entitled "In Memory of Aldo Leopold." He proposed that the AC create an Aldo Leopold Memorial Fund that would continue for five years, the money to be used for the purchase and planting of native Wisconsin species not yet represented in the Arboretum and for research into the interaction between native species and other established plant communities. "We do not propose a permanent endowment," he noted, because Leopold firmly believed that research money should be put to work as promptly as possible. The committee voted to create a Memorial Fund subcommittee, headed by Bunn, to manage the fund. The AC also discussed the possibility of putting together a Leopold memorial volume that would include a selection of his and his student's papers on Arboretum research and work projects. In July 1948, the AC approved, in principle, Bunn's suggestion that the committee create a revolving fund to facilitate the publication of the memorial volume.[86] The volume, however, appears never to have gotten beyond the planning stage. It's likely that the 1949 publication of *A Sand County Almanac* had something to do with the apparent scuttling of the plan.

By August 1950, the AC still had no definite plans in mind for a lasting and suitable Leopold memorial. There was disagreement on specifically what it should be. But the committee continued unanimously to agree that the memorial should be directly connected to something organic. On August 22, Gallistel approached Ernest Swift, director of the Wisconsin Conservation Department (WCD), with the proposal that the WCD rename the Flambeau River State Forest in north central Wisconsin as the Aldo Leopold Wilderness Area. Swift responded that since the Flambeau Forest was never intended to be a complete wilderness area and since large areas were intended for timber harvest, he could not approve the proposal. He suggested that something in game management would be more appropriate.[87]

In spite of the AC's initial objections to anything other than a natural memorial, in October 1951, the committee did accept an offer from Forrest Nagler, a Milwaukee engineer, for a memorial plaque. Dickson, Hickey, and McCabe were appointed to choose an appropriate inscription from Leopold's writings.[88] It is unclear whatever happened to this plaque. Finally, in March 1953, five years after the AC first proposed a Leopold memorial, Curtis insisted that the matter be resolved and a suitable memorial be established. He proposed naming the pine plantation in the Arboretum's southeast corner the Leopold Memorial Forest. The committee approved the proposal and suggested that the dedication ceremony be held at the May 1953 meeting of the University Board of Visitors.[89] The board, first organized in the early 1950s, consisted of local friends and supporters. It was a precursor of the current Friends of the Arboretum (FOA).

Prior to the May 23 dedication, the committee decided that the official name of the pines should be the Aldo Leopold Pines and that the event would be a twofold dedication to honor the memories of Leopold and Maurice McCaffrey, secretary of the Board of Regents, the man "on the inside," who had died in August 1947, eight months before Leopold. The pines were dedicated to Leopold's memory, and the Arboretum Road, what UW president E. B. Fred referred to metaphorically as the Arboretum's "backbone," was dedicated to McCaffrey and renamed McCaffrey Drive.

On the morning of May 23, the ceremony commenced in front of Longenecker's colorful lilacs. President Fred recalled fondly his memories of McCaffrey. Leopold was eulogized by Jackson and Arly William "Bill" Schorger, Leopold's friend and hunting partner. Jackson recounted the decisive series of events promoted by Dean Russell, College of Agriculture, in the late 1920s prior to the formation of the chair of game management and, ultimately, to Leopold's hiring. He also told of leaving for a two-week vacation in 1933 just before the pine seedlings for the southeastern corner were first planted. Jackson was told that, upon his return, a pine forest would be there to greet him. Returning, however, he was disappointed to see a field of seedlings, none taller than a tomato plant. He exclaimed to Leopold, "'Do you think I am going to be around here long enough to see those *little things* grow into *trees?*'" Leopold replied, with a "quizzical smile," "'Yes, I am *sure* you will.'" Jackson added that no one derived more satisfaction from those pines than Leopold. Bill Schorger spoke of his "admirable friend" who had devoted "the best years of his life . . . to the preservation of nature," who could not live "without wild things." And Schorger predicted that Leopold's *Sand County Almanac* would "live as a permanent part of our literature." The "Leopold Pines," he also predicted, "will long serve to keep his memory green."[90]

Dedicating the southeastern pine forest to Leopold's memory, first suggested by Curtis, was thoughtful and most appropriate for the ecologist who once wrote, "I love all trees, but I am in love with pines." The Leopold Pines continue to stand as a memorial to his contributions to the Arboretum and, on a much larger scale, to his efforts to instill in humankind a sacred and protective bond with the life force that Leopold—along with Thoreau and, I might add, Michael Olbrich—celebrated as "wildness."[91]

At some point in the early 1960s, an unknown source came across two documents in Fassett's files in the Botany Department Herbarium. One was a manuscript copy of Leopold's "Exit Orchis," written in 1940. The other, described "as a touching manuscript, written with many corrections on a faded piece of yellow paper," was Fassett's "An Appreciation of Aldo Leopold," carefully drafted in 1948 right after Leopold's death, with obviously intense concern for expressing just how deeply Leopold had touched his life. Fassett wrote: "I believe that each of us can, if he will, take something from the great men he has known (and I mean men of great spirit rather than men of wealth or power), and that by knowing such men I can make them a part of myself. And so I know my spirit has been strengthened and I have been made more nearly the man I would like to be, by years of association with Aldo Leopold. In spite of the grief and sense of loss so many of us have, there is that, which cannot be taken from me." Fassett sent a copy of the tribute to Estella, Leopold's wife.[92]

In 1989, a series of recollections written mostly by Leopold's former graduate students appeared under the title *Aldo Leopold: Mentor*. The edition was a compilation of memories collected in conjunction with the Aldo Leopold Centennial Symposium held in Madison on April 23 and 24, 1987. One of the accounts was Joe Hickey's recollection of a trip to the Arboretum with Leopold. The event was an Arboretum tree planting on one of the three Saturdays in April just prior to Leopold's death. The day was memorable, Hickey recalled, because "we gained some insight into his feeling for tree planting." As was common with so many Arboretum tree plantings during the 1940s, UW students were recruited to help plant the seedlings. As Hickey related the story:

> Each male student was given a shovel to dig shallow holes for seedlings, while the coeds did the actual planting. Late that afternoon, Bob McCabe . . . and I appeared back at the Department where "The Professor" was awaiting our report. All the trees . . . had been planted.
>
> "I have just one question," said AL at the conclusion of our report. "*Were the shovels sharp?*"
>
> "No, not particularly," we replied. "I'm sorry to hear that," said "The Professor." "I wish those youngsters could have had the satisfaction of feeling a *sharp* shovel cut cleanly through the sod."
>
> "He was quite a guy!"[93]

With Leopold gone, McCabe and Hickey, who had just returned to the campus in early 1948, were asked to take over the Department of Wildlife Management. Both would continue as members of the AC, McCabe mainly as an adviser.

In January 1949, Curtis was named coordinator of research. Leopold would remain the first director of research in Arboretum history. When Leopold assumed that title in 1933, the ill-defined position demanded extensive statewide travel, a far-reaching vision, and a trail-blazing but controlling sense of direction. Curtis, as coordinator of research, would stay closer to home with a far more clearly defined territory to manage, his sights distinctly on the growing diversity of the Arboretum rather than on state or federal concerns.

CHAPTER 8

❧

The 1950s

OUTSIDE THREATS, "THE BELTLINE COMETH"

By summer 1950, John Curtis, now the Arboretum coordinator of research, could claim, unreservedly, for the first time that it had become "abundantly evident that prairie restoration on abandoned agricultural land" was possible, no longer "merely a dream."[1] His March 1951 report on the progress of Arboretum prairies from 1935 to 1950, with detailed commentary on the early experiments of Fassett and Thomson, the early plantings of Sperry, and, later, of McCabe and Catenhusen in 1941 and 1942, and of Greene in 1943, was a testimonial to the careful work that had gone into the fifteen-year effort. His report also covered plans for future planting and recommended continued maintenance, including regularly prescribed burns (at least every three years), mowing and other control procedures, and the immediate control of runoff waters from the Beltline Highway. For the latter, he suggested a series of diversion dams placed at the heads of existing channels flowing from the highway. He also included a list of mammals trapped by Ruth Hine that were living on the prairies in 1950, including among the expected collection of shrews, ground squirrels, mice, and meadow voles an occasional American badger.[2]

Curtis's carefully detailed sixty-three-page report was indicative of his insistence on thorough scientific record keeping, something on which Leopold also insisted during his years as research director. But by 1950 the job of accurate record keeping had been complicated by the increasing number of research projects that required close scrutiny, in addition to the daily maintenance of extensive and expanding native and horticultural plant communities. Consequently, the Arboretum's Technical Research Committee (TRC), chaired by Curtis, advised hiring additional full-time personnel.

Hiring "Jake" Jacobson

Hiring an Arboretum superintendent who would also serve as the Arboretum horticulturist was the first challenge. To that end, Longenecker, Curtis, and Gallistel approached UW president E. B. Fred with a request for money. Since the Arboretum was being used more regularly as a biological teaching and research laboratory, they explained, it was essential to add staff members who could supervise *and* teach.[3] Consequently, the duties of the new full-time superintendent included the supervision of fieldwork and the maintaining of accurate records of materials planted and the use of assigned research project locations. Also, since the superintendent's work was developmental, it would facilitate both teaching and research requirements in the science departments. Curtis's hand in drafting the job description was clearly behind the inclusion of a coda emphasizing the need for careful, accurate long-term records, necessary because many of the plant communities and forest associations would take at least fifty years to mature. Thus, keeping adequate records was deemed an essential task for the new Arboretum superintendent.[4]

Once President Fred allocated the money, the search began. In January 1949, based on Fassett's recommendation, Longenecker offered Jacob R. "Jake" Jacobson the position of Arboretum superintendent at a starting salary of $3,600. Jacobson was a perfect match. He was a proficient field scientist and professional educator. After graduating with a bachelor's degree in science in 1922 from the UW–Madison, he spent twenty-six years in the Superior, Wisconsin, public school system teaching biology and served for three years on the State Science Curriculum Planning Committee. He also had long-term Arboretum connections. He served on Leopold's committee on deer investigation and gathered field data for Leopold on grouse, rabbit, and various carnivora in northwest Wisconsin.[5] He was also known personally to Thomson and Hasler as well as Fassett.

As a clear indication of where the Arboretum's emphasis was shifting as early as 1949, both Longenecker and Fassett, in support of Jacobson's application, told Gallistel that they wanted Jacobson because the position required a person who was comfortable with the public, with students, and with occasional short course teaching. What quickly becomes apparent reviewing the correspondence and committee reports in the days leading up to Jacobson's hiring is that the Arboretum Committee (AC) actually wanted more than just a full-time superintendent. They wanted a "super-worker" to oversee all the field and maintenance tasks. But they had the funds for only one full-time position. Consequently, not only was Jacobson expected to serve as both superintendent and Arboretum horticulturalist, but he also was expected to supervise the planting and all related fieldwork, including the prescribed annual prairie burns. He was also intended to be the record keeper for all field material planted—with locations. And, when needed, he was expected to aid Fassett and other science faculty in their courses in ecology, taxonomy, and conservation. He was also expected to assist students with Arboretum research projects and to be available for tours and field trips, involving field instruction in plants and nature study for all of the garden clubs, scout troops, bird study groups, nursery groups, recreational leaders, state groups, visiting conservation clubs, and any other group that requested the service.[6]

With all of that on his plate, it comes as no surprise that the Arboretum Committee, in an unintended ironic twist, thought that since Jacobson was going to be spending most of his time at the Arboretum, he should actually live *in* the Arboretum. And so, to enable him to stay close by, they offered him the "Arboretum caretaker's house" rent-free, "including fuel, water, electricity and telephone."[7] Built in 1932, shortly after the Arboretum came into existence, the caretaker's house needed a new furnace, a new pipe leading to the chimney, a new bathroom floor, and a new (or at least repaired) roof. Jacobson was assured, however, that the repairs would be made before he moved in. He and his wife would have the entire house to themselves, except for one small area set aside as an Arboretum office. The Arboretum Committee also arranged to have Jacobsen's position entered in the university records as a faculty appointment, since any teachers already in the State Retirement System automatically remained in the system if they moved to a faculty position.

The Jacobsons moved into their new residence on October 4, 1949. On October 6, A. F. Ahearn, UW superintendent of buildings and grounds, contacted Herman P. Kerl, sheriff of Dane County, and requested that since Jacobson's duties also involved protecting the Arboretum, he be appointed a deputy sheriff so that he could police the Arboretum with "authority to enforce the regulations."[8] And so Jacobson took on one more responsibility.

Henry Greene would acknowledge in an *Arboretum News* article on Jacobson's retirement in 1962 that he worked "far beyond any call of duty" and that owing to the persistent Arboretum need for additional help, the high incidence of success in Arboretum plantings during Jacobson's tenure was mainly the result of his "extra hours."[9] Nancy Sachse would say of Jacobson that "the stamp of his interest was everywhere."[10] And, indeed, Superintendent Jacobson would leave indelible marks all over the Arboretum. But as the record shows and as Greene stressed, given all the tasks Jacobson was assigned, how could he not? The unanimous sentiment around the Arboretum in June 1962, when he retired, was "that Jake would be missed."

A Busy Start: The Haen Property, Increased Research

With Superintendent Jacobson in place, the decade of the 1950s got off to a busy start. Acquiring the Haen property was a major accomplishment. The 27-acre property, contiguous to the east of the Grady Tract, belonged to Anthony J. Haen. Negotiations for the land began in fall 1951. By October 1952, the university had manipulated a property trade with Haen, who acquired a 42-acre parcel of university land on the southeast side of Fish Hatchery Road. The university acquired the 27-acre property south of the Beltline plus a single payment of $6,000. Gallistel, his eye always on the future, suggested that the money be set aside as a start toward the construction of a permanent office building in the Arboretum.[11] Acquiring the 27 acres also meant that the Arboretum could now extend the pine plantation in the northern part of the Grady Tract eastward parallel with the Beltline. The result was an attractive continuity of pines south of the Beltline. Another benefit emerging from the sale was the straightening of the east boundary of the Grady Tract, eliminating a "dogleg " that had created fencing difficulties.[12] All in all, the acquisition, the last major single land acquisition for the Arboretum, was in the best interests of both Haen and the university.

Still, one of the pressing concerns of the AC was the need for increased security. Plans were made for intensified patrols of the grounds. By April 1951, in anticipation of the annual rush of spring visitors, patrols were increased.[13] A surprising increase in research projects prompted Curtis in April 1950 to circulate a document to the faculty citing new rules and regulations for Arboretum research projects. All projects, for instance, now had to be registered with the AC and had to have some direct connection to the development and maintenance of the specific Arboretum area where the research was taking place. By April 1950, forty-five projects had been approved that included McCabe's (Wildlife Management) studies of house wrens, bluebirds, tree swallows, woodcocks, catbirds, yellow-headed blackbirds, mallards, and pheasants. McCabe was also working on the study of prairie species planted by direct seeding, the measurement of colony expansion in prairie indigo, rattlesnake master and transplanted trilliums, and the effects of nursery stock on tamarack reforestation. Restoration work on the 60-acre Arboretum prairie, by 1951 heading into its fifteenth year, continued under Curtis's close supervision. He also now underscored the claim that the Arboretum's prairie restoration effort, begun in 1936, was the nation's oldest.[14] Curtis was also working on the production of native lady's slippers, the antibiotic and autotoxic effect of native plant species, and trends in oak woods development on the Grady Tract. Other Arboretum research projects at the time included John Emlen's (Zoology) study of small mammal communities. A. J. Riker (Plant Pathology) was researching oak wilt infections in Wingra Woods, the resistance of red cedar to cedar-apple rust fungus, weed control in pine plantations, and the growth and development of rooted white pine cuttings. Longenecker was studying the varieties of flowering crab apples, the hardiness of lilac strains, and methods for oak reforestation by direct planting of acorns. Hasler continued to study minnows in small ponds on the Gardner Marsh. He was also researching wind effects on the microorganisms of ponds and the effects of carp removal on water plants. Robert J. Muckenhirn (Soils) was researching the long-term effects of plant cover on soil development. Sergei A. Wilde (Soils) was studying the depletion and methods of restoration of farm woodlot soils. Robert J. Dicke (Economic Entomology) was working on the biology of Arboretum mosquitoes. Royal A. Brink (Genetics) was doing research on forest-tree breeding, and C. A. Richards from the Madison Forest Products Laboratory was studying the resistance to decay of Douglas fir boards.

Indeed, throughout the 1950s, much to Curtis's satisfaction, the Arboretum bustled with research projects. In his April 1950 memo to the faculty, Curtis noted that the broad range of projects approved for the year was a commentary on the serious research commitment of the UW field science departments and a clear endorsement of the Arboretum's role in those research efforts. Although a budding interest in the possibility of educating local students and the public on the aims and purposes of the Arboretum was an increasing concern of Arboretum management, "research" was still the operative word in the 1950s, largely owing to Curtis's concerted efforts to sell the campus on the Arboretum research opportunities. In 1951, there would be another fifty-four projects approved. By 1953, the number was up to fifty-six.

Norman Fassett (1900–1954)

Curtis's memo to the faculty on the Arboretum's research program appeared in April 1950. In October 1950, Grant Cottam, in his second year as a full-time botany professor, was appointed to the AC as a replacement for Norman Fassett, who had taken a leave of absence. Cottam's appointment was temporary; he would serve until Fassett returned. In March 1952, Cottam was made a regular member of the AC, a permanent replacement for Fassett—who was not returning.

Owing to Fassett's reputation as an aquatic vegetation specialist, he traveled frequently to areas of the world where he could assist in the identity and study of water plants. According to John Thomson, a former student and good friend, Fassett was on a research trip to Guatemala, Honduras, and Nicaragua when an illness overtook him. It was December 1953; he was in Managua, Nicaragua, when he suddenly was hospitalized. After some weeks in the hospital, he returned to Madison for treatment and was told that he had a lesion on his brain, a mild one that would soon mend. On February 21, 1954, he had another attack, a hemorrhage, and was hospitalized again. On July 18, he wrote the following to Donald C. Peattie, his old Maine friend and Harvard classmate, who aptly described the July letter as truly indicative of "the fine, fighting, sporting spirit" that defined Norman Fassett:

The doctor said I might recover and walk again—you know how a poor old guy stumps along after a stroke. I now walk as well as ever, and yesterday I was climbing trees with a friend's children. On March 11 a tumor was removed just above the right ear. Five days later I staggered to my feet, grabbed a likely looking nurse, and polka'd down the corridor. Forgive my smarty boasting—I have to get what fun I can out of this. Since then I have been in and out of the hospital much of the time delirious or just plain unconscious. For a couple of weeks now I have been out of the hospital, and hard at work.

The cancer is coming back, and I am given a few months to live. I am working hard to finish a number of jobs. Just completed a monograph of Echinodorus. Now am starting a report on my recent trip to Central America. I have a lot of papers under way, and want to leave them so somebody can pick them up. I fear I shall not get to the Spring Flora of New England. . . . If you think of any likely candidates, let me know. Or, would make some nice theses, under proper direction.

My eyesight is still very bad, with the left upper quarter of the field of vision gone. Everything looks funny, and I feel as if a flash bulb had just gone off before my eyes. My hands tingle continuously. I get confused very easily. . . . Ever since you introduced me to Omar, on the banks of the Charles, I have lived by his philosophy, and it has paid off. Now I eat well, sleep well, and enjoy my likker. Wonder how it will strike me at the end—will there be a general decay of faculties as the cancer progresses, or will I have a grand hemorrhage that really slaps me down?

O.K. It's been fun.[15]

At some point in late August or early September 1954, Fassett traveled to Boothbay Harbor, Maine, where he spent a few happy weeks before he was hospitalized again—for the last time. He died on September 14. He was only fifty-four years old.

In November 1954, Peattie published a moving literary tribute to Fassett in *Rhodora: The Journal of the New England Botanical Club.* For some botanists, Peattie wrote:

> a living plant . . . may mean comparatively little. It must be brought into the laboratory and "controlled" before it yields up any intellectual satisfactions. . . . It is hard to believe that plants could ever have for such botanists any emotional content. A sentence from a letter which he wrote me in 1921 will show how different was Fassett's approach, even in these early days.
>   "I have fallen in love this summer," he wrote in 1921, "with the three Osmundas."
>   I can imagine a similar sentence (indeed there is one) dropping from the pen of the young Linnaeus, the young Theophrastus, the young Haller (poet-botanist of the flora of the Alps), William Hooker or Konrad Sprengel—all of them boni botanists. I cannot conceive that a Nigeli, and Ingen-Housz, a Correns, a von Mohl, or a Weissmann, would ever have fallen in love with a genus of ferns, or if such a weakness had overcome them in early youth, it would never have been admitted in writing![16]

Fassett would be missed. He had many friends, and his professional influence extended well beyond his publications. His UW colleagues, in memorials to honor his memory, spoke of the wide-ranging influence he had on his students. Colleagues teaching in departments other than botany frequently asked him to give opening and final lectures in one course or another. His lecturing style was that good, characterized by wit, humor, and a wealth of visual references. He brought the "salty air of the New England coast" to the Madison campus, one professor noted.[17]

His Arboretum connection was historical, deeply rooted, and productive. So was his connection to Wisconsin. As Peattie observed in his tribute, Fassett "loved plants as he found them, and where he found them," and he "looked with . . . distaste upon plants out of place." That's why, for most of his research, Fassett traveled to where the plants originated, the "wild" areas, and why he was delighted with Wisconsin's north woods. They had the "'right kind of trees,'" he would say.[18] One would expect that his dismay with the desire to plant "wild" things where they didn't belong was not lost upon his students.

Challenges in the Mid-1950s

The July 1953 meeting of the AC was held for the first time in the "new" Arboretum Lab, which was actually the "old" renovated CCC bathhouse. The minutes for the meeting record the AC's concern with a problem that would vex Arboretum management for years to come. Sewage from a sanitary sewer of the Madison Sewage District was draining into the Stevens Pond on Monroe Street. Gallistel contacted Madison officials, and the committee went on record as requesting an immediate correction to the problem.[19] The city responded with a directive that a new sanitary sewer would be constructed, but, owing to the heavy traffic on Monroe Street, it would be put through Arboretum land on the northwestern side of Lake Wingra. The AC formally registered its opposition, but to no avail.[20]

Movies were also on the minds of the AC members in summer 1953. Walt Disney Productions, in the spring of 1953, filmed prairie fires that were set in the Arboretum and at Faville Prairie for a Disney documentary entitled *The Prairie Story*. Curtis, after viewing the scenes, told Erwin Verity of Walt Disney Productions that he considered the shots "magnificent" and was planning to show the film to plant ecology classes in connection with their prairie studies.[21] In fact, Curtis was so impressed with the film's potential that he proposed putting together a complete documentary film showing all phases of prairie restoration work with examples from the Arboretum, including the Disney footage. Curtis argued that the return in publicity, emphasizing the Arboretum as the single worldwide leader in prairie restoration, would more than compensate for the cost. The AC unanimously agreed that the film should be made.[22]

By the mid-1950s, the minutes of the AC meetings began to tell a story. The Arboretum was changing. Gone were the matters integrally and primarily associated with growth, identity, and the desire for federal partnerships. The Arboretum had redefined itself. Even the official letterhead by 1954 had dropped the "Experimental Forest Preserve" label. The Arboretum was now known as the University of Wisconsin Arboretum and Wildlife Refuge. By the mid-1950s there could be no doubt that the focus was changing.

News and activities would now be generated mainly from without rather than from within the Arboretum. Even the increasing emphasis on research forced the AC to think outside the university for funding for projects that were never going to be funded internally. Curtis approached the TRC in November 1953 with a detailed report on possible ways to gain outside funding. He urged the TRC to study his report and to follow it up with applications. He warned the committee members that outside research funding would require time and effort, especially in the initial stages, a burden that the already hard-pressed Arboretum staff could not shoulder alone. Hence, he recommended that his report be submitted to the full Arboretum Committee and circulated among faculty with current or future Arboretum research plans.[23]

By the mid-1950s there were also increasing challenges and threatening developments from without that were encroaching on what once had been a relatively protected, insular environment. Gone for good were the days when Leopold and McCabe could expect to trap twenty-seven minks in one year, to round up and band four hundred pheasants, to attract migrating game birds to landing strips in the Gardner and Wingra Marshes, and to raise wood ducks that would return annually to the ponds. Gone also were the days when Jackson and Hasler could dream of turning Lake Wingra into a game fishery with, perhaps, a token population of lake trout. During the mid-1940s, the biggest threat to Arboretum security had been stray dogs raiding game traps, and an occasional trespasser disturbing Leopold's feeding stations. By the mid-1950s, the threats involved sewage and silted storm water from outside the Arboretum draining into Arboretum marshes and prairies, the construction of sanitary sewers cutting through Arboretum property, the loss of precious acreage to construction, the placement of radio towers and transmitters with the consequent but predictable damage to the ecosystem, and the looming danger of electrical power lines with a fire potential capable of wiping out the entire Arboretum.

Plane Crash in the Redwing Marsh

As if orchestrated by some sinister power, the mid-1950s threat from the outside was starkly demonstrated on a quiet Monday in November 1953, by a tragically ironic but emblematic intrusion. After dropping from the sky at a speed of four hundred miles an hour and then, out of control and with engines dead, making a ghostly silent vertical dive, a U.S. Air Force F89C Scorpion jet plane sliced into a bog in the Arboretum's Redwing Marsh on the southeastern edge of Lake Wingra and exploded, killing the pilot and his radar observer. An eyewitness noted that after the explosion, what was left of the plane was absorbed by the marsh. The plane left no gaping hole, "just a flat circle of mud about 40 feet in diameter." All that was left after the plane disappeared, according to the eyewitness, "was a small yellow flame near the center of the gray circle." Parts of the plane were later salvaged. The parts that were not recovered still remain buried in the mud at the bottom of the marsh.[24]

The officer piloting the plane, 1st Lieutenant John W. Schmidt, and the radar observer, Captain Glen E. Collins, were returning to Madison's Truax Air Force Base after an aborted routine training flight. They had been in the air for only thirty-five minutes when a sudden mechanical failure forced the return. They were approaching Madison and the Arboretum from the southwest when witnesses heard a mid-air explosion and the plane went into a steep dive. There is good reason to believe that the dive sent the plane downward in the direction of the St. Mary's Hospital neighborhood and that the pilot chose intentionally and heroically to ditch the plane on the western edge of Gardner Marsh in what is now known as Redwing Marsh. One eyewitness account recalled seeing the plane on a low-level path, heading northeast, directly in line with St. Mary's Hospital when it sharply banked westward, forcing the left wing down, and then it seemed to fall off to the west and to flash downward in a spiral vertical dive, smashing into the marsh.[25]

The tragic loss of the two air force officers was compounded by the fact that they both left young families behind. Damage to the Arboretum was minimal, but the November 1953 event was yet one more example of the increasing vulnerability of the Arboretum to potentially threatening and destructive outside forces. As fate would have it the pilot managed to ditch his plane in the wet soil of the marsh. Yet, as an item on the extreme 1953 drought in the *Arboretum News* for January 1954 pointed out, in Lieutenant Schmidt's attempt to steer his plane away from the St. Mary's neighborhood, he could just as easily, depending on his flight path, have ditched his plane, with nearly full fuel tanks, in one of the Arboretum's exceptionally dry forests, in which case the explosion and the inevitable, ensuing conflagration would have been catastrophic.[26]

How Much Is a Tree Worth?

The Arboretum was spared a destructive forest fire after the jet crash in 1953. It was not as lucky in 1954. Again, the threat to the Arboretum came from an outside source. On March 8, 1954, employees of the Chicago and North Western (C&NW) Railway were burning off grass and weeds from the right-of-way on the south side of the railroad tracks that paralleled the Arboretum's southern border and lost control of the fire. This was not a day when a fire should have been started, especially south of the Grady Tract. There was a strong southwestern wind blowing directly into the tract. The trees and grass were dry owing to an unusually dry winter. Large areas of dead vegetation were scattered along the floor of the oak forest. Nevertheless, the railroad crew started a fire. According to the crew, the fire was under control until shortly after 2:00 p.m., when the wind blew tufts of burning grass into the oak area. At 2:40 p.m. the UW vice president's office was told by university police that a fire was raging out of control in the Grady Tract. An emergency call went out to fire departments.

The fire spread over 70 acres and destroyed oak trees, some recently planted, before it was brought under control around 4:30 p.m. by firefighters who had to cut Arboretum fences in order to get their trucks positioned. The quick response kept the fire from spreading to the pines in the northern Grady Tract, which would have produced another major conflagration. The fire appeared to be out. Around 5:45 p.m., however, the fire flamed up again in an area adjacent to commercial dynamite storage sheds on private property next to the Grady Tract. The Dane County Sheriff's department broadcast a call for help on a police radio. The call was picked up and broadcast by radio and television stations. By 6:15, in a much appreciated show of concern, close to a thousand people arrived to help. Firefighters had already arrived, however, and had quelled the blaze without the need for any additional assistance.

The March 1954 fire threat was not over, however. About ten days after the Grady Tract fire, a potentially more serious blaze was narrowly averted by sheer good fortune. Curtis and some students just happened to be in the neighborhood of the Noe Woods on the western side of the Arboretum when they spotted a fire burning unattended in leaf litter directly behind a home in the Winslow Lane area, bordering Noe Woods on the west. Curtis alerted the fire department. Fire trucks, arriving promptly on the scene, extinguished the blaze. The Arboretum management, reviewing the incident, noted that if the fire had burned for only a few minutes longer, it likely would have spread to the pine plantings, which could have initiated a major inferno that would have destroyed most of the Arboretum timber north of the Beltline and seriously damaged homes in the Winslow neighborhood. The *Arboretum News* carried a report of the Winslow fire that ended with a condemnation of the "unthinking carelessness responsible for situations of this sort."[27]

The Arboretum's vulnerability to threats from outside became increasingly more apparent during the mid- to late 1950s. But the immediate problem, after the March 8, 1954, Grady Tract fire, was the question of reparation for damages done to the plants, the oak trees, undergrowth, and prairie. The ensuing legal debate over reparation focused on the controversial question of exactly how much monetary value one attaches to a plant, a tree, or a prairie. The problem was not easily resolved.

On March 10, 1954, Gallistel was advised by A. W. Peterson, UW vice president for business and finance, to sign a fire damage complaint against the C&NW Railway. On March 18, Gallistel asked Warren Resh of the Wisconsin attorney general's staff whether he should sign a complaint immediately or wait until the advent of the growing season in order to get a better assessment of actual tree and plant damage. In the meantime, he forwarded to Resh's office a report on tree damage that had been compiled by Albert D. Hamann, head of university police. Hamann estimated that 1,134 trees in the Grady Tract had been burned significantly enough to be considered lost. He counted 284 trees dead and 850 injured. He attempted a cost estimate of the damage based on the going 1954 price of oak for stumpage, logs, and lumber. He estimated the loss at 136,000 board feet or 10 cubic feet of wood per tree. He figured the loss for stumpage value at $3,400; the loss for log value at $5,440; and the loss for lumber at $9,520—for a total of $18,360. When he added to the total $1,172.60 for labor and materials to repair the fire damage, the final cost estimate amounted to $19,532.60.

Hamann's early damage assessment, however, by no means put the matter to rest. At the April 8, 1954, meeting of the AC, Gallistel reported that the question of a damage estimate, in spite of Hamann's efforts, was still unresolved. Curtis, underscoring the philosophical nature of the problem, questioned how anyone could actually determine the value of an established oak tree. The lumber value of the Grady Tract oaks was small, as Hamann had demonstrated. Intrinsic in Curtis's response, of course, was the much broader question of how one determines the value of any singularly beautiful work of nature and how that work is related to the overall value of the landscape of which it is a part, particularly if that landscape can be appraised not only on the basis of its material value but also its horticultural value.[28]

By June 1954, the argument for an assessment of the horticultural value of the fire damage was uppermost in the minds of the AC. Resh advised Gallistel that in order to recover damages, proving loss or injury to the property was all that was necessary.[29] Resh's advice left the matter of damage evaluation to Gallistel and the AC. After much debate, the argument that they decided to make was linked to the horticultural value of the trees *prior* to the fire. They based the damage costs on the amount of money it would take, not simply to "replace" the lost trees, but actually to "reestablish" them again in size and beauty. As with an insured home that burns, the insurance assessment would be based on the appraised value of the burned home, including damaged property surrounding the home. The objective was to provide enough money to reestablish what was lost—in terms of buildings, gardens, trees, and other natural landscaping.

The AC turned for guidance to a Milwaukee tree-moving company. The company advised that it would be impossible to move oak trees of the size damaged into the burned area, but elms of that size could be moved and replanted. The cost per elm was between $275 and $350 dollars. The AC decided, since oaks were so highly valued, to use the $350 dollar figure per oak.

Based on Hamann's earlier assessment of 1,134 trees lost, the number of lost trees multiplied by $350 amounted to $396,900 dollars. Adding to that figure an estimated cost of labor and materials for fence repair, tree disposal, and so forth and the cost charged by the various fire departments for their services, the grand estimated damage total amounted to $398,473 dollars. That was the figure that Gallistel sent to Vice President Peterson on June 1, 1954, with an accompanying explanation of the rationale behind the assessment.

The summer passed with apparently no action taken on the Arboretum's claim. Finally, in early September 1954, the C&NW Railway asked for additional information, including a list of projects that were being conducted in the burned area prior to the fire. The task of accounting damage done to locations where the projects, mostly faculty projects, were being conducted fell to Curtis, who by early October had drafted an eight-page report for the railroad. He argued that the fire had totally destroyed portions of the Grady Tract prairie and several century-old bur oaks that "had provided the nucleus of an excellent example of an oak opening community." He noted that "the fact the trees had survived all previous fires for 100 years" showed how exceedingly hot this ground fire was as it raged through tinder-like grasses, burning right through the ground layer of organic mulch. Owing mainly to the intense heat, research on the oaks, begun in 1940 as a long-term study, had been rendered useless. Ongoing studies of acorn production also had to be abandoned. He also explained how the loss of the oak canopy had destroyed other projects, many conducted by students on soils and fertilizers, vegetation, and insects that depended on the canopy. The great volume of scientific information gathered by the students, Curtis concluded, was now valuable as history but little else. The loss to the university's field science community had been severe.[30]

At the end of October, Gallistel sent a copy of Curtis's report to Resh. In early November, Resh sent a copy to Vern C. Smith, the district claim agent for the C&NW. Resh told Smith that he hoped he would agree that Curtis had prepared a careful analysis of damage to the trees and prairie and of the consequent loss of years of scientific research. Resh also told Smith that the university would cooperate fully in rendering any further assistance and that he hoped that a fair settlement could be reached.[31]

Apparently, at some point between June 1954 and October 1954, the AC's thoughtfully reasoned horticultural evaluation, set at slightly over $398,000 dollars, was dealt a touch of litigious legal realty that reduced the figure in the attorney general's office to a mere $15,000 dollars. Surely the oak trees, particularly the hundred-year-old bur oaks, were worth more than that, even sentimentally. But factored into the decision to ask for $15,000 dollars was also the economic reality that the C&NW was in financial trouble, and the Arboretum might actually consider itself lucky to be awarded $15,000.

After receiving his copy of Curtis's report in late October, Vern Smith contacted Gallistel on November 10. He didn't address the issue of the settlement. Instead, he asked for a copy of the map of the trees in the Grady Tract that covered about 58 percent of the burned area that had been mentioned in Gallistel's June 1, 1954, letter to Vice President Peterson. Either Smith was under the mistaken impression that the assessment of the tree loss was based on the map or he was stalling for time, hoping to frustrate the university's momentum by delaying proceedings. On December 13, 1954, Gallistel sent copies of the three-staged superimposed map to Smith. Each symbol on the map, made in 1953 by Cottam, represented the exact location of a tree. December came and went. So did January 1955. There appears to have been no follow-up on the matter of the map. Finally, in early February 1955, in a meeting with Smith, Resh was informed that the C&NW, "after considerable study and consideration," had authorized a check for $7,500 "in full settlement" of the claim. That was half of what the university and the Arboretum were expecting—$9,032 less than Hamann's March 1954 assessment; $391,500 less than what the AC had determined as the horticultural value of the destroyed oaks. Resh told Smith that there would no doubt be disappointment on the part of the Arboretum Committee and management.

Resh was correct. The AC and UW officials were disappointed and countered on February 28 with a compromise offer of $10,000. In early March, Smith contacted Resh with a counteroffer from the railroad of $9,000. Resh urged Gallistel and the AC to accept the offer. Smith had acted fairly, Resh observed, and he told Gallistel that owing to the railroad's desperate financial condition and considering the difficulties of establishing the extent of the loss "on a dollars and cents basis" should the matter go to court, the university should accept the $9,000 offer. Gallistel and the AC agreed to the terms. On March 21, 1955, a year and thirteen days after the date of the fire, Resh advised Smith that the Regents had accepted the offer of $9,000 as settlement for the Grady Tract fire damage. The matter was ended. The March 8, 1954, Grady Tract fire that destroyed 1,134 trees was history. The compensation for the loss demonstrates what each tree, even hundred-year-old bur oaks, was worth in the eyes of the C&NW Railway and, likely, in any court of law where the suit would have been tried. The figure—which one must assume, as Resh cautioned, was the consensual damage value attached to a tree in the mid-1950s "on a dollars and cents basis"—amounts to $7.94 per tree.

Needs in the Late 1950s

By the mid-1950s, the Arboretum management acknowledged that hiring Jake Jacobson in 1949 as the "super-working" live-in superintendent had been a truly wise move. Jacobson had taken over most of the responsibilities related to field and maintenance work, including the annual prairie burns, but his work schedule left little or no time for other tasks that initially had been expected of him, including guiding tours for Boy Scouts, Girl Scouts, and garden clubs. Neither did his busy schedule allow enough time to patrol the Arboretum, in his role as deputy sheriff. There was concern that Arboretum research projects needed to be safeguarded. As Longenecker pointed out to the AC in June 1954, the Arboretum was now more widely recognized than ever before and increasingly visited by school groups. Clearly, there was a need for at least one more position, perhaps two. The need was first addressed on June 19, 1954, when, in the course of an AC discussion on damage from visitors straying off the foot trails, the AC decided that funding for two ranger-naturalist positions should be included in the next budget and that the primary function of these ranger-naturalists should be public relations and the education of the public as to the proper use of the Arboretum.[32] This inclusion in the minutes, acknowledging formally the need to hire staff skilled in public relations specifically to educate the public, clearly speaks to an emerging new direction in Arboretum policy that the AC was considering by 1954, a direction that ran contrary to the protectionist vision that characterized the 1930s and most of the 1940s. The ranger-naturalist position, however, in spite of ongoing Arboretum needs, was not funded by the university until twelve years later, in 1965, when Dr. James H. Zimmerman (known locally as "Jim Zim") was hired as the first ranger-naturalist in the history of the Arboretum. But even then, in 1965, "owing to budgetary limitations," the position was only part-time.[33]

The increasing need for additional help, especially with tours, became a persistent complaint during the late 1950s. In 1955 a report on Arboretum use by UW and community groups noted that in 1955 alone, 650 students from thirteen university classes were making some use of Arboretum facilities, and although exact numbers were not available, guided tours were provided for Wisconsin garden clubs from Portage, Westby, Oconomowoc, Lake Geneva, and Racine. Tours were also provided for Boy Scouts from the Madison area and for third- and fourth-grade students from Spring Green. Many more school groups, not accommodated, toured the Arboretum on their own. The report concluded that it was the hope of the AC and Arboretum staff that the trend toward increasing use would continue.[34]

In May 1959, the sentiment was repeated by Curtis, now chair of the six-member TRC, who went on record with the AC appealing for an increase in tour availability for grade school children. He had recently conducted a tour for sixth graders from Nakoma School and was shocked to realize that the children, in spite of living only blocks from the Arboretum, had no idea what the Arboretum was about. Curtis argued that the lack of awareness on the part of the children cast a very unfavorable reflection on Arboretum public relations and provided additional evidence of the need for an educational program at the grade school level.[35]

"The Four-Lane Speedway"

The desire to educate the public on what the Arboretum was all about increased considerably in 1956, the result of shared civic consternation surrounding the widening of the Madison Beltline from a two-lane road to a four-lane highway. The original 1949 Beltline construction had cost the Arboretum 15 acres of pines. But the loss of the pines was minimal compared to the problems created by soil erosion and silting on Arboretum land owing to inadequate drainage. When Gallistel was informed of the Wisconsin State Highway Commission's (WSHC) plans for the widening, he was told that the construction of a new north lane would include a divided waterway that would ensure that water coming off the highway would not enter the Arboretum at a single point and, hence, should greatly improve the silting problem. He was also initially told, in 1955, that a pedestrian subway built under the Beltline enabling students to get safely from one side of the road to the other was a possibility.

Gallistel and the AC also wanted a vehicular subway built that would enable equipment to be moved easily from one side of the highway to the other. They also registered their concern over the possibility of cars rolling off the highway into the pine forests and starting fires. The WSHC assured Gallistel that a guard fence would protect the pines. He was also told in February 1956 that the WSHC would not approve the construction of an underpass for vehicular equipment. By March 1956, the commission was reneging on the construction of the pedestrian subway as well. Gallistel asked for an opportunity to present the Arboretum's case for the subway. In early April 1956, Gallistel, who was now referring to the Beltline as "the four-lane speedway," appealed to the Regents for a resolution endorsing the need for the pedestrian subway.

Gallistel and the AC would need all the help they could get. The WSHC district engineer, J. C. Jones, on April 9, told Gallistel that the subway was a bad idea: "it would be dark, difficult to police and keep clean." He also warned about potential graffiti, "unsightly drawings and worse," on the walls. Jones added that after consultation with the WSHC's Design Section, an alternative was devised. The designers had decided that the Arboretum should move its present private entrance on the north side of the Beltline westerly to a point directly across the Beltline from the gated south entrance. The new north side entrance then would be located about 1,500 feet east of Seminole Highway. Both vehicles and pedestrians "could wind their way through the Arboretum to this new private entrance and cross directly over the Beltline . . . to the entrance on the south side." In fact, the designers concluded, the whole situation would be helped now by the separation of opposing traffic lanes. How the separation of traffic lanes was going to help people and vehicles attempting to cross the Beltline at Seminole Highway is hard to fathom. Even more objectionable was the idea of vehicles and pedestrians "winding" their way through the Arboretum to the new north side entrance, especially since there was neither a road nor foot trail cut through at that westerly point. Installing an entrance on the north side near Seminole Highway would necessitate building a road and a foot trail from Arboretum Drive south through the Noe Woods, with the subsequent loss of valuable oaks, some over 150 years old. But a large loss of existing oak trees was apparently not at all on the minds of the members of the WSHC's Design Section when they devised the plan. For the Arboretum, of course, the plan was outrageous and, hence, was totally rejected. Gallistel was advised that if he still desired it, a conference with the Highway Commission could possibly be arranged. Gallistel requested a conference.

As it turned out, Gallistel's early April appeal to the Regents for an official resolution endorsing the pedestrian subway had been a sound move. On April 7 the Regents passed a resolution approving the Arboretum's request. The resolution with the request was forwarded to the WSHC. On April 30 Gallistel learned that he and the AC would represent the university at a hearing with the WSHC scheduled for May 3, 1956. Clarke Smith, secretary of the Board of Regents, advised in a letter to the WSHC that "favorable consideration by the Commission of [the] request will be very much appreciated." The support of the Board of Regents appears to have made the difference. At the May 3 meeting, the commission agreed to construct the subway. It would be five feet wide and seven and a half feet high. The commission also agreed to maintain the structure provided the university assume responsibility for maintenance and policing and construct fenced approaches between the right-of-way lines and the subway.[36]

For the AC, it was a victory of sorts. The Arboretum had its pedestrian subway, but having won the battle, they still were going to lose the war, and the AC knew it. In an April 1956 *Arboretum News* item on widening the Beltline, the author, likely Henry Greene, observed "whatever the worth of the new road as a highway, there can be no question that its construction is in the nature of a disaster to the Arboretum." The fence line on the north side of the Beltline would be moved at least sixty-six feet north of the present fence, resulting in the loss of many large white pines, nearly all the spruce adjacent to the Leopold Pines, most of the screen plantings above the Curtis Prairie, most of the sugar maples in the stand east of the screen plantings, and a number of old oaks. The soil would be compromised by the runoff of damaging claylike subsoil and silt that, as Greene pointed out, always accompanies road construction of the Beltline type. He added that the new highway, regardless of what the WHSC maintained, would also pose a fire threat to the pines since the road was too close to the pines. Greene's prognosis was disheartening but accurate. The water runoff from the Beltline following heavy rains still poses a threat to Arboretum plantings, particularly to the southern end of the Curtis Prairie, where reed canary grass continues to overtake the plants.

But Gallistel and the AC, in their battle with the WHSC, could take some comfort and satisfaction from the construction of that walkway under the road, which over the years has not only provided easy access to and from the Grady Tract but also has become the source of adventure for children and surely some adults who have found walking under the "four-lane speedway" a lot more appealing than walking across it.[37]

Robert McCabe's "Rabbit Shoots"

The Arboretum's perennial rabbit problem got worse in the mid-1950s. In 1955, McCabe came up with a solution. It was simple. With the approval of the AC, he arranged a "rabbit shoot." He organized a group of sharpshooters who gathered in the Arboretum on Sunday, December 11, 1955, for the season's first shoot. The effort met with success, McCabe reported. The only concern raised by the AC came from George W. Foster of the Law School faculty, who wondered about adverse publicity from shooting rabbits on Sunday owing to the increased number of Arboretum weekend visitors. McCabe explained that Sunday was the only day that his group could get together and that the public was informed about the hunt. The AC seemed satisfied with the answer and left the matter of scheduling future hunts up to McCabe.

By January 1956, however, when the monthly AC meeting rolled around, the matter of the Sunday winter rabbit shoots had taken a more serious turn. Not surprisingly, there were complaints—quite a number of them. Gallistel told the committee in January that he had just attended an open meeting of the Madison and Fitchburg town boards. The meeting, initially called to discuss a road issue, got seriously sidetracked, he reported, owing to rather vocal feelings "of bitterness among several residents of the two towns" over the Arboretum's weekend rabbit shoots. Residents of Fitchburg, living on the south side of the Arboretum, were particularly incensed. Some of the angry residents, unaware of the project's research focus, thought the hunt was some kind of decadent private sporting event. The AC discussed the matter, coming up with some thoughtful suggestions for countering the outcry. Hasler suggested, since Fitchburg residents were most angry and the Grady Tract bordered Fitchburg, that the shooting be restricted to the Arboretum area north of the Beltline. Dickson suggested that McCabe meet with school PTA's and explain the reasons for the hunt. Curtis suggested Arboretum guided tours during which the rationale for the shooting would be explained. By February 1956, the rabbit shoots were less frequent, and the controversy died down.

In summer 1956, McCabe, pleased with the success of the rabbit shoots, decided to target crows instead of rabbits. The crow shoot, designed to drive the crows out of the Leopold Pines, took place in June 1956. The shooters were members of the Dane County Sportsmen's League. In winter 1956–57, the rabbit shoots were renewed, accompanied by another round of complaints from nearby residents. This time, however, the complaints reached the Wisconsin Conservation Department (WCD), which promptly informed Arboretum management that the Arboretum, although designated as a state game refuge, was not an "open" refuge. Hunting of any kind in the Arboretum, therefore, was illegal and must stop.[38] The debate over whether the Arboretum was an "open" state game refuge with hunting privileges was an old issue that the AC had confronted before. In July 1936, as noted in chapter 4, the WCD had told Aldo Leopold that he could not shoot pheasants. Leopold wanted to study the palatability of grains that the pheasants had eaten. He applied to the WCD for a permit but was denied. Now, twenty years later in 1957, the WCD once again was telling the Arboretum management that hunting was prohibited on Arboretum land.

The Arboretum had first been designated as a state game refuge in 1934 by the WCD. This status, however, applied to the marsh areas and was intended for the protection of waterfowl, not for shooting them. In fact, the Arboretum was listed as a "closed" game refuge, and the discharging of any firearms, consequently, was prohibited, a legal directive reinforced in September 1938 by Order Number M-337 filed with the state, which made it "unlawful for any person . . . to take, catch, kill, hunt, trap, or pursue any species of wild animal or bird at any time, or to have in possession or under control any gun or rifle . . . upon . . . the University of Wisconsin Arboretum." The rabbit shoots would have to end.

This time, however, McCabe was directed by the AC to appeal. The rabbits were eating the Arboretum. In January 1957, McCabe directed a letter to the WCD requesting that the Arboretum be reclassified in order for the staff to control the animal populations. In February 1957, Gallistel asked for a hearing. In March, he was told by Harry Stroebe, supervisor of the WCD's Game Management Division, that if the Arboretum were approved as an "open game refuge," an order would be drawn up including a clause that would enable the Arboretum management "to operate the area for research purposes in wildlife management." The clause would permit the staff, "to take, catch, or kill birds or mammals where either control or research efforts" were necessary. At an April 12, 1957, meeting, the Game Management Division recommended that the Arboretum be officially designated a game refuge with hunting privileges.[39] McCabe was now free to carry on with the rabbit shoots. The rabbits, however, would continue to be an Arboretum problem.

Within a year, the AC would be apprised of another growing animal problem. For the first time in the Arboretum's history, serious concerns about damage to trees and shrubs from Arboretum deer began to surface. In November 1958, at Curtis's urging, the AC recommended that Gallistel approach the WCD once again, this time with a request for advice on the "control or elimination of the Arboretum 'herd.'" In December, Gallistel made no bones about Arboretum intentions when he told L. P. White, WCD director, that the AC "would appreciate having the herd eliminated." He asked for cooperation.[40] Eliminating the herd would doubtlessly solve the tree damage problem, but as we shall see, efforts to shoot Arboretum deer, given the angry response from the public, created a public relations problem that went well beyond the limited grumblings of Arboretum neighbors about McCabe's rabbit shoots.

Albert Gallistel Retires

The end of the decade also brought with it Gallistel's retirement and the end of his chairmanship of the AC. His service as AC chair went back twenty years to 1939. His service to the UW, however, went back fifty-two years to 1907, when, as Sachse put it, he "literally blew in to the Wisconsin campus," arriving during a raging February blizzard. He was young, fresh, and enthusiastic from studying art and architecture at the Chicago Art Institute. He had come to Madison directly from Chicago, where he had been working for the Chicago Board of Education. He had been recommended by the board to Arthur Peabody, the UW architect, as someone who could assist with Peabody's plans for the campus. In 1907, when Gallistel arrived, the campus enrollment numbered 3,051 students. Most faculty lived on or in the immediate area of the campus. Bascom Hall, sitting majestically atop its drumlin, was only fifty years old. Gallistel, a young architect influenced heavily by his study of John Ruskin's architectural works, set out to beautify the campus by planting trees and shrubs. He was fond of quoting Ruskin's lines from *Seven Lamps of Architecture* (1849): "When we build, let us think we build forever. Let it be not for present delight, nor for the present use alone; let it be such work as our descendants will thank us for."

Little wonder that Gallistel was enthusiastic about the future of the Arboretum and was involved from its beginnings in 1932. His influence and contributions touched every aspect of Arboretum development, from the formation of Arboretum governance to the design of its stone walls, its CCC camp, its main road, and its restored prairies. "Everybody knew Gallistel," Sachse wrote in a memorial tribute following his death on New Year's day, 1964, "not only for his professional distinctions" but also for his genial personality. Put Gallistel in a Santa Claus suit, someone had once observed, "'and you'll have perfect type casting.'"[41]

On September 17, 1959, Gallistel was the guest of honor at a ceremonial dinner in the Board Room of the UW Memorial Union that was attended by former and present AC members, including his old friend Bud Jackson, who recounted for the gathering the special qualifications that Gallistel brought to the AC chairmanship and the valuable contributions he had made to the Arboretum. After the dinner, Curtis announced that in Gallistel's honor, the old Camp Woods northeast of the Arboretum Headquarters had been officially renamed Gallistel Woods.[42] Although Gallistel retired from the university, he felt no compunction to retire from his commitment to the Arboretum. In subsequent years, he was available to Curtis as an AC adviser when Curtis needed him, and he was solidly behind Grant Cottam's efforts in the early 1960s to organize an ancillary support group that would be known as the Friends of the Arboretum and that would have a major impact on the Arboretum's future from 1962 to the present.

CHAPTER 9

❧

The 1960s

LAST LOST CITY LOTS, BUILDING A REPUTATION,
SHIFTING FOCUS

Albert Gallistel retired as Arboretum Committee (AC) chair in summer 1959. He was replaced by John Curtis, who gave up his position as Arboretum research coordinator in order to become the third chairman of the AC. The position of Arboretum research coordinator remained unfilled for the next fifteen years, a move that, in retrospect, was a harbinger of the move toward a new Arboretum focus on outreach and education that would characterize the transitional nature of the Arboretum into the 1960s and early 1970s.

At the end of the summer and beginning of the academic year 1959–1960, in response to a university-wide directive to reduce the size of UW committees, the AC complied by cutting its membership from fourteen to six. Curtis would serve as chair. Gallistel agreed to stay on through the year as an adviser. The remaining four members were Henry Greene, Arthur Hasler, G. William Longenecker, and Robert McCabe. Because of the reduction, the committee could no longer cover all the Arboretum's various commitments, so an Arboretum Advisory Committee (AAC) of twelve was formed to work with the six AC members. Curtis would turn to the AAC when necessary. All of the members of the AAC for 1959–1960 held UW professorial appointments.[1] The days of appointing advisers to the AC from outside the campus, like Colonel Bud Jackson, had passed. The Arboretum was conforming more and more to the internal tightening of university control over colleges and departments. Longenecker would remain as executive director, but there would be no director or coordinator of research for the 1959–1960 academic year.

Yet the subject of research was still very much on Curtis's mind. In March 1960 he told the AC that the lack of off-campus financial support and a lack of cooperation on the part of on-campus researchers working on existing projects were leading him to rethink the nature of the Arboretum's commitment to research. He was disillusioned. He had always strongly supported the Arboretum's research mission. Now he cautioned that some existing research projects, those of forestry pathology in particular, were using more Arboretum land than expected and were not filing required reports. He also registered his dissatisfaction with some of the faculty involved in Arboretum research for failing to give any acknowledgments to the Arboretum in their publications. In quite a turn-around for Curtis, he asked the AC to give serious consideration to what percentage of Arboretum money, land, and favorable treatment should be allocated now not only to research but also to teaching, adult education, and recreation. He acknowledged that research and major financial support went hand in hand, but in a foreshadowing of inevitable changes on the horizon, he asked the committee to come up with some ideas on what the principal function of the Arboretum should be in years to come.[2]

Not surprisingly, as Virginia Kline, Arboretum ecologist and Curtis's eventual successor as research program director (1975–1996), reminds us, it was Curtis, as noted in chapter 8, who in May 1959 first advocated the need in an AC meeting for the creation of some kind of education program for grade school children.[3] Eventually, it would happen, but Curtis would not live long enough to see this.

Curtis's Legacy, Cottam in Charge

Curtis was diagnosed with cancer in the late months of 1959. Around Christmas time he entered the hospital for exploratory surgery. According to Grant Cottam's account, Curtis spent most of the winter and spring 1960 in the hospital. And, typical of Curtis, he never told anyone what was wrong with him. "His silence," Cottam noted, "made things difficult for me. I taught his classes and assumed some of his committee roles" but had no "knowledge of what to expect." Finally, when Curtis was released from the hospital, he told Cottam what was wrong with him and that he thought he had five years to live. He began taking an experimental drug developed by UW medical researchers.[4]

By the end of the 1960 spring term, Curtis was losing strength. Yet, he felt well enough to meet with the AC and was intent on continuing to chair the meetings. In April 1960, in the first recorded reference to the idea, Curtis urged the AC to consider the possibility of hiring a managing director. In his mind, unknown to the AC, was the realization that he would not be able to continue. The AC and the AAC agreed at that April meeting to include in the next biennial budget request a full-time managing director's salary provision. A month later, in May 1960, Curtis told the AC that the duties attendant on running the Arboretum combined with chairing the AC were "bearing too heavily on him." He asked for and was promised help: assistance with labor matters came from McCabe, with land matters from Gallistel, with gifts and grants from Hasler, with publications from Greene, and with publicity, scheduling group meetings, maps, and blueprints from Longenecker. Curtis continued to assume responsibility for budgetary concerns and general Arboretum needs.[5]

Curtis managed to meet his classes through the fall 1960 semester. In spring 1961, he met his classes for the start of the semester. Cottam supervised the labs; Curtis delivered the lectures. In early March 1961, Curtis asked Cottam to take over his classes. He never returned to the classroom. He turned up for one of the class field trips but was unable to get out of the car. As Cottam recalled, "The last time I saw him, I went to his house to give him a new nature film. He gave me an anguished look and told me he was going to die." Curtis died a week later on June 7 at the age of forty-seven.

On a fall day in 1961, Cottam and Curtis's wife, Jane, scattered his ashes over the 60-acre Arboretum prairie that would be named for him a year later on October 14, 1962. As Cottam observed, the prairie very likely would never have survived had it not been for Curtis's interest and hard work. He is there "in his niche," Cottam added in tribute, "in the tall prairie grasses, the tiny herbs, the bacteria, and in the soil." Cottam hoped that both Curtis and his prairie would "be there forever."[6]

Curtis's legacy included more than the prairie he so loved and on which he spent so much time working. It was Curtis who pioneered a controlled burning technique that others would adopt and that enabled the Curtis Prairie to expand faster than ever and with more variety of prairie species in place. His legacy also involved the introduction of an Arboretum seed exchange program with arboretums and botanical gardens worldwide, an effort that did much to expand the Arboretum's global identity. His signature contribution to the exchange was the conviction that although routine seed offerings were a mainstay, a more desirable international exchange should involve plants native to Wisconsin and the Midwest, of which the Arboretum now had a bounty. As Greene observed, the "great demand over the years" for Curtis's seeds was a testimony to his energy and foresight. Indeed, in July 1961, the month after he died, the *Arboretum News* reported that the exchange mailing list had grown to 23 institutions in the United States and 132 in foreign countries. In 1961, some 75 institutions had requested seeds. A total of 552 packets were forwarded around the globe.[7] Curtis's final contribution to the Arboretum, made shortly before he died, was a carefully thought-out report for the AC and staff in which he offered his views on the Arboretum's future, especially on the direction that research should take.[8]

Thirty years after Curtis's death, Evelyn Howell and Forest Stearns, both of whom earned ecology doctorates at the UW (in 1975 and 1947, respectively), in a co-authored article published in 1993 on the lasting influence of Curtis and his students, suggested that Curtis's greatest legacy, in retrospect, was the systematic restoration of most of the Arboretum's native communities planted by the small army of graduate students who studied and worked with him during the late 1940s and the 1950s. The students were Curtis's protégés, forerunners of a new generation of ecologists interested in studying vegetation as a scientific continuum and in making certain that future restoration efforts would be based on the ecology of "communities" rather than random plantings.[9] That generation of students—Curtis's army—included Cottam, his successor, and David Archbald, who in 1962 would return from Indonesia to become the Arboretum's first managing director, the position that Curtis had designed.

Cottam and the "Great Treasure Hunt"

Grant Cottam was born on August 26, 1918, in Sandy, Utah, now a bedroom suburb thirteen miles south of Salt Lake City. But when Cottam lived there as a child, Sandy or Sandy City, as it was officially called, was a small town fortuitously located at the base of the Wasatch Mountains. Easy access to the mountains provided young Cottam with opportunities for extensive hiking and camping. As a young man he developed a deep interest in nature. In 1939, he earned a B.A. with a major in botany from the University of Utah. During the Second World War he served with the U.S. Army in the Pacific. He earned a field promotion to captain and was awarded Silver and Bronze Stars for gallantry in action and meritorious service.[10] In 1948, he earned a Ph.D. in ecology from the UW–Madison. Curtis directed his doctoral studies; in 1949, Cottam began working alongside Curtis in the Arboretum and as a colleague in the UW Botany Department, which eventually he would chair. Cottam collaborated with Curtis on a number of scientific research projects and subsequent publications. He retired from the University in 1986 and died on May 13, 2009 at the age of ninety-one.

When Cottam took over as chair of the AC in 1961, he inherited the same committee of six that Curtis had appointed in 1959. The only addition was Robert J. Dicke from the Department of Entomology, an "old hand" with Arboretum business, dating from the postwar 1940s. Dicke was a welcome

Professor Grant Cottam chaired the Arboretum Committee and directed
Arboretum policy from 1961 to 1971. (photo from Arboretum Photo Collection)

addition. Cottam must also have felt relieved to continue to have Gallistel
available as a resource. Gallistel had generously agreed to attend the meetings on an ex officio basis. Longenecker would continue as the Arboretum
executive director and as a member of the AC. Jake Jacobson was in charge of
day-to-day management. But, like Curtis before him, Cottam, as chair of the
powerful AC, was actually in charge of directing Arboretum policy.

One of Cottam's first major agenda items, one that was also close to Gallistel's heart, was the long anticipated formation of the Friends of the Arboretum. But before they could get to that serious item of business, Cottam,
Gallistel and the AC had to contend with a totally unexpected but potentially
dangerous exercise in absurdity.

The "Great Treasure Hunt," another threatening development from without, followed shortly upon Curtis's death. Early in the summer of 1961,
Madison Radio Station WISM, home of Rockin' Robb Steele and "top forty"
broadcasts of rock and roll tunes, decided to launch a promotion. Entering
into a most bizarre moneymaking pact with several undiscriminating local

merchants, WISM, no stranger to notoriety during those years, told its listeners of a "valuable" capsule containing a "treasure"—$1,000 dollars—that was buried somewhere south and west of the campus. The capsule's whereabouts was disclosed gradually in a daily series of broadcasts designed to keep listeners listening. But the upshot for the Arboretum was disastrous.

Without permission from or consultation with anyone officially representing the Arboretum, the WISM disc jockeys teased the listeners with clues, at first vague, but becoming more specific as the days passed. Unfortunately, the clues clearly began to send the "treasure seekers" in the Arboretum's direction. And that's when the trouble started. As the account in the *Arboretum News* recorded, "literally thousands of people roamed the area, trampling, poking, and probing, and overall doing considerable damage." They came by the carload armed with picks and shovels. Plants were trampled; rocks were overturned; holes were dug; sheds and garages were broken into. The stone shelter in Gallistel Woods was assaulted—shingles torn off the roof, the door forced open.

Finally, someone convinced WISM to redirect the mob—which they attempted, but with little success. The radio station started hinting that the treasure-seekers would be better served to abandon the west end of the Arboretum and to concentrate on the Lost City area to the east. Nevertheless, up to the last day of the hunt, the whole of the Arboretum continued to be overrun. Finally, the hunt ended when someone who had taken the WISM hint seriously discovered the capsule in the Lost City. The ordeal over, it was left to Jacobson's crew to repair the damage. The AC, hoping never again to suffer a similar comedy of errors, arranged for a consultation with representatives from the Wisconsin attorney general's office on possible legal steps that might prevent future invasions. The anonymous author of the article on the debacle in the *Arboretum News* held out lofty hope that in spite of the damage to plant communities and buildings, the enduring spirit of the Arboretum might somehow have filtered into the hearts and minds of treasure hunters who had been strangers to the Arboretum prior to the assault but now, having been unexpectedly introduced to the place, might have come "to appreciate the Arboretum for its own sake."[11]

The Friends of the Arboretum, Dedicating the Curtis Prairie and the Jackson Oak

With the 1961 summer's "Great Treasure Hunt" behind them, Cottam and the AC could get on with serious business. One of his first acts as chair, once the fall 1961 semester began, was to bring to the attention of the AC a matter that

Curtis had introduced years before.[12] The "Treasure Hunt," the latest example of seemingly interminable threats from the outside, led the Committee to debate the point that perhaps the Arboretum *could* benefit from better public relations and professionally directed public support. It was a point that Curtis had raised as early as spring 1960. Now Cottam, convinced of the necessity for additional public support, encouraged the AC to find a way to create an organization of concerned "friends" who would help with the promotion and protection of the Arboretum.[13] On the surface, Cottam explained the creation of the Friends of the Arboretum as a way to provide people interested in the Arboretum with "an opportunity to participate actively in its development." But he also intended the formation of the Friends, implicitly, as a defensive move.

In a memo he circulated to the AC, Cottam detailed his concern in no uncertain terms. He reminded the committee members of the importance of maintaining good public relations with the citizens of the entire state, not just the environs of Dane County, owing particularly to pressing dangers of exploitation or expropriation from sources that now extended well beyond the Madison metropolitan area. He had the following specifically in mind:

> The Wisconsin State Highway Department for its continued designs on expanding the Beltline Highway at the Arboretum's expense.
> Public utilities that continued to see the Arboretum as the shortest and cheapest route for electric, gas, water, and sewer lines.
> Local civil governments that periodically proposed that roads be cut through the Arboretum to connect segments of the city or township.
> Private real estate interests lusting to subdivide Arboretum land.[14]

Given the seriousness of impending threats, Cottam advised, the Arboretum needed all the "friends" it could get.

By October 1961, the AC had decided to establish a board that would govern the new Friends of the Arboretum organization. The committee also decided that given the statewide nature of Arboretum threats, members of the Friends board should be drawn from Madison and other parts of the state. By March 1962, the AC, in consultation with the AAC, had come up with proposed by-laws and a constitution, most of which was drawn up by Jacob H. Beuscher of the University Law School. Beuscher, also a member of the AAC, would work closely with Cottam on the organization of the Friends throughout the spring and summer of 1962.

The formation of the Friends of the Arboretum was formally announced in the *Arboretum News* for April–July 1962. "Our budget is hardly sufficient to carry out our primary functions, and the many things we could be doing to make the Arboretum more attractive and useful," the announcement noted. Nevertheless, the announcement continued, "we are establishing an organization to be called *Friends of the University of Wisconsin Arboretum*. Its aims are twofold. The first is to provide funds primarily for the purpose of improving facilities for the public utilization of the Arboretum, to include the preparation of trail guides, the marking of trails, and the organization of occasional field trips for the members of the *Friends of the Arboretum*. The second aim is to . . . help protect the Arboretum from the demands made by the expanding urban area that now . . . surrounds us, for roads, utility rights-of-way, highways, and even real estate development."[15]

On the afternoon of Sunday, October 14, 1962, the Friends of the Arboretum (FOA) held its first official meeting outdoors at the entrance to the 60-acre Arboretum Prairie where a glacial boulder, found on the prairie and inscribed with Curtis's name, had been placed. The first FOA meeting, appropriately, was held in conjunction with the dedication of the prairie to Curtis, who first officially proposed the formation of the Friends. At the meeting, Beuscher presented a constitution and bylaws that were unanimously adopted. The first elected president was A. W. Peterson. Lowell Frautschi was elected vice-president and Jane Curtis, John Curtis's widow, was named secretary-treasurer. An eight-member board of directors was appointed that included Peterson, Frautschi, and six other members, including Gallistel, who had been a strong supporter of efforts to form the Friends from the time of Curtis's proposal. An advisory committee that would operate as a liaison between the FOA and Arboretum management was also formed. It consisted of Cottam, Longenecker, Beuscher, and David Archbald, who had been hired as the first Arboretum managing director in June 1962.

Following the morning business meeting, Cottam gave a short introductory speech in which he noted that over the years Curtis had applied himself to all of the Arboretum's plant communities, but the prairie was his favorite. "He loved to wander through it," Cottam continued, "noting the success or failure" of the planting efforts and burning experiments "applied under his direction." The prairie "in all its beauty and complexity," Cottam went on, was Curtis's gift to us. Consequently, "we are proud," as a gift to him, "to be able to give it his name." After his introductory remarks, Cottam introduced Professor Robert P. McIntosh of Notre Dame University, another of Curtis's former students, who gave the principal dedicatory address. McIntosh spoke of Curtis's scientific contributions, particularly to the science of ecology. After the dedication, the crowd in attendance divided into groups for a tour of the

"Curtis Prairie" led by members of the AC, the Botany Department, and a contingent of Curtis's former graduate students.[16]

The FOA had gotten off to a very promising start. By early November 1962, less than a month after the dedicatory fall meeting, the membership was at 180 and counting. As planned, many of the Wisconsin members were from outside Madison, some from outside the state.[17] On May 18, 1963, another important dedicatory meeting of the FOA coincided with yet another historical Arboretum event. This time it was Colonel Bud Jackson's turn to be honored with a dedication, and in his honor the FOA met at the western end of the Curtis Prairie under a large white oak with Jackson in attendance. Following lunch and a brief business meeting, the dedication ceremonies began.

On that cool but sunny May afternoon in 1963, the large white oak that stood at the west end of the Curtis Prairie was to be officially dedicated to Jackson and would be known from that day forward as the Jackson Oak. As he had done six months earlier during the Curtis Prairie dedication, Cottam offered some introductory remarks. He acknowledged Jackson's important role in the history of the Arboretum, referring to him as the "White Knight" who "rescued us" from oblivion after Olbrich's unexpected death in 1929. He also related the story of how the large white oak came to be chosen for the memorial. Jackson had asked for it, Cottam revealed. The great oak "was here when the Indians roamed this country. Before the changes wrought by the white man it was growing . . . in a grass land. . . . When the early settlers changed the prairie into a pasture, cows and horses rested in its shade. Now it is a prairie again." Cottam noted that he thought it most appropriate "that this tree bear Colonel Jackson's name." Cottam was followed by Lowell Frautschi, FOA vice-president and a long-time friend of Jackson. He spoke of Jackson as a "rare bird" with "rare talents," intelligent and imaginative, "persistent . . . beyond the point of stubbornness, persuasive, with a flair for words and oratory." We are "the *Friends of the Arboretum*," Frautschi added, but Jackson was "surely the foremost among its friends, and in a sense its creator." Frautschi was right. Without Jackson, the idea of an Arboretum emerging out of the 1920s likely never would have become a reality. "Olbrich was the prophet of the Arboretum," Frautschi noted; "Colonel Jackson became its sword."

Jackson, sitting in the audience, must have been pleased. He was eighty-five years old, and he went on to live six more years. Actually, Jackson, in 1963, and Gallistel, in 1959, had set a precedent, being the only major players in the thirty-one-year history of the Arboretum who had been honored for their contributions while still alive. McCaffrey, Olbrich, Leopold, and Curtis had not been so fortunate. Jackson took advantage of the moment to address the audience, as expected, with "humor and dignity." His bearing, "his firm voice and straight posture belied his years," the *Arboretum News* reported, "giving evidence of his strong, determined character." He was in fine form," noted Nancy Sachse, another old friend, in a personal journal excerpt published years later. "He made a sonorous and very neat speech," she added, "to the relief of all who feared he would go on forever."[18] Two inscribed photographs of the great white oak were presented to Jackson.

When the ceremonies ended, Cottam announced that guided tours of Gallistel Woods would be available for the people in attendance. He gave a brief account of the salient features of the Gallistel Woods and of future plans for its development. Most took the guided tour in groups of fifteen to twenty, led by various persons associated with the Arboretum. One tour guide eminently familiar with the woods was Jim Zimmerman, "a remarkable young man," Sachse would observe, who knew "every bug and berry on all 1218 acres."[19] In spring 1965, Zimmerman would become the first official Arboretum ranger-naturalist. After the tours, a few lingered at the "watchman's trailer" that now housed a young couple, Botany Department graduate students Robert and Kathy Riehm, who, as Arboretum custodians, a new security measure, were given the privilege of living in the Arboretum watchman's trailer rent free. The trailer was located where the old security residence once stood. The Riehms had a pet, a rescued red-tailed hawk, who entertained the stragglers that remained with an exhibit of impressive responses to Bob Riehm's falconry commands.

The FOA members in charge of arrangements, the *Arboretum News* reported afterward, judged the meeting a great success and expressed appreciation for "the interest and cooperation of the many *Friends of the Arboretum* who made it so."[20] Immediately following that memorable meeting, the FOA could boast that the membership increased to 252, that $4,500 dollars in dues had been collected, and that the organization was moving forward. And, indeed, it was. In less than a year and a half, activities sponsored by the FOA would include the publication of trail guides, the placing of markers along the trails, and the creation of a new Arboretum map mailed to members and, eventually, made available to the public. The trail guides helped with identification of plant species and eventually would provide useful information on particular features of every major Arboretum trail. The new colored map, prepared by UW artist John H. Wilde, displayed all the Arboretum trails on one side and on the other side showed the location of specific plant communities.[21]

Even more ambitious than the trail guides and the colored map, however, was a movie designed to tell the story of the Arboretum "through the seasons."[22] On the surface, the film was a colorful artistic tour of the Arboretum's natural beauties. On a less explicit, more subtle level, however, it was a reminder for viewers of the Arboretum's ecological contributions to the community, both local and statewide.[23] The film, which lasted thirty-two minutes, premiered on September 30, 1964, to a packed house in the Wisconsin Union Theater. Attendance totaled over 1,300.[24] Following the premier, the *Arboretum News* acclaimed it both an artistic and educational success and noted that owing to its timely conservation message, it would be added to a list of other nationally recognized "Conservation Films" for 1965.[25]

From the time of its inception, the FOA was a force for progress and expansion. The membership, numbering 252 in 1963, increased to approximately 700 by 1970, and by the end of the century had catapulted to over 3,000. It was the FOA, as discussed below in more detail, that provided funds in 1963 for the acquisition of the last Lost City lots. Through the years, the FOA assisted with the funding of Arboretum publications, the Arboretum Tour Guide Program, the volunteer program, a variety of educational initiatives, and some of the most important events designed for public entertainment and education. Truly, the early 1960s was a time of tremendous expansion for the University of Wisconsin that necessitated "a whole new operating system." Without any organized outside support, the Arboretum ran the risk of becoming totally a UW "stepchild," a bleak project, since the whole Arboretum dream had depended historically on the work of many people, such as Olbrich, Stark, Jackson, and Louis Gardner, who had no direct UW connections. Consequently, in the 1960s, there was a public participation gap that needed to be filled—and it was filled and it continues to be filled, to the great benefit of the Arboretum—by the FOA.[26]

David Archbald, First Managing Director

In late spring 1962, approximately four months before that first memorable FOA meeting in October, David Archbald was appointed Arboretum managing director. The appointment was concurrent with the retirement of Jake Jacobson as Arboretum superintendent. Eugene D. "Gene" Moran, an Arboretum staff member since 1955, succeeded Jacobson, but Moran's title was foreman of the labor crew. Moran worked almost exclusively outdoors. Jacobson's responsibilities as superintendent had included many of the managerial demands that Archbald now assumed as the first managing director. Hence, the position of superintendent was phased out. Also, Jacobson's position, as a faculty line item initially, was now transferred to the line item that Archbald filled, and Archbald was hired at the faculty rank of assistant professor.

As early as April 1960, while Curtis was still alive, the AC had approved a salary provision for a managing director. The salary line was included in the budget for 1961 to 1963.[27] In September 1961, only three months after Curtis's death, Cottam told the AC that Archbald earlier had assured Curtis that he would accept the position and now, as an official candidate, would be available by April 1, 1962.[28] Archbald returned to Madison early in 1962 and was appointed managing director and assistant professor of botany in late spring 1962.

Archbald, born in 1925, was a native of Buffalo, New York. In 1943, at the age of eighteen, he enrolled in the Naval Air Program at Cornell. He served in the Navy from 1943 to 1946. After the war, he earned a bachelor's degree from the University of Buffalo. In 1948, he enrolled in the graduate program in the UW's Botany Department. He came to the UW specifically to work on wildlife management with Leopold, but upon his arrival in 1949 he discovered that Leopold had recently died. Eventually, he met Curtis and told him he was interested in "applied ecology" rather than "theoretical ecology." Curtis agreed to take him on as one of his botany graduate students. For the next six years, from 1949 to 1954, Archbald worked with Curtis, mainly on the Arboretum prairie. In 1949, as Archbald recalled, the prairie still had large sections of invasive bluegrass that Curtis wanted replaced with native prairie grasses. Following Curtis's advice, Archbald used Ted Sperry's maps to identify the original plantings. Plants were still in clusters, he noted, "locked in with the blue grass." Sperry had planted in blocks, which was anathema to Curtis, who insisted on planting communities with a variety of native species. Curtis's graduate students collected and planted seeds from relic prairies. At times, they also brought in sods. Often they put the sods in car trunks, wet them down, and then planted them. And the planting did not stop with the prairie, Archbald noted. "We planted . . . the maple undercover in the Gallistel woods. . . . We put in the Ponderosa pines out there."[29] While working on his doctorate, Archbald also held the position of Arboretum botanist, a perk that had traditionally been awarded to promising graduate students.[30]

Archbald received his Ph.D. in spring 1954. From 1955 to 1961, he worked as a researcher on natural rubber production for U.S. Rubber (Uniroyal) in the Sumatra Plantations at Kisaran, Indonesia. In 1958, he was appointed director of plantation research, and in a development that anticipated the direction in which his professional interests were moving, even as early as the late 1950s, he devised and set up an organization through which interested Americans living in the Kisaran area could volunteer for local social and educational programs. Archbald's growing concern with environmental protection issues would become even more acute as his tenure as Arboretum managing director played out in the 1960s. "The abiding concern was what we were doing to the environment. We were destroying everything," he was quoted as saying, when asked if his earlier field ecology efforts had been based on any specific underlying principles. He also lauded Leopold's habitat rehabilitation efforts in the 1940s as an early example of a desire for environmental land reform and renewal of wildlife and plant species. Leopold wanted the Arboretum to be a "generating factory" of animals and plants that would "go out into the community," Archbald recalled. The Arboretum would be a "generator of wild things" in the pursuit of preservation and restoration, a commitment largely approved by Archbald and one on which he hoped to build as managing director.[31]

Archbald's return to the Arboretum in early 1962 was in time to help Cottam and the AC launch the Friends of the Arboretum. Archbald was also the first director to have an official Arboretum office in room 329 of the Botany Department's Birge Hall. The new office had a secretary, John Curtis's widow, Jane Curtis, who worked part-time.[32] Archbald was also in time to see the last of the Lost City lots acquired by the University, thanks to FOA funding. The Arboretum and the University had spent decades acquiring the lots. Along the way, the fascinating story of Bernard "Bernie" Chapman and the Lake Forest Land Company, of the demise of the Lake Forest Development, of Victor Arnold's treachery, of the financial loss to land purchasers and investors, and the tragic end to a dream city tagged in the 1920s as the "Venice of the North" had become, and still is, a tantalizing piece of Madison folklore. With all the lots now the property of the Arboretum, the story and the lure of the Lost City became history and continues to peak the curiosity of Arboretum visitors who regularly ask just exactly what was or is that large southeastern portion of the Arboretum called the Lost City Forest.

The Purchase of the Last Lost City Lots

Archbald was present at the meeting of the AC on March 11, 1963, when Cottam announced the happy news that "the last three outstanding 'Lost City' lots" had been acquired. The Regents approved the purchase on February 8.[33] It had taken thirty years to acquire all the property in the 108-acre Lost City Forest that now covers most of the Arboretum's southeastern edge—thirty very long years, with a myriad of recorded problems and long-term aggravations. Olbrich, Stark, Gilbert, Leopold, and McCaffrey would have been pleased. And without doubt, the remaining surviving pioneers—Longenecker, Gallistel, and Jackson—had to have felt a mixture of sanguine relief and merited triumph when they got the news—especially Jackson, whose persistent efforts to find donors, from the beginning, made possible the purchase of most of the Lake Forest land.[34]

The history of the ill-fated Lost City began a hundred years ago in 1911 when Chandler Bernard "Bernie" Chapman, Leonard Gay, and E. J. B. Schubring, Madison contractor-realtors, organized the Lake Forest Land Company and devised a plan to build "The City Beautiful," a grandiose suburb on the south shore of Lake Wingra on land that had previously been owned by the Gay Dairy Company. The suburb would be called Lake Forest. The three entrepreneurs had high hopes and grand expectations for the model community that was advertised as the future "Venice of the North": lagoons would be dug in an effort to lower the water table, some homes would be built on lots extending to the Lake Wingra shoreline, canals would be dredged providing boat owners in homes on the lagoons with a water route to the lake. A drive to be called Marshall Parkway, extending two and a half blocks across the west end of the development, would lead to the Lake Wingra shoreline, affording access to the lake for the rest of the residents. Access to the lake was guaranteed.

The neatly platted lots, most designed with small bungalows or cottages in mind, averaged only 40 by 105 feet. Lot prices ranged from $600 to $2,000 and were acquired for as little as $10 down. The lots were arranged in a design similar to the old colonial village green concept with arterial roads leading to residential neighborhood blocks radiating out from a large circle or hub, similar to Pierre Charles L'Enfant's eighteenth-century design proposals for Washington, D.C. A large imposing obelisk was planned for the center of the circle. The circle itself, proclaimed metaphorically by its Milwaukee designers as "'a man-made pearl . . . dropped on a bed of organic velvet,'" was intended to remind viewers of the piazza in front of the Vatican's St. Peter's Basilica. The designers, from the firm of Hegemann and Peets, also intended the circle to be the community's commercial and civic center. The plan also called for a nearby mall with a community center, shops, public buildings, and a park.

A double boulevard, built across what is now Gardner Marsh, would be called Capitol Avenue and would extend in a straight line from the Civic Center Circle to South Mills Street, offering travelers a clear sight of the State Capitol. The plans also called for a trackless trolley running from South Mills Street across Murphy's Creek (now Wingra Creek), and down Capitol Avenue to the Civic Center Circle. A swimming beach was planned on the eastern shore of Lake Wingra just opposite Vilas Park. There was also a plan for a nine-hole golf course along Fish Hatchery Road.[35]

In April 1916, the Lake Forest Land Company, originally a *holding* company, was reorganized as the Lake Forest Company, a committed *development* company with Bernie Chapman as president and project contractor. To finance the construction of the new subdivision, all of the Lake Forest land, under a deed of trust, was put in the hands of a loan company called the Madison Bond Company, which issued and sold Lake Forest Company bonds at rates of 10 percent down and 1 percent per month. The president of the Madison Bond Company was Victor H. Arnold. He had formerly been the president of a Chicago-based bond firm called the Victor H. Arnold Company. According to a Chicago datelined news story, cited in John Newhouse's 1967 account of "Madison's 'Lost City,'" Arnold left Chicago several years before his financial swindles were discovered and built a flourishing bond house in Madison.[36]

And so, with $500,000 in capital stock, Bernie Chapman, in 1916, began work on the massive construction project. Two dredges began digging canals that drained water from the low-lying land. Some land was filled in. When Chapman thought the land was firm enough, concrete for streets and sidewalks was poured. According to John C. "Jack" McKenna Jr., whose father had worked as a sales manager for the Lake Forest Company, after the Lake Forest canals and lagoons were dug, Chapman, judged as "a fine man" by McKenna, used the dredges to improve the shoreline of Lake Wingra, which was much shallower in 1916. Chapman also put locks in Murphy's Creek near the Vilas Park Zoo in order to raise the lake's water level. He started work on Capitol Avenue, originally intended to be a double boulevard. However, when it was discovered that the road bed would require more fill than anticipated, the avenue was limited to a single double-lane road. But as promised in the original plan, Capitol Avenue extended in a direct line toward the State Capitol and Capitol Square. "It was a good dream," McKenna observed. "It had every chance of success." As a child growing up in a house on one corner of the Lake Forest property (now Irwin Place), he remembered fine lawns and shrubberies and "streets and sidewalks [that] were on good drained ground."[37] In summer 1921, the projected nine-hole golf course, designed by Guy Martin, the Maple Bluff Country Club professional, opened for business.[38]

Of the 840 total acres in the Lake Forest plat, only 200 acres were planned for the housing subdivision. Approximately 800 lots were up for sale.[39] World War I temporarily slowed sales, so by 1918, only 48 lots had been purchased. Many of the buyers were from outside Madison and thought of Madison,

with its four lakes, as a potential retirement spot. At the time, Madison was
served by three railway systems, nine railroad lines, carrying traffic in and out
of the city daily. Madison was also considered an attractive vacation spot by
out-of-town tourists, including many from Chicago and northern Illinois.[40]
Cottages and other short-time rental properties were being built in subdivisions around and close by the lakes. People on holiday came from out of town
and stayed for a week or two. Some stayed longer. By 1920 there were almost
forty thousand people living in Madison, and Chapman was convinced that
the population would increase, and so would the sale of his lots. But the lots
continued to move slowly. Many were being bought by land speculators hoping for a substantial increase eventually in the property value. Surprisingly, the
speculators were encouraged by the Lake Forest Company through a semi-
monthly newsletter called the *Lake Forester* that carried news and photos of
the development's progress. Unfortunately, encouraging people to buy and
then to wait to make a "killing" through resale turned out to be a fatal mistake
since speculators never intended to build houses on the lots. And houses were
what Chapman really wanted. By 1920, four years into the project, 61 lots had
been sold, but only one house had been built. In 1922, the company's best year,
73 lots were sold but only six houses built.

The six houses were located east of the Civic Center Circle on the side
streets connecting Carver and Martin Streets and are still standing. So is the
seventh house, but it's located on the high bluff east and north of the Civic
Center Circle in the area now known as Forest Park, the only part of the
original Lake Forest development that contained marketable properties after
the Lake Forest Company's financial collapse in 1922. The seventh house, at
2601 Balden Street, was built in 1922 by William Twenhofel, UW professor of
geology.

There is also an account and a photograph in the *Lake Forester* for July 15,
1921, of the first house that was under construction west of the Civic Center
Circle in what was known at the time as the Burr Oak section of the Lake
Forest development. By September 1921, the house was for sale, according to
the *Lake Forester*. "Drive out there some of these fine fall days," the public
was advised. "There are open skies and fresh air, beautiful oak trees and per-
fumed wild flowers in Lake Forest. It is no strain on the imagination now to
see there in our latest section the homes of happy children, the open hearths
of contented families, the attractive nests of newly-weds. Some of them are
already there."[41] Three other houses planned for the Burr Oak section were to
be located between the newly constructed house and the street directly east
known as The Mall. All four of the new houses would face Capitol Avenue.

How many of the four were constructed is not known. Very likely, two were built, or at least the second foundation was put in. Also, very likely, the houses that were constructed were never lived in, including the one for sale in September 1921. Two ghostly foundations still lay under the tangle of overgrown vegetation in the Lost City Forest. They are all that remains of those two Burr Oak houses and have for years been the spooky destination for Arboretum guided tours during Halloween.[42]

Included among the first six houses that were built between 1920 and 1922 east of the Civic Center Circle, on the side streets bounded by Carver Street on the north and Martin Street on the south, was a large stucco home purchased by the McMurry family in 1920, which has the distinction of being the first house in the Lake Forest development that was lived in. The house, at 2005 Baird Place (now 2025 Dickson Place), was built by Karl McMurry, a UW professor of business administration.[43] The McMurry house was followed by the McKenna house, a stucco bungalow, located on Arvin Place (now 2030 Irwin Place), directly behind the McMurry house. Three of the other four houses belonged to the Erd family, whose house (its "handsome chimney suggesting a cozy fireplace within") was located on Arvin Place (now 2013 Irwin Place); the Gibson family, whose white frame colonial was located on Baird Place (now 2009 Dickson Place); and the Burroughs family, whose house was also located on the 2000 block of Baird Place. The sixth house (original owner unknown), also built between 1920 and 1922, was located on the corner of Martin Street and Dunn Place, the side street just west of what had been Baird Place.[44]

Elmer McMurry, Karl's son, who was eleven years old in 1920 when the family moved into the house on Baird Place, recalled, in a 1998 interview, that the neighborhood seemed isolated during the early years. For one thing, there was no electricity. The McMurrys were obliged for several years to use kerosene lamps; perishable food was kept in an icebox.[45] McMurry's account squares with unmistakable signs of impending problems for the Lake Forest Company as early as 1921. Aside from no electricity, there was little fire or police protection. There was no gas line, no nearby school, no sewage system. The land in the low lying areas was a soft marshy mixture of peat and marl extending down through the subsoil for hundreds of feet. Because the soft ground made the construction of Capitol Avenue as a double boulevard impossible, the company had settled for a single road, but the drain tile and sand fill failed to prevent the land beneath the road from settling. Other streets in the low-level development also began to settle and crack. Concrete buckled; streets shifted. Weeds started growing between cracks in the concrete. The canals, partially dredged, were clogged with mud, runoff, and deadfalls. Concrete was poured endlessly on the streets in the low subdivisions, but to no avail. They would never be solid.[46]

Indeed, as early as 1921 the dream suburb was in trouble. But the coup de grace came early in 1922 when the news broke that the Madison Bond Company holding the Lake Forest mortgages had failed. The Lake Forest Company, consequently, was bankrupt. Victor H. Arnold, bond company president, had squandered the money and was "on the lam," nowhere to be found. Interest payments on the mortgage bonds were refused. Failure of the bond company left the Lake Forest Company without funds and hopelessly in debt. Without adequate capital, Chapman could no longer fund the construction. Losses were estimated as high as $500,000 to $750,000. Arnold, now a fugitive, had fled Madison and by spring of 1922 was in New York.

The Devil and the Lost City

Several years before the collapse of the Madison Bond Company in 1922, Arnold had fled Chicago, also leaving behind his Chicago-based bond company that had failed, bankrupting many of the investors. Arnold moved to Madison, where he resurrected himself and became a local celebrity. Frank Custer, retired *Capital Times* reporter, remembered Arnold as "a golden-tongued, junk bond salesman, a preacher-financier, and a flamboyant big-spender." He had a home on fashionable Sherman Avenue, owned three cars—a Pierce Arrow, a Buick, and a chauffeur-driven Rolls Royce. And he had a reputation for associating with silent movie stars and other theatrical icons. In fact, Arnold, like a character out of a Dickens novel, for whom "nose pincher glasses . . . his cane, his greenish brown slouch hat and tan oxfords polished to perfection were trademarks," was a theatrical icon himself. He once rented a Madison high school auditorium and "lectured on brotherly love and the example set by Jesus Christ" before a packed house. But his "biggest caper," according to Custer, was a Christmas party he threw for the Madison poor in December 1921, just before he fled to New York. Basking in the theatrical limelight one final time before leaving Madison, Arnold placed a notice in the *Capital Times* inviting poor Madisonians to visit any Madison store and to buy food, clothing, and fuel (up to two tons of coal), on him—it was his Christmas present. Little did anyone expect, however, that Arnold paid the bill, over $25,000, with embezzled money before leaving town.[47]

Arnold's story took an ironic twist, however, about a year later. In November 1922, the *New York Times* began reporting on Arnold's exemplary religious mission to assist New York's poor. The *Times*, unaware of his criminal dealings in Madison, lionized him as the "banker-preacher," and the November 6 story's headline, in boldface capital letters, read: "BANKER IN SERMON CALLS POOR TO HIM." Arnold, once again in the limelight, had begun to offer a series of Sunday morning religious services in Town Hall, 43rd Street, east of Broadway, where, the headline continued, "RENT EXCEEDS COLLECTION." Arnold, it seems, had magnanimously decided during the service to "dispense with a collection," but the appreciative audience, in a gesture of support, voted to make an offering anyway. Happy with the way his preaching effort had begun, Arnold observed, "I would rather have the services start off modestly and gradually grow to a crowded house than to start with a big crowd and have the movement simmer down."

So, by the end of 1922, the flimflam man who had bilked thousands in Madison and Chicago had become in New York an evangelizing Christian hero, a street preacher. "If you know of any poor or sick please give me their names and addresses and I will visit them and try to help," he was quoted as saying. Arnold, the *Times* story reported, "retired from business" in order to devote "all his time to Christian work." For the Sunday morning Town Hall services, he traveled to New York with his wife and four daughters from their sumptuous estate on 5½ acres in Bayside, Long Island. He also had opened a studio on Fifth Avenue.[48] The Sunday morning Town Hall services were based on the theme "The Spirit of Truth." For seven consecutive Sundays in November and December 1922, Arnold, ironically, preached to his followers on "The Spirit of Truth."

When he was arrested in late December, he claimed that he was the victim of a frame-up originating in Madison with Madison Bond Company officials who took over after he resigned as president in 1919 and who were responsible for the company's collapse. In fact, he told the federal authorities who arrested him that he had gone to Washington, D.C., to urge the U.S. attorney general to issue arrest warrants for those errant Madison bond company officials. On December 27, 1922, the morning after his arrest, his wife was quoted as insisting he had been victimized by a Madison "political ring" that was jealous "and was doing all in its power to undo him." She told the *Times* reporters that she was in the process of trying to raise bail to get Arnold out of jail, "but we are penniless." And then she added, as an afterthought so ludicrous that it's hard to believe that she may actually have meant it: "My husband telephoned me from the jail tonight," she said, to tell her "that perhaps he can do some good while there." The fact that Arnold thought he could continue on in New York as the evangelizing "Banker-Preacher" who could do only " good" and who

had done no wrong suggests that with federal authorities hot on his trail and
soon to arrest him, he was fabricating a defense. Perhaps, he sincerely thought
that his evangelizing efforts would exonerate him and keep him out of trouble.
It was the 1920s. One might say that he had aspirations to be like celebrated
evangelist Billy Sunday. And although he was no Billy Sunday, he had a comparable amount of theatrical flair.[49]

Arnold was arrested on a federal warrant that listed thirty-one charges, including mail fraud, embezzlement, and taking flight to avoid arrest. He was
held on $25,000 bail, which he couldn't raise. In his six years in Madison, he
had sold approximately $14,000,000 worth of bonds. At one time, he owned
eleven automobiles. Now he was broke. Early in 1923, he was extradited to
Wisconsin to face federal fraud charges. In October 1923, he was convicted
by the U.S. District court in Superior, Wisconsin, of using the mails to de-
fraud the public. He was fined $10,000 and sentenced to five years imprison-
ment in Leavenworth.[50]

Jackson's Personal Mission

With no money to finance the construction work, the Lake Forest Company's
dream of a Venetian-style housing development on Lake Wingra's shore faded.
As early as March 1934, the year of the Arboretum's formal dedication, Colonel
Bud Jackson, sensing an opportunity to add the Lake Forest property to the
Arboretum acreage, approached Chapman with a deal. He told him that he
expected the university to acquire a sizable amount of money from a federal
loan, and the money would be available for land purchases. He told Chapman that if he wanted to do something to help everyone, he should "figure
out a price for *everything* that the Lake Forest Company has left—lots, acres
and all." Citing a price for everything, Jackson added, might encourage the
university to make a deal.[51] But Chapman, in typical bulldog fashion, would
not budge. He was determined to hold on to whatever he could in spite of
the outcome. Jackson, as usual, was relentless in his pursuit of the Lake Forest
acreage. In fact, Jackson had a longtime personal connection with the Lake
Forest land that went back at least to the late 1920s. Bernice Elver Gesteland,
who lived with her husband and son in a Lake Forest home on Dickson Place,
recalled seeing Jackson often in the late 1920s strolling through the "Lake Forest woods and meadows with friends." He often visited with the Gestelands.
He was "vitally interested in the area," she recalled, and often came hiking, and
"always with a cane."[52]

Jackson's deep personal interest in the Lake Forest land helps to explain why he worked so assiduously for thirty years to acquire it. From 1947 until 1963, as the archival record shows, he personally handled most of the correspondence and conducted most of the business involved in settling the sales arrangements. As the university bought up the lots, streets were sealed off and torn up. Some holdout lot owners complained and threatened lawsuits. Some negotiated with Jackson and the AC for years before selling. As late as 1962, Jackson was still at it, at times, sending handwritten notes to deed holders. In February 1962, he could report to Gallistel that there were only six lots left in private ownership, and he was bent on acquiring them. By March 6, 1962, three of the six—lots 35, 36, and 37, all fronting on McCaffrey Drive—were donated as gifts. The university would cover the cost of 1962 taxes including delinquent interest payments.[53] By March 11, 1963, with financial help from the FOA, deeds to the last three lots—lot 14, Block 17, and lots 14 and 15, Block 12—were officially accepted by the Board of Regents. After almost fifty years, all of the original land that was part of the projected "Lake Forest . . . City Beautiful" belonged to the Arboretum and would now take up a position in Madison folklore, in perpetuity, as the mysterious Lost City Forest.

And Jackson himself? He had lived to see the purchase of all of the lots. He had outlived two generations of Arboretum pioneers. Cottam was not being facetious when he told Jackson just prior to the acquisition of the last of the Lake Forest properties that he had "only recently begun to realize the magnitude of your contribution to the Arboretum."[54] In fact, one might say that the tall elderly gentleman, well into his eighties by the early 1960s, who wore a fedora and walked with his cane through the Arboretum lands, now including the entire Lost City Forest, had been around so long that, like the scenery, he was often just simply taken for granted, as if he had always been there and always would be. On May 23, 1969, Colonel Bud Jackson died at age ninety. As Nancy Sachse stated, "Without him there would have been no Arboretum for his story is bound up with the land like the roots of the trees."[55] And she was right. Jackson was that important.

Jim Zim, Rosemary Fleming, and the Guides

On New Year's Day 1964, Albert F. Gallistel, known around campus as the "walking blueprint" of the UW, died. His long history of service to the university dated from 1907. In 1963, the year leading up to his death, he was still actively involved with the FOA and was still serving as an adviser to the Arboretum Committee. Gallistel always enjoyed talking about the less hectic past when the UW campus and the enrollment were smaller and it was easier for

UW people to get around and to get to know each other—to watch President Charles Van Hise, for instance, taking a leisurely horseback ride down the lake drive. But Gallistel never let his fondness for nostalgic history get in the way of what he judged was in the best interest of the Arboretum's future.[56] He did not live long enough to see the Arboretum's successful tour guide program in place, but given his interest in making the Arboretum increasingly more accessible to the public, he likely would have approved.

The Arboretum tour program, one might say, unofficially began in 1965 with the hiring of Jim Zimmerman as the first Arboretum ranger-naturalist. Although his position was part-time, "Jim Zim," as he was known around Madison, would prove to be both the inspiration and the source of an experimental program that eventually mushroomed into a full-blown Arboretum tour-guide program. Zimmerman's background suited him well for guided walks through nature. He had a Ph.D. in botany, with a plant taxonomy concentration, and he also had a strong practical grasp of the principles of nature education. He was excellent in the classroom and in the field. Jim Zim was a pleasant anachronism, an Old World gentleman-scholar, too unconventional to worry about the vagaries of job markets. And his world was nature. When Zimmerman was hired in 1965, he was also simultaneously teaching an extremely popular Madison Area Technical College (MATC) night class entitled "Reading the Landscape." His course became the recruiting source for future Arboretum naturalists who would assist him with many of his Arboretum projects.

Zimmerman wasted no time after he was appointed to the Arboretum staff. As a harbinger of the direction in which his influence would inform the Arboretum guide program, he fashioned a mimeographed trail guide to the Curtis Prairie, likely the first interpretive guide to the prairie expressly intended for the general user. He distributed his home-made guide to those who joined him for two-hour walking tours offered on several weekends during summer 1965. By the end of summer, over 470 people had attended. The tours were so popular that the Arboretum administration decided to repeat them in the fall.[57]

Zimmerman's tours were much more than simple "walkabouts." They were interpretive, designed to attract people with at least a threshold interest in nature. And whether he knew it or not, he was planting pedagogical seeds for a future training program.[58] He was a gentleman-scholar by avocation. But he was first and foremost a teacher. His "Reading the Landscape" course enrolled a cadre of committed student-admirers who continued to take courses from him, even when course material was duplicated. They eventually found

Rosemary Fleming in front of a derelict set of concrete steps abutting one of the
abandoned foundations in the tangled depths of the Lost City Forest. (photo by
Mary Knoll, Arboretum Photo Collection)

themselves studying with him in small informal groups. Hence, the fact that
the original Arboretum "guide group" that gathered around Zimmerman in
1965 became recognized as a formal "study group" comes as no surprise. Nor
should it be a surprise that among the original guide group were two giants—
Rosemary Fleming and Virginia "Gina" Kline, who would become key figures
eventually in the Arboretum's educational program. Being part of the original
guide group meant, on many occasions, instant teaching experiences. When
he announced a tour, Zimmerman would have as many as two hundred people
show up. Unable to shepherd all of them by himself, he recruited—on the
spot—talented and knowledgeable people he trusted from his "Reading the
Landscape" course and from the Arboretum study group. Rosemary Fleming
and Gina Kline were often pressed into service. Both recalled the challenges
and how, oftentimes, they got lost with their tour groups on the Arboretum's
unmarked trails.[59]

   And so, facing more people on tours than he could handle alone, Zimmerman asked Fleming to work as his assistant at some point between 1965 and
1966. She accepted and with the endorsement of the AC was hired in 1966 as
Arboretum "tour coordinator," a position that was generously funded for six

months by the FOA. She worked with Zimmerman but not directly under his supervision. She was also charged by the FOA to develop a workable program. In order to do that, she drew heavily on her past experience in public service and as a leader in local Girl Scout activities. Upon her retirement from the Arboretum, fourteen years later, it would be said of Fleming that "she was concerned with bringing an appreciation of the natural world to people of all ages, and she showed a particular genius for working with children."[60] During her six-month probationary period in 1966, she projected a vision that was slightly different from Zimmerman's, one geared more to younger visitors and to a pedagogical approach that was basically experiential or, as she put it in her first full-year report, one that placed "greater emphasis on multisensory, direct-approach learning."[61] One might say of the difference between them that Zimmerman "taught the teachers" how to be knowledgeable naturalists, and Fleming "taught the teacher-naturalists" how to be effective guides. But she and Jim Zim, the tour-coordinator and the professor, apparently worked very well together, both in the Arboretum and with neighborhood civic and public school environmental science projects, such as the Dudgeon School's coordinated effort to keep the nature trails around the Arboretum's Ho-nee-um Pond, opposite the school on Monroe Street, trimmed and accessible to visitors.

Zimmerman was an intellectual, and a rather Bohemian intellectual, at that. He rode a bicycle to MATC for the first class he ever taught there, and he continued to use his bicycle for transportation. As one account has it: "he wore tennis shoes. There were debates as to whether he owned a necktie. He got haircuts, now and then, when they did not interfere with the more important things in the world of nature."[62] He saw the Arboretum as he saw most of nature, primarily as a vast and wonderful laboratory for study and research. And, like Leopold, he wanted it protected. He completed work for his Ph.D. in 1958. But his inspiration as a scientist researcher had come much earlier in his life. At the age of fifteen and living in Madison, according to a 1968 *Wisconsin State Journal* feature that dubbed him "The Pied Piper of Madison," he "met the first of two great men who would mold his life": Professor Norman Fassett, of the UW Botany Department. "'He was not particularly visible,'" Zimmerman was quoted as saying, "'but he was great.'" The other great man, whom Zimmerman met a few years later after he had enrolled in the university, was Aldo Leopold.[63] Zimmerman's vision of nature as a wonderful wildlife laboratory to be investigated, studied, protected, and preserved reflected the influence of both Fassett and Leopold on him.

Zimmerman published a news article in 1971 expressing frustration with the

direction in which he believed the Arboretum was heading. He saw it moving
into a "passive, 'caretakership,'" and he faulted it for having "no problem-
oriented staff" actively seeking grants and attracting researchers. The vital
need, he thought, echoing Leopold, Jackson, Gilbert, Curtis, McCabe, and
other Arboretum pioneers, was for "government funding of a permanent research and development staff."[64] Once again, funding had become the issue.
The university, from the inception of the Arboretum, had failed to fund it
adequately. Federal funding was still out of reach. It was an old story, and
Zimmerman was repeating the mantra of frustration, felt so acutely earlier
by Leopold. As Jackson repeatedly reminded people, the Arboretum was put
together WCTU—"without cost to the university." But in 1966, without adequate funding, all hope of a workable guide program was about to go down
the drain. Funding for Fleming's position was essential, and Archbald, as managing director, was determined to get it, in spite of the university's refusal to
support the position.

Although the documentation is slight on this development, it seems that
shortly before Fleming's six months as Arboretum tour coordinator were up,
Archbald took his concern directly to the Dane County Board. As he explained later in a 1989 interview, he made a strong case at the meeting with the
board in 1966. He told the board members that the Arboretum was, and had
been for years, understaffed and underfunded; that Dane County schools were
regularly sending busloads of students to the Arboretum for tours; and that the
increasing number of tours was more than the Arboretum could handle. He
argued that providing Dane County public school children with educationally
oriented guided tours was not a fair use of the Arboretum staff's time. "We
are paid by the state," he reminded the board, "so why work for the county?"
Educating county school children was not the responsibility of the UW
Arboretum. He appealed to the board to either help fund the tours or to
restrict the number of students coming into the Arboretum. As a welcome
compromise, the board agreed to fund the Arboretum tour coordinator position and to hire Rosemary Fleming, but her title would be "Dane County
naturalist." She would be paid by Dane County, but technically affiliated with
the Arboretum. In a sense, she was stationed there.[65] Her position as a full-
time county naturalist was the first in the nation.

By 1967, Fleming had a full-blown guide training program in operation.
She had begun to put together the first guide group in 1966 during her six-
month probationary period. She initially recruited twelve guides. By 1967–68,
she had fifteen. Some were friends, many were from Zimmerman's "Reading
the Landscape" class (a requirement for prospective guides lacking an adequate

natural science background), and some were old Girl Scout acquaintances. Most of the guides—housewives with time to volunteer or women working on degrees—came with varied skills, varied backgrounds, and generally some experience in group leadership. They were available when she needed them, often on short notice. They were paid small amounts for guiding tours, funding that was, and still is, provided by the FOA.

Fleming's program initially involved weekly training sessions that often had the additional advantage of instruction from UW faculty. The guides shared learning experiences with each other in indoor and outdoor study sessions and learned the value of dry runs, dress rehearsals, or, "pre-flights," as they called them. The program was a remarkable success. In 1968, after only two years, tour demands doubled, and then doubled again between 1968 and 1971. In the first five months of 1969, the guide program conducted more than four hundred tours for close to seven thousand people.[66] In August 1969, the *Capital Times* reported on a summer program supervised by Marion Sutherland, one of Fleming's guides, who offered an Arboretum program for children from the fourth through the sixth grades. The program, sponsored by the School-Community Recreation Department under the direction of the Madison Community Center, was a conscious public outreach effort by the Arboretum. The young students met three times a week for three weeks, during which Sutherland chose a different part of the Arboretum for exploration. The children learned about nature but also about the Native American tribes that lived around Lake Wingra, about the Lost City, and about characteristic differences between prairies, woods, and other plant communities. The summer program was an immense success, filled to capacity with young students, with a waiting list of seventy to eighty students. Sutherland told the *Capital Times* reporter that she hoped to expand the program for summer 1970 to include older students and, of necessity, additional guide personnel.[67] There would be a corresponding increase in training sessions for the guides.

By January 1971, Fleming's guide training program had expanded to include a series of courses designed to train outdoor educators who would be eligible, upon completion, for seasonal part-time employment as nature interpreters at the Arboretum and other Dane County nature centers, including the Madison School Forest in Verona and the Cherokee Marsh Outdoor Education Area on the northern edge of the Madison city limits. The 1971 program enrolled sixty people and included orientation meetings and over a hundred formal class sessions at the Arboretum taught by Arboretum instructors. Students in the program were also required to take Zimmerman's "Reading the Landscape" course which, by 1972, was being taught by Gina Kline.[68] In 1971, Fleming logged over 750 tours accommodating over 16,000 people.[69] By 1974, Fleming's guides had conducted 4,940 tours that had attracted 89,175.[70] The numbers increased exponentially between 1974 and Fleming's retirement in 1981.

Following Fleming's retirement, however, according to Katharine Bradley, the Arboretum director in 1981, the tour program experienced a major setback. In 1982, Bradley's last year as director, the Dane County Board, no longer interested in supporting an Arboretum naturalist, stopped funding the position. Bradley retired in 1983, leaving the matter in the hands of her successor, Greg Armstrong, who fortunately managed to acquire funding for a combination tour and education director who took over responsibilities that had been Fleming's for fourteen years.[71]

By the early 1990s, the Arboretum guide program had recovered and started to expand. By 1993, more than three thousand public school students annually were coming to the Arboretum for guided tours. The number of participating nature guides had risen to twenty-four. Arboretum training sessions once again accommodated not only Arboretum guides but also naturalists from other Wisconsin city and outdoor education centers.[72] The personnel changed over the years, but the mission remained the same. As Fleming put it in 1968, "as public interest in the Arboretum mounts, understanding of measures necessary to maintain ecologically sound plant-animal communities grows. . . . Capitalizing upon public interest, guides gain an excellent opportunity to . . . interpret Arboretum purposes while encouraging proper use." Her overriding point? "The best insurance for securing and maintaining Arboretum goals is an educated and concerned public."[73]

An Edenic Oasis

By the end of the 1960s, the Arboretum's growing reputation as a good place to visit for educational tours, for seasonal changes, for spotting a hawk or a pileated woodpecker or even a spiraling timberdoodle, as well as for taking quiet hikes through the prairies or the woods had begun to gain well-earned national publicity. The Arboretum was described in glowing terms. "A world-famous outdoor laboratory of nature" was how one news service referred to it.[74] Arthur Godfrey came to town in May 1967, visited the Arboretum one Sunday afternoon, and was so impressed and stayed so long, he almost missed his matinee performance. When he returned to his radio show, he bragged to the "world" about the Arboretum's beauty. "Somebody way back," he told his listeners, "had a good head on his shoulders and set aside this acreage." And what happens while you're there? "You should see the birds. . . . I counted personally 12 different species that I have never seen before. They're only in the wilderness." The place was "beautiful," with "exotic birds." It was "amazing . . . fascinating . . . wonderful to walk in there, to sit down and watch, to be quiet. Pretty soon the wildlife starts to come around. . . . It's fascinating, just fascinating."[75]

Eight months after Godfrey's visit, in January 1968, an editorial in the *Wisconsin State Journal* lauded with similar hyperbolic praise what it called "The Arboretum Idea," an obvious allusion to the legendary outreach mission conceived in 1904 by then UW president Charles Van Hise. "The Arboretum Idea," the editorial continued, was the inspiration for the new UW–Parkside campus in Kenosha. And "only in Wisconsin" was this phenomenon possible, because only the UW–Madison had the now famous "outdoor laboratory of nature." The new Parkside campus, by embracing Madison's "Arboretum Idea," was also projecting the Arboretum dream of Jackson one more "step toward immortality."[76]

The dream of the Arboretum as something "immortal" was a novel claim. But who would now argue with the possibility, as Godfrey also implied during his visit, that, by the late 1960s, the Arboretum, in the eyes of many, had evolved into something larger than life, something "immortal," other-worldly, Edenic—a "wilderness" with "exotic birds;" a wonderful place to walk in, to sit down in, to "watch;" a place, where, if one sat quietly enough, as Godfrey suggested, "the wildlife starts to come around." For a moment in time in May 1967, Arthur Godfrey, in true 1960s fashion, had gotten himself "back to the garden." He had a delightful time; the Arboretum left an indelible impression on him. "There's a warm place in my heart for Madison," he told his audience.

Unrest in Eden: A Concrete Disaster

Godfrey's exuberant response to the Arboretum is frozen in time. The Arboretum in 1967 seemed Edenic to many—a wooded oasis, a sanctuary, a garden of delight, in the midst of urban sprawl, where if one sat quietly enough in the surrounding silence, the wonders of nature unfolded. But there was growing unrest in what had become—by spring 1968—a very troubled Eden. The "concrete monster" on the Arboretum's south border was stirring again. Zimmerman and Cottam sounded the alarm in a notice in the *Arboretum News* emphasizing the need to "Protect the Arboretum" from "progress" perpetuated under the banner of the public good. "The scientific problems involved in making the Arboretum the collection of natural communities it is becoming," Zimmerman and Cottam wrote, "are large and complex. Until now, we have devoted most of our efforts to solving these problems. It appears that from now on more of our efforts will be devoted to the people problem, to rectifying the damage caused by the works of man."[77]

The Beltline Highway (combining U.S. highways 12, 14, 18, and 151 where it cuts through the Arboretum), which Cottam, leveling his own hyperboles, referred to as the "cement desert," the "noisy, stinking, inanimate ogre," the "great catastrophe that has plagued us from the day it was conceived," was scheduled to expand in 1969 into a six-lane highway with an underpass and interchange on Seminole Highway and with a two-lane frontage road on the south side that would cost the Arboretum 3.77 acres of fertile upland and approximately 1,500 trees. It would also result in that ugly straight-line-edging effect that comes from clear cutting a forest on its boundary. But the effect the expansion would have on the Curtis Prairie would be even worse.

"These are troubled times," Cottam wrote. He acknowledged that he had received complaints and threats from people who questioned why the AC and the university had agreed to allow the Wisconsin Department of Transportation to take more land and destroy more trees. We've been labeled as "traitors, thieves, or fools," Cottam said. Some have even suggested that we "sold our birthright for a pile of money." Nevertheless, we made the decision, he added; and under the circumstances existing at the time, we thought it was proper. If we were to do it over, we would have second thoughts.

Deliberations over the Beltline expansion began in 1968. Also in 1968, the Wisconsin highway death toll had reached a record high. Governor Warren Knowles, in response, launched a statewide campaign for highway safety. In 1968, Arboretum representatives met with the Wisconsin State Highway Commission (WSHC) and, according to Cottam, came away believing that the WSHC plan included some details favorable to the Arboretum. For one thing, the frontage road on the north side of the Beltline would terminate before it reached Arboretum property. For another, Cottam added, Arboretum representatives were told that the WSHC would design and build erosion control structures, "including two concrete flumes and a water-retention, desilting pond to hold and clarify the water so that we could let it run across the prairie in the condition and quantity it would have had were the Beltline not there." In return for state-built control structures, the university agreed to sell the state a total of 3.77 acres, which included land on both the north and south sides of the Beltline. The acreage also included approximately 1,000 large pine trees and 500 small spruces. The Arboretum representatives were not pressed to agree hastily to the plan, but as Cottam observed at the time, to have objected to the plan likely would have resulted in a state-initiated condemnation or action by the state legislature, particularly given the statewide concern over highway safety and the record level of traffic deaths in 1968 and the general lack of concern on the state level for environmental problems. Neither did the Arboretum representatives want a legal fight in the courts or a battle with the state legislature. Nor did they want some bureaucratic state agency making the final decision for them.

"The subsequent negotiations were the most traumatic I have ever experienced, and I emerged from them a sadder and wiser man with some firm convictions about what one must do when negotiating with a bureaucracy." Cottam's response captured the mood of the Arboretum *after* the construction had begun. The consensus was, in retrospect, that the AC had made a mistake by not negotiating a clearer and more effective erosion control structure with guarantees. As Cottam explained it, the Arboretum got $60,000 for the acreage, including a $5,000 state contribution toward the Arboretum's erosion control structures. Out of the $60,000, the Arboretum used $42,000 for the erosion control facilities. The additional $18,000 went toward the purchase of the 3.6 acre Keepman property, a weedy, low-lying swampy area off Fish Hatchery Road. In addition to the loss of 3.77 acres of upland and about 1,500 trees, in the final account, the Arboretum also gave up an additional 3 acres of prairie that were dug out for the desilting pond that, as Cottam explained, "was . . . necessary largely because the highway changed the water flow characteristics and caused flooding and erosion in our pine and spruce forests, as well as the oak opening and the prairie."

Initially, as noted above, the WSHC had convinced the AC that the water-retention, desilting pond would "clarify" the water so that it would be left to flow "across the prairie in the condition and quantity" it originally retained—as if the Beltline expansion had never happened. That was the hope; the reality turned out differently. Archbald had predicted that the desilting pond wouldn't work, that without yearly dredging, it would have about a five-year life span. Cottam concluded that the Arboretum had come up short on all counts. A "disaster" in all respects. So where was the "virtue" in the massive highway project, the benefit that had made it all worthwhile? Well, "as a highway safety measure," Cottam sardonically opined, "it is said to have some value." His challenge to those in the future who would find themselves in similar no-win situations was to figure out exactly with whom to negotiate. Should we negotiate with the WHSC or should we "let the legislature decide on the relative merits of highways versus trees?"[78]

Closures

The year 1969 was marked by closures. G. William Longenecker, the Arboretum's executive director and the last of the Arboretum pioneers, died on February 25, 1969. He had retired in 1966. In 1967, the Arboretum's horticultural area was named after him. Another unexpected development in 1969 occurred in late spring. David Archbald announced that he was planning to resign from the Arboretum managing directorship. He had accomplished some impressive objectives that would shape the future of the Arboretum, including planting a maple undercover in Gallistel Woods, planting spruce trees in the Grady Tract, adding key personnel (Moran, Zimmerman, Fleming) to the staff, and, most important, establishing a guide program that continues to be an Arboretum public outreach centerpiece. As noted above, it was Archbald who convinced the Dane County Board to mandate Rosemary Fleming's county naturalist position, an appointment that led to a shift in Arboretum policy from an exclusive preoccupation with teaching UW students and encouraging faculty research projects to an additional focus on public outreach and education.

By 1969, Archbald was ready for a career change or, at least, a change in the direction in which he thought his talents as an ecologist should lead. He had been "politicized;" one might say, "radically politicized." He announced his decision to resign at a June 1969 meeting of the FOA. He told the Friends he had been caught by surprise "by the remarkable rate of increase" in global environmental degradation. "I feel I've just got to counterattack in a massive way," he added. On the surface and for the record, he explained that he was leaving in order to become director of the Man-Environment-Communications Center (MEC), an organization with a global focus promoting a combined school- and community-based approach to environmental education.[79] But off the record, he was disillusioned, frustrated with what he deemed was the parochial nature of his Arboretum appointment, and, subsequently, his life. He felt inhibited by a "global bias . . . going to hell in a basket" at the Arboretum. Aside from the fledgling tour guide program that he had initiated, the Arboretum was still basically in the business of teaching and research, not political activism. He was disillusioned. He wanted to devote his energies to environmental education on a broad political scale, particularly to young people.[80] But left unsaid was the fact that by spring 1969 he no longer had the support of an AC now pressuring him for his resignation.

In some measure, Archbald's frustration had anticipated Zimmerman's 1971 criticism of the Arboretum as lacking a sense of purpose and promoting, what Zimmerman referred to as a "passive 'caretaker' concept" of operation. For

Zimmerman, the basic problem was the Arboretum's inability to rally any sizable community-wide political support. He saw the problem also as a failure in leadership (a direct criticism of Cottam and the AC) as well as a lack of adequate funding. These troubling times required more, he argued; specifically, a more active involvement in countering threats to the environment, both in the Arboretum and beyond the Arboretum.[81] The tenor of the times—it was the end of the 1960s—demanded that one become informed on the issues and express concern publicly—and protest publicly, if necessary.

It *was* the late 1960s, the era of social protest, the extension of a cultural movement that had developed nationally during the late 1950s and early 1960s as a reaction against what campus activists, in particular, deemed was a "do nothing" or, borrowing Zimmerman's word, a "caretaker" attitude toward life, especially toward issues like the Vietnam War, civil rights, and threats to the environment endangering both world security and global well-being. The issues needed to be addressed. And how were they to be addressed? Initially, one had to become informed, as Archbald advised in a heated article in the *Wisconsin State Journal* on the "pollution explosion" and public apathy. "Become informed," he told his readers. "There are no quick-fix environmental cures." He urged them to express their concern to their elected officials. Readers should challenge them; ask them "what they are doing to set things right."

During his last year at the Arboretum, Archbald became more stridently political, particularly in his attacks on global pollution. He began to appear more and more—particularly, in print—like a very angry and radical environmental activist. The signs were there throughout 1969 and into 1970. On March 30, 1969, the *Wisconsin State Journal* carried a full-page feature on Archbald with a banner headline reading "We Are on a Collision Course with Nature." In the article, Archbald cited the National Research Council's warning "that by the mid-1980s, the increase in sewage would demand ALL of the dissolved oxygen in our lakes and rivers." He noted that the "green house effect" was trapping heat in the atmosphere, that the warming of waters had stimulated algal growth, that Lake Michigan was inundated with DDT, and Lake Erie, owing to industrial pollutants, was now a "sewer." The answer to the problem, he urged, was "a greatly aroused public, putting pressures on the government, correcting as much of the damage as yet can be corrected before it is too late." He also stridently advocated world population control. "Saying that a population explosion [was] a problem" only in underdeveloped nations," he challenged, was like "saying to a fellow passenger, 'your end of the boat is sinking.'"[82]

On April 6, 1969, the *Milwaukee Journal* published an article under the headline "UW Charts Describe Dangers of Pollution." Apparently, a series of charts with statistics on the environmental impact of the continuing "population-pollution spiral" that was put together by Archbald was on public display at the Arboretum. He was quoted as saying: "I don't think we'll be able to stand it in another 30 years. We'll have a bunch of sewer pipes for a river system." He was also quoted as saying that he found "the present 'birthquake'" particularly frightening.[83] He had used the term *birthquake* for an earlier article in which he recommended support for a bill revoking tax exemptions for dependent children. In fact, he argued that Congress should enact a bill imposing taxes on people *with* children. The article, appearing in the *Capital Times* on April 6, 1969, carried the inflammatory headline, "UW Ecologist's View: Would Tax Babies to Save Nature."

Advocating taxing babies to save nature was not the kind of suggestion, even in the subversive 1960s, that curried much favor, especially among Cottam and the AC as well as among UW faculty and administrators alarmed already by their Arboretum managing director's strident radicalism. But by summer 1969, the year oil from an offshore rig coated the beaches of Santa Barbara, Lake Erie was declared "dead," and Cleveland's Cuyahoga River caught fire, Archbald was offering no apologies. Throughout the remainder of 1969 and into 1970, identified as a UW ecologist, he authored a controversial column on global environmental issues entitled "Our Environment" for the *Wisconsin State Journal.*

At the end of the AC meeting on January 14, 1970, Cottam excused Archbald and Zimmerman from the meeting in order "to permit a discussion of personnel matters" by the remaining members of the committee. On February 18, the last item on Cottam's agenda for the AC meeting was another couched discussion of a personnel matter. Zimmerman was absent for the February meeting. At the March 11 meeting of the AC, Archbald submitted his letter of resignation effective July 1, 1970. The committee, after accepting his resignation, announced that it had already at hand a list of potential candidates for the position, compiled even *before* Archbald had officially resigned. Also at that March 11 meeting, Zimmerman informed the AC that he and his wife were planning to write a nature series column for the *Wisconsin State Journal,* and that he was now increasingly troubled, as he tactfully observed, by its "possible relationship to the Arboretum." He asked for a response to his concern from the committee. A discussion followed, at the end of which it was decided that the AC considered the column "an independent project."[84] Yet, the fact that Zimmerman felt obliged to ask the AC for a judgment, a consensus, on whether he and his wife should publish nature articles in the

State Journal suggests that fallout from Archbald's inflammatory articles was very much on the minds of Cottam and the AC in 1969.

On July 1, 1970, Archbald left his position as managing director of the Arboretum. "While it is with regret that I leave my official position with the Arboretum," he wrote in the summer 1970 issue of *Arboretum News*, "it is also combined with a sense of urgency and of the mission of applying ecology to a formidable challenge—helping to maintain a quality environment."[85]

In July 1970, Roger C. Anderson succeeded Archbald as managing director. The AC welcomed Anderson in an unsigned article in the fall 1970 *Arboretum News*. The article acknowledged that Anderson was "admirably equipped to manage the Arboretum." In addition, the article affirmed, "he is interested in research on the Arboretum and research is our most important function."[86] Anderson, however, would resign the managing directorship only a short three years later in 1973. In his letter of resignation sent to UW Chancellor Edwin Young, copied to the AC, he observed: "It should be of concern to the University that in a period of three years the Arboretum has had two Managing Directors and both have resigned. The first has been vilified. . . . But there must have been other circumstances which caused him to lose interest in his work."[87] Anderson's resignation would underscore a conflict of leadership at the Arboretum that would not be resolved until Katharine Bradley, one very strong-minded administrator, assumed the directorship in 1974.

❧

CHAPTER 10

The 1970s

WORLD FAMOUS PRAIRIES, CONFLICTS IN LEADERSHIP,
NEW APPOINTMENTS AND A VISITOR CENTER

Professor Roger Clark Anderson, a Wisconsin native, was born in Wausau. After receiving his bachelor's degree in 1963 from Wisconsin State University–La Crosse, he entered the graduate program in botany at the UW–Madison. In 1968, upon completion of requirements for a Ph.D. in botany, he accepted a position as assistant professor in the Botany Department at Southern Illinois University in Carbondale. He was working at SIU when he was contacted by the Arboretum Committee (AC) in early 1970 and offered the managing directorship. Subsequently, he requested and received a one-quarter teaching appointment in the UW Botany Department at the rank of assistant professor, effective spring 1971. He explained to the AC before accepting the position that he also wanted to teach and to direct graduate students. He was only twenty-nine years old.[1] He would go on, after his three-year tenure at the Arboretum, to work as a prairie restoration ecologist, eventually becoming a Distinguished Professor of Ecology at Illinois State University.

"We can't be all things to all people"

In summer 1970, Anderson returned to Madison and the UW with high hopes. Like Archbald before him, he was a committed campus activist who campaigned while at SIU to save Lusk Creek, a canyon stream surrounded by native vegetation in the Shawnee National Forest. The Forest Service planned to dam the stream to create a lake but under pressure abandoned the effort. Anderson also saw in his new Arboretum position an opportunity to help build in the Arboretum a teaching and research facility—an actual building—a possibility that had been explored in some detail by the AC prior to his arrival. Anderson also saw the teaching end of his appointment as an opportunity to spread the environmental message to a wide spectrum of Arboretum visitors and supporters. He was quoted as saying, "I like to teach." When he first started at SIU, he noted, he taught ecology with eighty students in his class. By the time ecology had become "a national issue," his class had grown to four hundred. The nation needed "a change in attitudes and ideals," he noted. "We have to reach people starting with grade school. Our greatest hope lies with the young people."

So Anderson was particularly pleased with Rosemary Fleming's guide program and the fact that 120 young people would soon, shortly after his arrival, be touring the Arboretum. In a show of support for Fleming's program in 1971, he endorsed enthusiastically, in the face of open opposition from influential AC members, the program's alliance with a new extensive county-wide training program.[2]

When asked after he arrived in summer 1970 how he thought the UW had responded over the years to environmental threats and the need to preserve natural resources, Anderson responded, "very well." He felt Wisconsin had always had an interest in conservation—"John Muir, Aldo Leopold, Norman Fassett—and today we have some of the top people in the field."[3] Perhaps motivated by his own desire, in the tradition of Muir, Leopold, and Fassett, to promote conservation and protect Wisconsin's natural environment, Anderson initiated, within a year, a public relations campaign aimed at state of Wisconsin pollutants and polluters—particularly Arboretum polluters. Following the lead of Zimmerman, who by September 1971 had given up his ranger duties and, by request, went to a half-time appointment as Arboretum naturalist, Anderson launched his own environmental campaign through a series of newspaper and magazine interviews. The impending "urban crush" on traditional Arboretum commitments was his major concern. In one article, he acknowledged that the urge to get away from the noise and distractions of urban life was understandable and accounted for the large increase in visitors to the Arboretum in the early 1970s. For many people in the Madison area, he observed, the Arboretum was "the only open space" where they could experience a natural environment. The problem, as always, was how to keep the Arboretum's research function intact in the midst of the ever-increasing number of visitors, because research, he added, "is our official mission. We can't be all things to all people."[4]

Indeed, by 1971, the Arboretum could boast of a long, impressive record of productive field research. Since the early 1930s, well over 127 scientific papers and 87 theses, all products of Arboretum research, had been produced. In spring 1971 alone, 20 professors were conducting 32 separate Arboretum projects, all with global significance. Lake Wingra, in 1971, was the site of an ongoing International Biological Program on the study of land and water interactions. Yet, Anderson was well aware of the ironic position he and the AC were in at the time, because the increase in visitors was also, in large part, owing to increased public relations efforts to make the Arboretum better known and appreciated by the public for its commitment to research and its developing international reputation.

As recently as 1969, he noted, before the word *environment* had become commonplace, a lot of Wisconsinites wondered why 1,200 acres of old fields, wetlands, and woods in the middle of a city were left wild and undeveloped. Better to use the land for a public park or a housing development, they thought. The wonderment, it seems, extended as well to representatives of WHA Radio (Wisconsin Public Radio) who, as early as 1958, requested permission to erect a three-hundred-foot radio tower in the middle of what they referred to as an Arboretum field "grown to weeds" and sitting idle. They were referring to the Curtis Prairie, no less, the oldest restored tall grass prairie in the world, on which the Arboretum staff had been working since 1936—twenty-two years. The WHA officials, like many other Wisconsinites, were totally unaware of the prairie's history or its significance.[5]

The solution was education. "In the long run," Anderson remarked, "the Arboretum's survival will depend on the public's sensitivity to what it is and what it is not." The guided tour program, which attracted more than 16,000 people in 1970, raised some public awareness of the Arboretum's fragile ecosystem and, as he put it, had "sensitized many to the intricacies of nature's handiwork."[6] But, ironically, even the tours and the guide program itself had strained Arboretum use, and there were new pollutants in the early 1970s that threatened the Arboretum's sensitive plant communities. There were also snowmobiles in the Arboretum in 1971. Silt from nearby housing developments polluted the marshes. McCaffrey Drive had become a speedway. Cyclists were riding bikes on foot trails and in other planted areas where bike riding was posted as prohibited. Noise pollution from the six-lane Beltline Highway dividing the Arboretum discouraged the presence of animals and birds, especially migratory species. Automobile exhaust polluted the air and threatened trees bordering the highway. In 1972, Dane County considered widening Fish Hatchery Road on the Arboretum's extreme east boundary. The Arboretum wetlands and wildlife refuge with four experimental ponds would have been directly affected.

By 1972, pressure to protect the Arboretum, its natural resources, and its increasing research potential from outside threats persuaded UW administrators that it was time for the university to look for solutions. Chancellor H. Edwin Young told a gathering of environmentalists in March that the university's "growth era" had ended, and that in the future the university would be paying less attention to constructing buildings and parking lots and more attention to environmental issues, including mass transit and the protection of the Arboretum. Young noted that much of the pressure to address pollution problems was now coming from UW students, especially students in the biological sciences who were demanding that more attention be given to the effects of university activities on the environment. And one of the most pressing environmental concerns, he added, was the future of the Arboretum.[7]

To Anderson and the AC, the best available resource in the struggle to keep the Arboretum vital, yet relevant and accessible, *was* the university itself. Finally, after almost forty years, the Arboretum was able to engage in a closer partnership with an admittedly supportive, commiserating university that guaranteed guidance, increased security, and the availability of facilities. Coincidentally, a well-advertised and well-attended conference on the subject of tall grass prairie restoration, where the Arboretum's prairies were center stage and celebrated internationally, just happened to have been scheduled for the UW campus around the same time Anderson was appointed director.

A National Prairie Restoration Movement

Anderson assumed his duties as managing director in September 1970, just in time for the dedication on September 18 and 19 of the UW's Biotron (a controlled environment facility) and for an important three-day meeting on the campus, from September 18 to 20, of the second Midwest Prairie Conference. In time, the Midwest Prairie Conference would evolve into the prestigious, more widely recognized North American Prairie Conference. The first Midwest Prairie Conference, a "Symposium on Prairie and Prairie Restoration," was held at Knox College in Galesburg, Illinois, in 1968. The UW Arboretum was well represented. David Archbald, who was still managing the Arboretum in 1968, was the keynote speaker. Another conference speaker, Ray Schulenberg, a prairie ecologist from the Morton Arboretum in Lisle, Illinois, presented a history of "Morton Arboretum Prairie Restoration Work" in which he noted that "the precedent, inspiration, and basic procedural information for prairie restoration had come to us from the University of Wisconsin Arboretum at Madison." Schulenberg, in 1973, would team up with Robert F. Betz, professor of biology at Northeastern Illinois University in Chicago and a pioneer in the preservation of Illinois relic prairies, on a prairie restoration project at Fermilab, a U.S. government atomic accelerator located on 6,800 acres of farmland in Batavia, Illinois. Betz and Schulenberg, admittedly influenced by the work done thirty years earlier at the UW Arboretum, began the restoration efforts on approximately 1,000 acres in Fermilab's main ring, a project that continues today with the help of Fermilab employees, community teachers, local students, professional ecologists, and volunteers.[8]

Given the Arboretum's reputation by the late 1960s as the place where prairie restoration began, it was not surprising that the UW–Madison along with UW–Parkside, another pioneer in prairie preservation, and the Wisconsin Department of Natural Resources (DNR) would be invited to host the second Midwest Prairie Conference. As Vivian Hone, a *Capital Times* nature writer covering the conference, aptly put it, the "preservation of whispering blue stem and golden Indian grass" has "achieved such esteem" that lovers of the prairie have come together now at the UW–Madison to celebrate their shared interest. The choice of Wisconsin was overdue, she stated. Before Wisconsin's rich soil had been broken with plows, over 2 million acres of treeless prairie covered the state—close to 7.5 million acres if one included prairie with oak openings. And the choice of Wisconsin was also appropriate, Hone suggested, given the University's "world famous Arboretum" and the global importance of its two pioneering restored prairies: the Curtis Prairie, "started from scratch" with "transplanted sod," and the Greene Prairie, planted almost single-handedly by its creator. "Today," she continued, "more than 300 native herb species are . . . found in the Curtis and Greene restorations . . . world-renowned replicas and models of a Midwestern botanical past." "Researchers come . . . great distances to study and view them," she noted. The Arboretum prairies have also been the "featured players," she added, in two widely circulating films, the Disney Corporation's award-winning documentary, *The Vanishing Prairie*, in 1954 and, more recently, an *Encyclopedia Britannica* educational film on America's Midwest prairies.

Weather-wise, autumn 1970 turned out to be a particularly good time for enjoying prairies. At UW–Parkside, the Chiwaukee Prairie, a highly prized native prairie growing on sand ridges and rich with fall color, was a highlight of the conference program and the focus of a well-attended field trip. Three field trips, also conference highlights, took place on the UW Arboretum prairies. At the conference banquet, the Arboretum was presented an award for "outstanding management of natural lands used for educational and research purposes" by the Soil Conservation Society of America.[9]

Prairie restoration efforts were gaining extensive national attention in the early 1970s. In October 1970, the *New York Times* published a lengthy syndicated feature by Pulitzer Prize–winning journalist John Noble Wilford on national efforts, particularly in Wisconsin and Illinois, to preserve and restore "vanishing grasslands." There are people who love the prairie the way "others love the sea," Wilford wrote. "They love the feel of its black, spongy soil, the splendor of its many wild flowers and the sweep of its tall grass." With Willa Cather, they feel that "'the grass was the country,' and that its roots ran deep and shaped the heartland of a nation." But native remnant prairies were still very difficult to find in the 1970s. One looked for patches of grass on rocky hillsides or south-facing unplowed ridges, or for forgotten cemeteries and railroad rights-of-way. Nevertheless, Wilford noted, efforts were increasing nationwide to find and preserve remnants. He cited, in particular, efforts supported by the Nature Conservancy, the Prairie Restoration Society, the Sierra Club, and other land conservation groups to preserve the few examples of virgin prairie that survived and to restore other salvageable grasslands. Across the nation, land was being bought up before real estate developers could get to it. Relic prairies were being fenced in. Native prairies gone to pasture were being restored, and there was increasing interest in establishing national prairie parks. Significantly, the Arboretum's Roger Anderson was quoted in Wilford's feature as observing at the Wisconsin Prairie Conference that "man had probably misused the earth's grasslands more than he had misused any other plant environment" on earth.[10]

Indeed, the history of relic prairie preservation in the Midwest in the 1960s and 1970s was characterized by the discovery of small untouched areas. But, Wilford cautioned, the discoveries were minimal. Hence, ecologists and conservationists who attended the second Midwest Prairie Conference emphasized the need to concentrate more efforts on restoration. The aim in the 1970s was to recover land that had been lightly grazed and to restore it by planting local, native prairie grasses and prairie flowers. And there was a prototype, a model, for that type of restoration, Wilford acknowledged. It was the "65-acre Curtis Prairie, developed on the outskirts of Madison by the University of Wisconsin"—the oldest restored tall grass prairie in the world. Only now, however, after more than thirty years, he added, can the Arboretum's prairie, so diligently worked on by early ecologists, be considered a reasonably authentic example of a wild, native prairie. It takes that long to build a prairie, because "the process of restoration is . . . slow and difficult."[11]

Prior to the publication of his article, Wilford had visited the Arboretum. Grant Cottam, while being interviewed during a prairie walk with Wilford, had explained to him that thirty years was actually a reasonable expectation. If the field had not been a pasture and had been cultivated for any length of time, Cottam added, restoration would take a century or more. One needed to plant and wait patiently, and one needed to burn regularly in order to kill off non-native competition. Wilford's syndicated October 1970 *New York Times* feature circulated widely. The timing was perfect. The second Midwest Prairie Conference had just taken place the month before. The Arboretum, particularly its prairies, had figured largely in the conference programming and in Wilford's influential article. The pioneering role of Arboretum prairies in the history of prairie restoration worldwide had now been catalogued as a seminal influence on the development of international interest in prairie restoration. Lloyd Hulbert, a Kansas State University prairie ecologist, was quoted by Wilford as observing that people have traveled long distances to see glaciers and mountains. Perhaps, some of them might now consider traveling to see what a native prairie looks like. Prairies are "part of our heritage," he added.[12]

The sentiment would be echoed by many conservationists, including Douglas E. Wade, professor of outdoor education at Northern Illinois University and a major force behind restoration efforts throughout northern Illinois in the 1960s and 1970s, and Arthur H. Ode, prominent national horticulturist who supervised prairie restoration at the Boerner Botanical Gardens in Milwaukee.[13]

Land Acquisition, Cottam Resigns, the McKay Bequest

Even before Roger Anderson's arrival on the UW campus, the AC had been engaged in an attempt to acquire funding for additional Arboretum buildings. The committee had unsuccessfully applied a number of times for National Science Foundation grants to subsidize the building of a teaching and environmental research facility that would provide classroom and laboratory space and, finally, adequate office space for the staff. Into early 1971, the AC still was having no luck finding building money.

Compounding the funding difficulty, Anderson, with the support of the FOA, initiated a land acquisition program in the summer of 1971 that was directed, in spite of zero funding, at acquiring "buffer properties" surrounding the Arboretum. Anderson had ambitious hopes of raising $560,000 to purchase land around the Arboretum that he thought would create a buffer zone protecting it from what he termed "urban creep," that "relentless opponent of wilderness." He wanted to buy seven properties in particular totaling approximately 40 acres.[14] Anderson observed, "the Arboretum foresees no way to finish this land acquisition other than through private gifts." And "therein lay the rub." There appears to have been little response to the request and little accompanying support from the campus. Contributors were asked to send contributions directly to the Arboretum and to designate the gift for land purchase.[15] In spite of Anderson's hopes and efforts, no buffer properties were acquired, but Anderson's efforts, continuing well into the early months of 1973, did convince some members of the AC of the benefit of acquiring buffer properties even after he had resigned his directorship.

Far more successful were the efforts to find funding for the classroom and laboratory research building. The AC started the 1971–72 academic year with some new members. Cottam, after ten years as committee chair, resigned at the end of summer 1971. The new AC chair was Robert J. Dicke, for ten years the chair of the Entomology Department and a ten-year member of the AC. Three new faculty members were also appointed (Gerald C. Gerloff, Harold C. Jordahl, and John W. Thomson). Anderson, now officially listed as "Consultant," was the ninth member of the committee.[16] In the early months of 1972, Anderson and a selection of AC members met with the UW Foundation to discuss a memorial bequest of $250,000 from the W. G. McKay Foundation for the proposed Arboretum classroom and research laboratory facility. The result of the meeting was both encouraging and unexpectedly sobering. The McKay representatives made it clear at once that the McKay Foundation money favored "a . . . public education facility," but not "a research facility."[17] Hence, the AC's long hoped-for "research laboratory facility" was not going to be realized, at least not with McKay money. But the committee had at its disposal up to $250,000, and even the most ardent of researchers could hardly turn it down.

The McKay Foundation bequest scuttled the dream of having solely a research center. But it also created some concerns that would consume a sizable measure of AC time. Two pressing questions needed to be addressed: where would the new facility be located, and if not a research center, what exactly would be its design and purpose? By the end of February 1972, the AC had considered locating what they now referred to as the Arboretum Interpretive Education Center on one of four possible sites, including two on the edge of Anderson's buffer zone. They were also considering property north of Carver Street. Anderson and the committee, worried about the potential increase in interior Arboretum traffic, favored locations on the Arboretum's outlying boundaries.

By summer 1972, the AC had decided on another name for the proposed facility—the McKay Nature Education Center. When the committee met in September 1972, it had a new member, Emily H. Earley, a professor of environmental studies and the first woman to serve on the AC. In February 1973,

Dicke, the AC chair, appointed a Site Selection Committee, an ad hoc committee of three AC members charged with developing criteria on which to base a location for what was now, in 1973, being called the Arboretum Nature Center. Anderson was not included on the Site Selection Committee.[18] In April 1973, the Selection Committee reported its conclusions to the AC: the chosen site should offer easy access to the facility; the site should not be "in a sensitive or intensively used area of the Arboretum"; and, most important, the location should not increase the flow of traffic into the Arboretum's interior. Hence, the committee continued to look primarily at outlying areas and locations on the Arboretum boundaries. Eventually, the area the Selection Committee chose as the location for the future Arboretum Nature Center, the location that would become the nerve center also for Arboretum business and Arboretum educational pursuits and outreach efforts, was the Duck Pond (Spring Trail Pond), or that portion of the Arboretum near the pond that bordered Manitou Way.[19] On July 20, 1973, the AC voted on yet another name (the fourth) for the facility—the McKay Nature Awareness Center.

In March 1974, the AC reactivated the Site Selection Committee because the Duck Pond had been rejected owing to unsuitable substrata. After examining the alternatives, this time the committee recommended locating the building in the headquarters area, in spite of the possibility of increased traffic into the Arboretum's interior. The committee also recommended shifting the function of the building from solely a nature education center to a multipurpose structure providing facilities for a visitor reception center, accommodations for work rooms, and offices for Arboretum staff and researchers. By March 1974, the project cost had also increased to over $286,000. Additional money would have to be raised.[20]

Frustrated, Anderson Resigns

At some point in late October or early November 1972, the AC chairman Robert Dicke met with Chancellor Young in order to clarify, for the benefit of the AC and Anderson, questions about Arboretum governance. At the AC meeting that immediately followed his meeting with Chancellor Young, Dicke asked that the following sentences be recorded in the minutes:

> 1. The Arboretum Committee is the Chancellor's governing body and the committee chairman serves as the liaison between the Committee and Chancellor.
> 2. The Managing Director is appointed by the [Arboretum] Committee and is responsible to the Committee.[21]

In his monthly report in April 1973, under the heading "Items for the Committee's Consideration," Anderson queried the AC about input into committee plans for the proposed McKay Nature Awareness Center. He had two particular questions:

> 1. At what point during the development of the Nature Center will the Arboretum staff have input into the program?
>
> 2. What will the Arboretum Director's role be in the administration of the Center?

In May 1973, a four-member Building Committee was created, with Edward Hasselkus (Landscape Architecture) as chair. Anderson was appointed, but as an ex-officio member—he would have *no vote.*[22] In June 1973, Anderson, out of patience with the AC for failing to include him in Arboretum policy decisions, resigned his position as managing director, effective July 27, 1973. By July, Anderson had accepted a teaching position at Central State University in Edmond, Oklahoma.

In his undated letter of resignation sent to Chancellor Young and the AC (cited in the discussion of David Archbald's resignation in chapter 9), Anderson acknowledged that he did not want to generalize about Archbald's experiences prior to his 1970 resignation, but that he wanted to outline what he saw as current problems. He addressed particularly what he called "the administrative structure of the Arboretum." He noted that, contrary to popular belief, demonstrable and enduring progress at the Arboretum had been achieved by the "dedicated pioneers," the legendary figures who were once central to the Arboretum's history but who were now gone. They included, primarily, "Jackson, Gallistel, Longenecker, Leopold, Curtis and . . . a few others."

Anderson faulted the AC for promoting in recent years a "patriarchal system," making the Arboretum managing director subservient to the Arboretum Committee. The committee, however, had neither the time nor the energy, he argued, to tend properly to Arboretum interests. And that lack of involvement was essentially the problem. The early Arboretum pioneers assumed leadership roles and got things done. The current AC, by contrast, was ineffectual. He recommended restructuring it to include, once again, "individuals of distinguished stature in the University," dedicated conservationists with a deep appreciation of the environment.

Unless responsibility for the Arboretum was assumed by distinguished campus leaders, he argued, and "unless the utmost care and concern [are] exercised by those responsible for the [Arboretum's] future, it will slowly erode away." He noted that hardly any effort was exerted by the AC to raise funds to purchase the critical buffer zone property adjacent to the Arboretum that he deemed essential. In fact, he was discouraged from trying to raise funds. Yet several AC members fought vigorously, he added, to give away Arboretum land to WHA (Wisconsin Public Radio) for a radio tower. He also took the AC to task for what he considered a jaundiced attitude toward the public. The committee should realize, he noted, that "the public is . . . interested in developments that take place in the Arboretum," and should be aware that public assistance is essential if the Arboretum is going to achieve its goals. The committee also should realize that the director has more at stake in the Arboretum than the AC members. Consequently, the AC should hire a director they can trust and then "give him his head." The director is full-time, while the committee members are part-time. His final thought—he hoped his resignation would stimulate thinking about serious administrative problems and would bring about a better, more effective governance system.[23]

Anderson must have felt somewhat vindicated in January 1974, only five months after his resignation, when the AC appointed a subcommittee, chaired by Gerald Gerloff of the Botany Department, charged with overseeing the "Long Range Acquisition and Sales of Arboretum Property."[24]

Like Archbald before him, Anderson left frustrated and discouraged. But during his time as managing director, he and the AC did make some gains. The Monroe Street bike trail through the Ho-nee-um Pond area was completed. Following Zimmerman's resignation from the ranger position, Anderson managed to wrench an allocation from the chancellor's office that temporarily funded a new ranger position. The Duck Pond (Spring Trail Pond) was temporarily restored and dredged; the northern shoreline was no longer eroding. With the assistance of the Wisconsin DNR, close to three hundred ducks from the exploding Duck Pond population were snared, banded, crated, and transferred for release to a spot along the Lake Michigan shore, leaving a population of approximately fifty ducks at the Duck Pond for the neighboring children to feed.[25] Repairs to the Duck Pond shoreline continued into 1975, with a concentration on the reconstruction of the south bank. The ever-expanding duck population was reported to have taken the disturbance in stride. Also during Anderson's tenure, in December 1972, the first effort was made to block off McCaffrey Drive as an east-west commuter's shortcut into town by placing a chain across the road at the Wingra parking lot and then dividing the parking lot with a barrier. It was clear by the end of 1972 that something had to be done about the drive, now dubbed the "speedway" by local newspapers.[26] By January 1974, the AC went on record recommending the closing of McCaffrey Drive to through traffic.

When Anderson left at the end of July 1973, the Arboretum professional office staff, excluding the field-workers, included only two other people: Rosemary Fleming, Dane County naturalist, and Jim Zimmerman, half-time Arboretum naturalist. But waiting in the wings, although it would take four months to convince her to accept the position, not as "managing director" but now as "director," as Chancellor Young insisted, was Katharine Bradley, who would serve from 1974 to 1983—ten years during which she would increase the size of the professional office staff, shift administrative control of an impressive portion of Arboretum business to the director, and establish precedents that would shape the direction in which the Arboretum would move for the remainder of the century.[27] But first, she had to come to terms with Grant Cottam, who had resigned as chair of the AC in 1971 but who was still, in 1973, the most powerful influence on Arboretum decision making.

Katharine T. Bradley

In a 1991 interview, Katharine Bradley told the story of her hiring as Arboretum director. Anderson left in July. The position remained open and unfilled well into the winter of 1973. In September 1973, the Search and Screen Committee reported that notices that had been sent out had elicited little response. After forcing two resignations within three years, the AC rather ingenuously concluded that the lack of response obviously had something to do with the "time of year."[28]

At some point between September and November, Bradley decided to apply. Some of her "learned friends," as she put it, aware of her background as a trained botanist, had urged it. She had grown up in the Hudson Highlands near West Point, where her forester father managed a large experimental forest. She earned an A.B. degree from Vassar College in 1943, a master's in botany from the University of Minnesota in 1944, and a Ph.D. from the University of Michigan in 1953. While working on her dissertation she taught botany and biology at Wellesley College. She also worked in New York art galleries, the Bettmann Archive, and the Rockefeller Foundation's Reference Service. Fluent in French, she spent a postwar summer working with children at a French camp for young war victims in the Alps foothills. She also had some elementary school teaching experience and a year teaching biology at Bowling Green State University prior to beginning doctoral work at Michigan.

Katharine T. Bradley was Arboretum director from 1974 to 1983. Her time as director was marked by significant changes in Arboretum governance. (photo by Tom Rust, UW Photographic Media Center, Arboretum Photo Collection)

She came to the UW–Madison in 1954 to do cancer research as an American Cancer Society Fellow. She subsequently worked in the Botany Department until, as she put it, she decided at age thirty-seven that it was time to retire and to raise "offspring." Now, in 1973, at age fifty-three and with two children in their late teens, she decided that she might enjoy returning to the professional world—not because she needed the job, but because, as she put it, "I like to work" and to "get things done."[29]

Initially, however, she had serious reservations about applying for the job. For one thing, as she astutely observed, the "job was ill-defined." John Curtis had created the position of managing director in 1960 while he was still AC chair. Hence, she added, it followed that the managing director originally had been intended to be "a flunky for Curtis." Curtis had been the mainstay, and after him, she noted, it was Cottam who became the "guiding spirit" for the Arboretum. "Cottam . . . was dedicated to the same school of ecological development." And even though Cottam had resigned as AC chair in 1971, he remained the dominant influence in decision making during Anderson's last two years. Bradley knew that in order to be an effective director, it was essential to have his support.[30]

So when she considered taking the job, one of the first things she did was to call Grant Cottam. "She had known him," she noted, and she could "hear his chin hit his chest when she told him" of her plans. It was clear, she added, "Cottam considered her the last person he would have thought of in the position." Yet, sometime later when she met Cottam and his wife at a gathering, they both persuaded her to apply. If he wrote a letter guaranteeing his support, Bradley told him, she would apply. Cottam, the Arboretum's "guiding spirit," wrote the letter, and she applied.

Bradley knew that without Cottam's unconditional backing, she could suffer the fate of both Archbald and Anderson. And she was not about to become the AC's third "flunky." She skirted the Search Committee and sent her letter of application directly to Dicke, the AC chair. She had "spontaneous support," she observed in a 1991 interview, including the symbolic support of Estella Leopold, Aldo Leopold's wife, whom she had known for years. She liked "running things," she admitted. She did not want the AC "running things" for her. Nor did she want to be under the thumb of what she determined at the time was the increasingly powerful and intimidating Friends of the Arboretum (FOA).

When Bradley took over the directorship in 1974, the FOA, as she quipped, was "muddling" along. Some members of the FOA Board of Directors thought that they "ran the Arboretum," she observed. The FOA had been behind the funding for successful, influential Arboretum programs, principally the guide program. They also had begun publishing their own newsletter, entitled *Friends of the Arboretum*, in 1975.[31] The Arboretum budget, however, was part of the UW budget. The FOA board didn't seem to realize that, Bradley noted. And so, she decided that she would try to encourage the FOA to "develop a functional working philosophy which was consistent with their real role." The FOA could do with a professional director, she opined. With funding provided on short notice by the Downtown Rotary Club, she eventually hired Gene Glover as the new FOA director. Glover's first task, as Bradley wryly put it, was to help the Friends to "accept" a professional director. Glover, Bradley observed, "made the operation go" and eventually, as had been the idea originally, the FOA paid her salary. The Rotary Club provided only enough money to get the position started.[32]

Bradley successfully outflanked the ambitious FOA board members, but reaching a workable compromise with the powerful AC presented a more daunting challenge. She realized early on that the AC was the governing body she had to convince about Arboretum needs if she was ever to accomplish anything. The basic problem all along during the Archbald and Anderson years, she thought, was a lack of definition: "We needed to define exactly what the committee did and exactly what the director did." Essentially, she liked the AC and found them helpful. When she took over as director in January 1974, Dicke was still the chair. By the fall of 1974, however, Dicke resigned and was replaced by Robert Ellarson, a professor of wildlife ecology who had studied and gone hunting with Leopold and who had five previous years of AC service. Ellarson had a long, close relationship with the Arboretum and recalled fond memories of his personal experiences, dating back to the days when Jacobson was superintendent. Bradley had a healthy respect for Ellarson, and they had a healthy and productive working relationship right up to his retirement in fall 1978, when he was replaced by Botany professor Gerald Gerloff.

Bradley was an administrative wizard. She had a keen sense of what motivated people and of how to bring them around to her way of thinking. If past problems with the AC had been the result largely of a lack of awareness and accurate information on the part of the AC members, she was prepared to remedy the deficiencies. What the committee needed to know in order, as she put it, "to make the kind of realistic, long-range plans that have never been made before" was relevant and timely information about the Arboretum's realistic needs. And since the AC met with the director only once a month, she decided to write and distribute a monthly report to the committee members prior to the meetings in order to keep them "current" regarding what was going on around the Arboretum, what questions needed to be addressed, and, subsequently, what items could be included on a possible agenda for the AC meetings that would enable her to get the advice she needed and, also, in the process, what she wanted. The move was brilliant—whoever controls the agenda controls the meeting. No one objected to her generous offer to spend time writing lengthy monthly reports for all the AC members to read and mull over.[33]

A New Direction

As time passed, Bradley and the AC defined their respective roles. The committee was the "policy formulating body"; the director was responsible for implementing the policy—*policy* in this case meaning "broad operational directions." In a January 1974 newspaper interview, Bradley explained the relationship this way: "The committee makes policy decisions. The nitty gritty is mine. . . . If I think their policy is wrong, it's up to me to try to persuade them—but I had better know what I'm talking about! They're all authorities."[34] Elsewhere, she wrote that she was "not entirely without influence in the formulation of those policies, but the fact remains—I don't make them."[35] Her charge was to execute and manage policy decisions. In order to do that effectively, she realized that she first had to acquire a solid knowledge of the operational construct that was now her charge. She knew the Arboretum, but not well. She spent the first month of her directorship walking the grounds. Around springtime 1974, she persuaded Harriet Irwin, who had just finished an M.A. thesis on the natural history of the East Marsh, to take her out to the Gardner Marsh and to tell her all about it.[36]

The first pressing operational matter she faced was, as she bluntly put it, "no staff." The crew chief, Gene Moran, was helpful, but he had hardly any professional staff. Bradley inherited both Zimmerman and Fleming, but by summer 1975 Zimmerman was no longer the Arboretum naturalist. He assumed the less conspicuous role of "consultant." In winter 1977 he left the Arboretum completely to accept an appointment in the UW Department of Landscape Architecture. In 1991 he died unexpectedly while cutting grass on a Cambridge, Wisconsin, prairie. It would correctly be said of "Jim Zim" that his death marked "the end of an era and a great loss for environmental studies."[37]

Staffing Needs: A Ranger and Virginia Kline

What little "staff" the Arboretum had between 1974 and 1977 worked out of the "security residence," with one big room for an office. The rest of the house was occupied by the security residents who lived on the property. By the spring of 1974, Bradley had a full-time, seven-day-a-week ranger, Wayne Pauly, who was assured of a salary thanks to a generous three-year renewable $30,000 grant from the Rennebohm Foundation. She counted on Pauly for help with the educational program as well as for security. "His walking patrol [was] essential," she observed.[38] But even a full-time ranger was not enough. "We needed a staff ecologist," Bradley recalled in her 1991 interview, someone "to plan the care and feeding of the ecological communities. We had no one who could do that." She wanted someone who had come out of the John Curtis school, a student of plant ecology dedicated to rebuilding an understanding of vegetation based on the continuum concept and one who, like Curtis and Cottam, was prepared to gather large amounts of quantitative data. Accurate record keeping by faculty researchers using the Arboretum had been

Virginia "Gina" Kline, Arboretum prairie ecologist from 1975 to 1996. (photo by Melinda Bailey, Arboretum Photo Collection)

a persistent problem that Bradley identified almost as soon as she took over the directorship. There are "researchers and teachers," she observed, "who . . . regard any efforts at record keeping . . . as a personal affront. How to persuade these scholars that records of activity . . . are an essential basis both for future zoning and for requests for support . . . is . . . part of the Director's job."[39]

She found her proponent of the Curtis school of record keeping and sampling methods in Virginia "Gina" Kline, whose Arboretum connection went back to 1965, to the early days of Zimmerman, Fleming, and the nascent Arboretum guide program. When Kline was appointed Bradley's Arboretum staff ecologist in 1975, she was finishing her doctoral dissertation, "The Terrestrial Vegetation of the Kickapoo Valley." Her director was Grant Cottam, the most visible symbol of the Curtis school on the UW faculty. Kline was nearly fifty at the time. Bradley was happy with the choice.

Kline's experience in the mid 1960s with Zimmerman's famous "Reading the Landscape" course and her experience with the Arboretum's early guide program provided her with credentials that enabled her in 1964 to get hired as a

guide and then as a trainer of guides in the newly established Madison Metro-
politan School District's outdoor education program at the Madison School
Forest, a 307-acre hilly terrain in Wisconsin's unglaciated Driftless Area. The
forest is also known as the Jackson School Forest in honor of the Arboretum's
Colonel Bud Jackson who, not surprisingly, was instrumental in acquiring
the land. In 1966, Kline was named coordinator of the Outdoor Education
Program for the Madison School Forest. In 1972, she also started teaching
Zimmerman's "Reading the Landscape" course at MATC. When she came
to the Arboretum in 1975, she was officially appointed as the first Arboretum
ecologist, working three-quarter time at the Arboretum and one-quarter time
as a lecturer in the Botany Department. She taught "Vegetation of Wisconsin"
jointly with Cottam until his retirement in 1986.

Bradley was quoted as saying, "getting Gina out there in the Arboretum as
a professional ecologist was very important because it meant that we had professional direction for the management of the plant communities." A major
concern was the increasing spread of invasive plants and weeds. Kline came
up with a burning schedule for the prairies. She also started screening plant
poisons. "We didn't want to use them," Bradley observed, but cutting the
weeds only made them worse. And so, reluctantly, the Arboretum began a
weed-killing program. When Kline was asked in 1977 about the weed killing,
she replied that "a biological control" would have been "wonderful, but we
haven't found it." She noted also that she and her staff were "studying methods
other than spraying."[40]

Kline became Bradley's "multitasker." Eventually she worked full-time in
the Arboretum. She was invaluable. She worked on thinning the Leopold
Pines and other pine plantings along the Beltline. She engineered the planting
of other trees along the Beltline that would better withstand the pollution.
She worked on eliminating woody invasions in the prairies and wetlands. "I'm
really concerned about the wetlands," she observed. "They're a natural heritage
here," and they "need study and emphasis." She was particularly concerned
about the lowered water levels.[41] As the 1970s gave way to the 1980s and the
1990s, she would be the key influence on expanding a corps of volunteers
committed to weekly work parties with Arboretum staff members. Bradley
observed in 1991 that Kline just kept getting "busier and busier."[42]

By 1991, Kline had also completed a time-consuming, comprehensive 180-
page Arboretum Master Plan for all the ecological communities, including the
wetlands, the Grady Tract, and the Lost City Forest, areas that were not included in Curtis's earlier, pioneering 1951 Master Development Plan. Kline and
Bradley conceived the idea for the master plan as early as January 1977. They

acknowledged at the time that the work would take decades. And it did.[43] The plan, which began circulating in 1991, is still in use. It provided a long-range vision for the development and management of native Wisconsin communities as well as a projection of ways in which the Arboretum could better serve faculty and students, non-academic field science professionals, and the general public.

A shy, modest person who never intentionally pursued special recognition, Gina Kline nevertheless ended her career with an established international reputation that included, among other honors, invitations to address the International Botanical Congress in Berlin, the Society for Ecological Restoration, the American Association of Arboreta and Botanical Gardens, and the Wisconsin Association of Science Teachers. She also lectured publicly over the years at Longwood Gardens in Pennsylvania, the Wisconsin State Garden Club convention, and Madison General Hospital as an Osler lecturer. She also worked as a consultant with thirty-four groups in sixteen states and five countries, including the Canadian Forest Service, the U.S. Forest Service, the San Diego Zoo, Washington's National Zoo, and the Chicago Botanic Garden. In 1994 she received the Robert Heideman Award for Excellence in Public Service. She retired in 1996. "I think I'd like to do something with young people," she was quoted as saying at the time.[44] She died on February 23, 2003 at the age of seventy-seven.

"Putting Humpty Back Together": William R. Jordan III and Restoration Ecology

Another pressing policy matter that Bradley faced was finding someone to manage operations for the newly constructed McKay Nature Awareness Center. The building, which took two years to construct, was available for occupancy by mid-summer 1977.[45] She went hunting for money to hire staff. She was having little luck when fortune smiled, and one of the Brittingham family members sent her a letter informing her that, by chance, the Brittingham Foundation had learned that she was looking for funding. She consulted Chancellor Young, who told her to apply for a Brittingham Grant, and she likely would get it. The Brittingham Foundation agreed to provide $15,000 a year for three years to support a public services coordinator for the McKay Center. The position, as Bradley envisioned it early in 1977, would be demanding. It would involve initiating work policies for the building's operation, recruiting and educating a cadre of volunteers to assist visitors, soliciting contributions for and editing the *Arboretum News*, keeping a record of guided tours in order to lighten Rosemary Fleming's work load, attending guide training sessions with an eye toward eventually working also as a guide when needed, and even finding space in the new building for organizing an in-house library with reference books, theses, dissertations, and other titles that had been donated to the Arboretum.[46]

Bradley began advertising the coordinator position nationally in late 1976. By mid-January 1977 she had received thirty-five applications and interviewed three applicants. In February she offered the job to William R. Jordan III. Jordan was well qualified. He had spent a portion of his youth in the midst of Wisconsin woodlands. His father, a district state forester, worked in the northern section of the Kettle Moraine Forest. Jordan later majored in developmental biology at Marquette University and eventually earned a Ph.D. in botany from the UW–Madison, after which he worked in the Biology Division of the Oak Ridge National Laboratory in Tennessee. Unsure of a future commitment solely in the sciences, he returned to the UW and finished an M.A. in journalism. He also worked for a time as an editor for the American Chemical Society in Washington, D.C. In a 1979 interview he explained that he "wanted to work with words, as an editor and writer, and with people in . . . parks or an arboretum."[47] Consequently, he was thrilled when the UW Arboretum position became available, and by mid-February 1977 he was on the payroll as the full-time public services coordinator.

Between March and July, Jordan busied himself with preparations for the McKay Center opening. He had in mind a series of slide shows featuring Arboretum highlights, such as the prairies, and set about in April putting together an extensive collection of slides that would be the basis for an extant Arboretum slide collection that numbers well into the thousands. "The important thing," he observed in his April 1977 report to Bradley, was "to have something to show in the Center when it opens." He insisted that the Arboretum could benefit from a "well organized slide library." To that end, he prepared guidelines for labeling the slides. He also worked with Fleming on a "library" of laminated Arboretum plants that could be displayed with the seasons and a set of plant murals that would be left on permanent display. He and Fleming also planned to have groupings of major prairie grasses on display in the entryway. By May 1977, he reported that he had enlisted thirteen volunteers.[48] Yet he was still worried.

Jordan was concerned mainly about attendance when the Center opened. The public might "love the Arboretum to death outside," but inside at a new, unfamiliar visitor center interest might be limited to a few visitors looking for restrooms. He was right to worry. The McKay Nature Awareness Center opened to the public on Wednesday, July 6, 1977, at 8:30 a.m. The first volunteer receptionist, one of the original thirteen recruited, was Mildred Lloyd.

The first visitor was a Madison man, looking for the restroom, who wrote his name in the guest register—"for all posterity," as Jordan put it—but his handwriting was so bad that no one could read it. Jordan was disappointed. He noted in his monthly report to Bradley that it might be a good idea to put a notice or two about the Center's opening in the local newspapers. He also told her that he thought "the long run solution" was to get something in the center that people wanted to see.[49]

By January 1978, he had in place the plant exhibits and a variety of slide show programs accompanied by recorded, taped narratives. And there was more to come. A "Longenecker Gardens in the Spring" slide show was being prepared by Ken Wood, Arboretum guide, and Gar Dawson, Arboretum horticulturalist. Sue Bridson and Clifford Dennis were working on a slide show on Arboretum history. "A Winter in the Arboretum" slide show was also planned. For the exhibits, a plant display was ready for viewing. A bird exhibit would soon be finished; "components" were arriving, Jordan noted, from "various quarters of the world." He, Bradley, and Gina Kline had also been working on a map exhibit that he hoped would be ready for showing at the McKay Center dedication scheduled for April 21, 1978.[50]

Jordan had been hired in February 1977, but since the McKay Center did not open until July, for four months he was free of daily, pressing administrative chores. As a result, he and Bradley had ample time to mull over new directions, particularly a new direction for the *Arboretum News,* which was now under his stewardship. They discussed a general shift in editorial focus away from Arboretum internal activities and toward ideas with wider interest and wider audience appeal. It occurred to Jordan, Bradley noted in her March 1977 Director's Report, that an *Arboretum News* focus on plant community restoration might encourage "financial support from companies." Jordan had in mind companies connected to heavy construction, the Army Corps of Engineers, and strip mining owners, for instance—companies that were legally obliged to provide land restoration and had no clear idea of how to proceed. Bradley was supportive but worried about journal production costs and the possibility that such an ambitious project might require additional staff.

In early April 1977, only a brief few weeks after his discussion with Bradley about the journal, Jordan was promoting a brand-new in-house publication, something beyond the parochial *Arboretum News;* a publication dealing with ecological restoration and land reclamation that could attract funding and "push the Arboretum's research and education missions to a new level of achievement and significance." He told Bradley in a report dated April 7, 1977: "I intend . . . to prepare a proposal for a magazine that will bring this kind of [restoration] information together in a well-edited form that those interested . . . will be able to rely on for thoughtful, up-to-date reporting of research, goings-on and thinking in this area. As far as I can tell, this stuff is now being published willy nilly in all sorts of places."[51]

Jordan wanted to publish, at the Arboretum, a journal that would provide a forum for the exchange of information on plant community restoration and land reclamation, mainly among ecologists and people working in land repair. The journal, he suggested, would also appeal to park managers, naturalists, landscape architects, and others interested in the restoration of plant and animal ecosystems. At the time, there were no professional journals being published that focused exclusively on restoration ecology. There was a market, and the Arboretum was the appropriate location, the "center"—the "research center"—he argued, from which the journal would originate. Jordan's design was without a precedent at the Arboretum, but by no means was he the first to have recognized and to have acclaimed publicly the historical significance of the Arboretum in the history of restoration ecology. In fact, Jordan was echoing sentiments that had been expressed four years earlier by Jim Zimmerman and his wife, Elizabeth, in an article that appeared in the *Wisconsin State Journal* on January 21, 1973, entitled "UW Arboretum Tackles Ecosystem Restoration."

In the article, the Zimmermans claimed that although "few" were doing anything seriously in 1973 about restoring environmental communities, the Arboretum, for decades, had been actively engaged in "restoration." They called the program Ecosystem Restoration. It had begun, they claimed, "with Aldo Leopold's belief in learning from nature, G. William Longenecker's discovery that beautiful landscapes also are ecologically correct, and the expertise of John T. Curtis who studied the small bits of undisturbed natural vegetation." For over the past forty years, the Zimmermans continued, the Arboretum has been "trying to put Humpty back together again." And they cautioned that "letting lands go back to nature" did not ensure that the total ecosystem would "return, at least not in a lifetime." Arboretum efforts to restore the five major ecosystems of the Midwest had been a series of hits and misses, trials and errors. But the pursuit, they suggested, if well publicized, can "become a popular hobby wherein everyone works outdoors and learns much ecology through making mistakes," just as the Arboretum had. As examples, they cited what had been learned over the years through the following:

Arboretum efforts to restore conifer forests "where dense one-species planting" had made "the soil too dry and infertile for appropriate undergrowth."

Arboretum efforts to restore deciduous forests with native trees that when

"left alone" would "come back naturally" and where oaks were not replaced because the original native environment could not be recreated.
Arboretum efforts to restore wetlands where "the deepwater marsh can be
rescued from ditching and siltation by shoreline gradualization" and the
restoration of "natural waterlevel cycles."[52]

Zimmerman was correct in his assessment of extensive Arboretum restoration efforts dating from the 1930s when work on the Curtis Prairie began.
Restoration, even prairie restoration, as discussed above, was by no means a
new ecological idea in the 1970s, although interest increased markedly in the
1970s. In northern Illinois, for instance, counties once covered by tall grass
prairies had been promoting prairie preservation *and* prairie restoration since
the 1940s. DeKalb County, sixty miles west of Chicago, was on record as
creating the DeKalb County Forest Preserve District in 1940 in order "to provide recreational opportunities, to preserve *and restore* our natural plant and
animal life for both public enjoyment and education, and for the protection
of our natural heritage."[53]
In November 1954, George Ward and Paul Shepard, two biology professors
from Knox College in Galesburg, Illinois, had traveled to the UW Arboretum
to tour the Curtis and Greene Prairies. After the visit, they went back to Galesburg and began planting a tall grass prairie at Knox College's Green Oaks Field
Study Center, a 760-acre field station located twenty miles east of Galesburg.
Their objective, inspired by the Arboretum prairies, was a "complete restoration: the establishment of a group of species in abundances and proportions
similar to those in natural communities." That was in 1954.[54] Clearly, prairie
restoration, particularly in the corn and soy bean corridor in northern Illinois
and southern Wisconsin, was nothing new in the late 1970s, and it somehow
seems fitting that southern Wisconsin's UW Arboretum should emerge during the 1980s as the definitive center for the global study and propagation of
restoration ecology. And most of the credit for that 1980s distinction belongs
to William Jordan's persistent efforts to create a forum for the study of the
history and significance of restoration ecology at least from the 1930s, the time
of the Arboretum's founding and also the time of the ecologically devastating
dust bowls.

❧

CHAPTER II

The 1980s and Beyond

A RESTORATION ECOLOGY LEGACY,
"EARTH PARTNERSHIPS," REFLECTIONS

By the end of 1977, William R Jordan III, the Arboretum's public service coordinator, was giving serious thought to the publication of a journal for people interested in the reclamation of plant and animal ecosystems. Within a year, he would explain to Katharine Bradley, Arboretum director, that land repair was "big business," and that since the Arboretum had pioneered land restoration, he believed it was forty years ahead of anyone else. A lot of money is being spent, he suggested, to learn "about just the sorts of things we have been finding out here since the early days." A perfect example of the kind of commercial land reclamation to which he was referring had been addressed in a summer 1978 *Arboretum News* article, written by Darrel Morrison, UW professor of landscape architecture. In the article, entitled "Native Plants for Man-made Moonscapes," Morrison described the 1976 prairie restoration efforts of Phil Dinsmoor, one of his graduate students, at the only active iron mine in Wisconsin, an operation near Black River Falls owned by the Jackson County Iron Company.[1]

"Look at what we have to offer mining companies being pushed to solve reclamation problems," Jordan told Bradley. Consequently, "we have to seize opportunities . . . as vigorously and as imaginatively as Olbrich and Stark and Jackson and the early planners seized the Depression-era opportunities to acquire cheap land and free labor. Perhaps the continued well being . . . of the Arboretum will depend on this." To that end, he suggested, "publications" would be important in a program promoting research on land reclamation connected with the Arboretum's history. In early 1979, he proposed three future publication possibilities in a report to Bradley. The first would be a one-time illustrated book on landscape design with native plants that would include articles by authorities in the field. The second would be "a technical

William R. Jordan III promoted restoration ecology as a new direction for the Arboretum. (photo by Melinda Bailey, Arboretum Photo Collection)

journal on restoration and maintenance." The third would be a popular magazine for gardeners. He told Bradley, "once we got something like this off the ground we wouldn't have to spend so much time explaining to our public that the Arboretum [was] not a park." Bradley, skeptical, yet interested in the possibility and impressed with Jordan's contagious enthusiasm, agreed to support the effort. It was a wise decision.[2]

In July 1980, Gerald Gerloff, the AC chair, invited Jordan to attend the July AC meeting in order to provide the members with an update on the new ecological newsletter that he was proposing. Jordan told the committee that the newsletter, an occasional publication, would include short summaries of ongoing research in the "restoration and management of natural ecosystems in the central and northeastern U.S." It would enable people involved in restoration work to see what others in the field were doing.[3] In October 1980, Jordan asked Grant Cottam if he thought developing a center for restoration at the Arboretum combined with a supportive journal might be "far-fetched." Cottam, diplomatic as usual, was supportive. "The horses are here," he said; "we don't have to go outside."[4] Encouraged by support from Bradley, Gerloff,

Cottam, and others, Jordan, in the 1980 fall issue of *Arboretum News*, announced a new Arboretum publication, entitled *Restoration and Management Notes*.

Originally scheduled for publication in March 1981, *Restoration and Management Notes* (*R&MN*) appeared three months later in June. Initial response was reassuring. Jordan reported receiving notes and submissions "from ecologists, naturalists, land reclamationists, landscape architects, foresters, and managers of parks, wildlife preserves and other natural areas."[5] Bradley would say of the journal ten years later that by 1991 it had evolved into a publication that was appreciated all over the country, not just regionally, as Jordan had first anticipated.[6] In early 1984, the University of Wisconsin Press took over the journal's publication, including subscriptions, promotion, and financial accounting. The Arboretum retained editorial control. A publication surplus of $5,000 was transferred by the press to the Arboretum in late summer 1984.[7] By 1999, *R&MN*, which originally published nonrefereed articles on an occasional basis intended mainly for a central and northeastern U.S. and Canadian audience, evolved into *Ecological Restoration*, a peer-reviewed quarterly with an international advisory board and a worldwide focus, publishing articles and notes on the recovery, repair, and restoration of ecosystems ranging from grasslands in Illinois to fen meadows in Switzerland to degraded mountain land in South Africa.[8]

Jordan, with the support of national and international restoration ecologists, *R&MN* subscribers, and the Arboretum professional staff, also sponsored numerous influential meetings, conferences, symposia, and workshops on restoration ecology throughout the 1980s and early 1990s. In 1988, Jordan, the Arboretum, and the journal had a hand in the formation of the Society for Ecological Restoration (SER), an international organization currently with members in thirty-seven countries and all fifty U.S. states. In May 1988, SER was given office space in the Arboretum Security Office.

Jordan's ambitious 1977–78 projections, drafted during the first years of his appointment, became, by the mid 1990s, the mainstay of the Arboretum's identity as a center for the advancement of restoration ecology. The Arboretum's commitment to the recovery of land, plant, and animal ecosystems also was increasingly viewed by the 1980s as a logical extension of the "land ethic" that Aldo Leopold, the Arboretum's first director of research, first advanced during the 1930s dust bowl years. As Leopold explained in his 1947 essay, "The Ecological Conscience": "The practice of conservation must spring from a conviction of what is ethically and esthetically right. . . . A thing is right only when it tends to preserve the integrity, stability, and beauty of the community, and the community includes the soil, waters, fauna, and flora, as well as people."[9] His position on conservation ethics, on the need to "heal the land," resonated throughout the century not only through recovery and land reclamation efforts but also in the promotion, dear to Leopold's heart, of farms as natural game habitats that would promote the "land ethic" by putting nature back in the farm, an endeavor involving Wisconsin farmers that consumed much of his time during his Arboretum years (1933 to 1948).

Katharine Bradley Retires

When the first issue of *R&MN* came out in June 1981, Katharine Bradley still had two years of tenure left as Arboretum director. Jordan had been at the Arboretum for four years. When Bradley retired in summer 1983, she was pleased to have been a part of the genesis of Jordan's journal in spite of some obvious concerns, owing to his time-consuming outside commitments, about his position within the Arboretum professional staff. She had expanded the office hiring, but she was still seriously understaffed. Jordan and Gina Kline represented two-thirds of her professional staff. She depended heavily on them, especially between 1979 and 1980 when the prospect developed for the purchase of vacant lowland properties on the Arboretum's southeast boundary. She wrote in 1983 in her "Goodbye" to the staff and Arboretum supporters that of all the things she and the staff had accomplished during her tenure, she thought one of the most important was "to have acquired all the vacant lowland on the southeast, down to the Chicago and Northwestern Railroad." That entire area, she added, "can now be managed properly as a unit, and we have a readily recognizable boundary there without major irregularities or inholdings."[10] She was referring to the Fisher and the Selix properties acquired in 1979 and 1980, respectively.

The Fisher property, bought with financial help from the Friends of the Arboretum (FOA), was a marshy triangle of wetland and sedge meadow, covered with native wetland species and with pre-settlement bur oaks. The property's location on the southeast edge of the Arboretum provided filtering protection for the Gardner Marsh to the north and ultimately for Wingra Creek and the stream flow east into Lake Monona. In December 1979, Jordan published an article on the marsh. He noted that the property was "one of the few remaining open wetlands in the Madison area." No road "touches the area," and "no trails cross it." He also noted that Kline had visions of turning the property into "a microcosm of pre-settlement Dane County" with an oak grove restored to savanna, "carpeted with prairie grasses and flowers, and giving way to oak and maple forests in the Old Lost City area to the west." During a visit, Kline and Keith Wendt, Arboretum ranger, waded into the marsh to identify plants. They found marsh aster, turtlehead, marsh fern, sedges, and blue joint grass, and on one small spot, Wendt found a single specimen of fen goldenrod.[11] Bradley also turned to Jordan, Kline, and Wendt in 1980 for help with the acquisition of the 9.8 acre Selix property on the Arboretum's southeast border. The Selix property, another wetland, extended south and east to abandoned Chicago and North Western Railway tracks.

The acquisition of the Fisher property in 1979 and the Selix property in 1980 was followed in 1981 by a gift from the Sinaiko family and a gift from the estate of Louis Gardner, who had recently died in December 1979. The money enabled the Arboretum to acquire a remaining open wetland parcel on the southeast, just north of the railroad tracks.[12]

A final pressing challenge that Bradley, her staff, and the AC were grudgingly obliged to face in the last years of her tenure was the lingering aggravating problem of what to do with "the road"—McCaffrey Drive, the main road through the Arboretum. Bradley worried that as time passed, the biggest problem facing the Arboretum would be less about speeders than about the increase in commuters using the drive as a shortcut to and from Madison's West Side. If the number of commuters using the road continued to increase, then the road could conceivably be removed from Arboretum and UW control. In this case, she argued, it would become a public road, a city or state road, perhaps, and it could be widened and turned into a major thoroughfare at the obvious expense of precious Arboretum land and everything that went with it.

And so, the solution somehow involved closing the road to forestall the inevitable commuter traffic increase. In December 1978, Madison police chief David C. Couper, concerned particularly about cars on the drive endangering runners, suggested to Bradley that through traffic could be eliminated by turning the two entrances to the drive into a cul-de-sac. Motorists could drive to the center of the Arboretum from each entrance, but they would be unable to continue beyond that point.[13] The suggestion, a sensible one and one that would prevail, met with little support from the AC. In 1980, Arboretum field crew foreman Gene Moran proposed a similar plan to detour traffic coming from the east and west into the McKay Center parking lot by putting up a barrier across McCaffrey Drive at the Longenecker Drive junction. The detour would also discourage commuter traffic. For the entire month of July 1980, as an experiment, the drive was dead-ended to through traffic by chains set up in the Wingra parking lot.

Reaction was mixed. Many who opposed closing the road, including some members of the AC, argued that closing it denied access to elderly motorists who traveled the drive for pleasure. By 1981, however, in spite of objections, a permanent dead-end barrier was set up in the Wingra parking lot. It was essentially the same plan that police chief David Couper had suggested to Bradley in 1978. The end result, however, Bradley observed, was malicious vandalism—smashed locks, smashed gates, fences torn down, and damaged barricades. In April 1984, Gregory Armstrong, after taking over the director-ship, decided to open the barrier once a week on Sunday afternoons from 12:00 to 6:00, a policy that continues today and that Bradley approved. "If I had thought of it," she later observed, "I would have done it."[14]

In 1991, Armstrong would have the McCaffrey Drive gate removed from the Wingra parking lot and a new gate installed in the McKay Visitor Center park-ing lot, where it remains today. The barricade at Longenecker and McCaffrey Drives detoured all traffic moving west and east into the McKay Center park-ing area—exactly what Gene Moran had recommended eleven years earlier in 1980.

Before she left the directorship, Bradley also obtained grants from the Evjue Foundation for guide program support, money that helped to offset user costs for guided tours. The fees were low, she later observed, "and the schools got used to it."[15] Owing to the regretted loss of Susan Folley, Rosemary Fleming's successor as tour guide coordinator, along with her position and salary as the Dane County naturalist, Jordan's position as public services coordinator was revised in November 1982 to include the tour program responsibilities. Bradley had no one else to turn to. Gradually, however, Jordan was relieved of working with volunteers and of administering the McKay Center. Gene Glover took over both responsibilities, and she was "good at it," Bradley went on record as saying.[16]

Jordan was never totally content with the quotidian administrative demands of supervising a visitor center and a volunteer program, and once relieved of those duties, he had more time to work on *R&MN* and the Arboretum's role in the expanding new discipline, now officially tagged *restoration ecology*. As Bradley observed, Jordan was "an intellectual . . . bright, very much interested in ideas," whose "skills" ran contrary to "administering volunteers."[17] When Armstrong took over as director, he revised Jordan's position even more radi-cally in order to provide Jordan with more free time to work on editorial and public relations promotions intended specifically to advance the Arboretum as an international center for research on the restoration of land, plant, and animal ecosystems.

During her 1991 interview, when asked what her major contribution to the

Arboretum had been, Bradley responded that the most important things she had done were related to "the evolution of the working relationship between the Arboretum Committee, the Arboretum Director and the development of staff personnel into able professionals who could contribute in so many various ways to the development of programming."[18] Bill Hilsenhoff, AC chair, and Emily Earley, an AC member, in tributes to Bradley, endorsed her assessment of her greatest accomplishment. They referred to it as "the organization of the organization." As they put it, before she took over the directorship, "the Arboretum had a long history of being run by academics and, while the operation was far from being a shambles, there was plenty of room for improvement in administrative efficiency." Bradley had guided Arboretum governance in a new direction. She had solved the dilemma. As Hilsenhoff and Earley observed, she had "put things together and made them work. . . . She had put us in touch with the administration down on campus in a way we never were before, and [had] given the Arboretum a voice the people on campus respect and pay attention to. She had done the same thing for us in city, town and state government. . . . Such accomplishments need to be pointed out . . . because they are invisible. Yet they are important and are something we are sure Katharine's successor will appreciate."[19]

And for Greg Armstrong, her successor, Bradley asked the following favor from her supporters: "Please be kind to my successor, and treat him as well as you have treated me."[20]

Gregory D. Armstrong, Clarifying the Arboretum's Mission

Gregory D. Armstrong became the Arboretum director in August 1983. He was well prepared for the position, having since 1970 been director of the Botanic Garden at Smith College, Northampton, Massachusetts. While at Smith he managed an arboretum, a range of greenhouses, and several gardens. Armstrong was a Wisconsin native who grew up in the small Rock County community of Cooksville, where his father, Miles, a retired teacher, ran the legendary Cooksville General Store for a period after World War II. His mother, Beth, was also a teacher. Both of his parents eventually taught in the Evansville School District.

After finishing high school, Armstrong attended the UW–Madison, majoring in horticulture. Ed Hasselkus, the Arboretum's talented garden curator who was still on the Arboretum staff when Armstrong arrived, was one of his favorite professors. While at the UW, Armstrong worked summers in the Milwaukee County Parks Department, at Longwood Gardens in Kennet Square,

During Gregory Armstrong's tenure as director from 1983 to 2004, endowments increased, new staff was added, and new public education programs were pioneered. (photo by William R. Jordan III, Arboretum Photo Collection)

Pennsylvania, and at the Wisconsin governor's mansion. In 1967, he completed requirements for his bachelor's degree. In 1970, after working for three years as a student gardener at the Royal Botanic Gardens in Kew, Richmond, England, he earned a Kew Diploma. During that period he also traveled in Europe, spending a summer at the Jardin des Plantes in Paris. In 1980, he received a master's degree in botany from Smith College.

When Armstrong accepted the Arboretum directorship, he knew he was stepping into an experience that would be different from the Smith College Botanic Garden. The 125-acre Smith Botanic Garden was much smaller, and unlike the Arboretum, it was not separate from the campus. In fact, it *was* the campus. Its collection was a more traditional variety of cultivated trees, shrubs, and gardens than the Arboretum's vast 1,260 acres of native plants and natural ecological communities. Armstrong looked forward to the challenge of directing the Arboretum and early on expressed his particular interest in drafting a much needed statement of purpose that would clarify the Arboretum's future.

In March 1984, in anticipation of the fiftieth anniversary celebration of the Arboretum's dedication scheduled for June 17, 1984, Armstrong included in his "Director's Report" to the AC a three-page document entitled "Our Purpose," in which he emphasized the Arboretum's unique historical character as a facility that was indeed to some extent a collection of trees and shrubs of horticultural interest but that was also, and to a larger extent, "a collection of restored ecological communities representing the major ecosystem types of pre-settlement Wisconsin."[21] The observation, of course, was a partial para-phrase of Leopold's now oft-quoted dictum from his 1934 dedication speech on the Arboretum "idea, in a nutshell," being the reconstruction "primarily for the use of the University," of "a sample of original Wisconsin—a sample of what Dane County looked like when our ancestors arrived here during the 1840s."[22]

On October 11 and 12, 1984, Armstrong and Jordan sponsored a well-attended international symposium on "Restoration Ecology: Theory and Practice." The term *restoration ecology* referred, at its simplest and as Leo-pold advised, to restoration development "as a technique for basic research."[23] Armstrong's and Jordan's assessment of the Arboretum's mission was by 1984 heavily influenced by Leopold's writings, enough so, in fact, that prior to the fiftieth anniversary celebration keynote address, given by Peter Shaw Ashton, director of Harvard's Arnold Arboretum, Nina Leopold Bradley was invited to read aloud her father's 1934 speech, "What Is the Arboretum?"

In his "statement of purpose," Armstrong drew more from Leopold, how-ever, than just a recognition of the Arboretum as, partially, an ecological sample of pre-settlement Wisconsin. He also essentially reiterated Leopold's observation that it would take "50 years to do this thing" and that the recon-struction would be "done for research rather than for amusement, and for use by the University, rather than for use by the town." As Armstrong paraphrased it, "The idea here is that the value lies in the actual restoration and manage-ment of the ecological communities as opposed to an alternative notion that the end product or completely restored communities are the objective. This idea," he added, "that the actual restoration *process* provides an important op-portunity for basic ecological research has been a part of the Arboretum ef-forts for some time." The end product of research (a restored community, for instance) was neither the research nor the teaching objective. In fact, the process of land and plant restoration and its management was or could be more productive both for research and for teaching than the ready availability of natural communities. Given this position, Armstrong, in fact, was making a landmark case for a consideration of the Arboretum's purpose as an ongoing center for researching and for teaching the process of restoration ecology, a process rooted in Leopold's earlier thoughts on "healing the land"—that is, "bringing nature back to the farm." At this point, Armstrong was speaking directly to the future, in anticipation of the current eminently successful Arboretum "Earth Partnership" programs.[24]

Old Problems: Storm Water Silt and Sediment

The Armstrong years, twenty of them, from August 1983 to July 2004, were years of relative tranquility and steady growth with the occasional intrusion of old problems that—being old problems—invariably kept surfacing (and continue to surface to this day!), including particularly the need to contain and filter the silt and sediment runoffs from adjacent storm sewers, the Beltline, and nearby commercial development in order to keep the runoffs from inundating the Arboretum wetlands, prairies, and eventually Lake Wingra. Without the Arboretum filtering polluted water, Lake Wingra would have become a sewer. As early as May 1978 a "first-of-its kind study showing that the desilting pond in the Arboretum's Curtis Prairie" retained about 75 percent of the phosphorous entering it from storm water runoffs led to the construction of three similar ponds in the Arboretum in an effort to improve water quality in Lake Wingra.[25] By 1981, it was evident that the desilting ponds also decreased the buildup of weed growth choking Arboretum wetlands. In late August 1981, after a summer of heavy rains, 30 acres of vegetation, including trees and brush bordering the Arboretum's West Marsh (east of Monroe Street), were destroyed by silt and sand deposits from city and suburban storm sewers.[26] By January 1983, a new desilting pond, the fifth, was under construction east of Monroe Street at the foot of Glenway Street. Five acres of trees had to be cleared at the spot where the large pond went in. Its banks were planted with prairie grass and plants.[27] A sixth pond, located off Manitou Way, was finished early in 1984.

But even six desilting ponds surrounding the Arboretum were not enough. By the 1990s, the prairies and some wetlands were seriously endangered by an increasingly pernicious invasion of reed canary grass. By 1996, silt in the Greene Prairie, largely the result of storm runoff from the imposing housing development immediately south of the Arboretum boundary, created an irreparable situation. The encroachment of reed canary grass into the prairie was unstoppable.

By the end of the century, in June 1999, Professor Joy Zedler, newly appointed to the Leopold Chair of Restoration Ecology and also the new Arboretum director of research, began publishing a series of articles in *NewsLeaf*, the Arboretum's in-house newsletter, on the decade-old threat to the survival of the prairies and wetlands, among other Arboretum water issues. She argued that the wetlands, which at one time were rich in native species, had become over time "poor in native species and 'rich' in invasive plants." Gardner Marsh, for example, in twenty-five years, had suffered an increase in destructive shrub cover and a noticeable decrease in the diversity of its vegetation. Large cattails dominated the marsh in 1999 and threatened to displace the sparse remnant of sedge meadow that remained from what was once the largest sedge meadow in the Arboretum. Taller and denser than native sedge meadow plants, the cattails shaded out light essential to their survival.

Zedler also called attention to the fact that the flume designed to carry storm water runoff from the Beltline to the Curtis desilting pond had crumbled and that the pond was filling with silt, which was flowing into the prairie. The result was an increase in the population of willows and reed canary grass, both invasive.

By the late 1990s, the Arboretum had become "an urban wilderness at the mercy of all the land that surrounded it." The solution necessitated finding ways to infiltrate the runoff at its source. Ken Potter, a UW engineering professor, suggested creating ponds on higher ground closer to the highway and replacing the ditch that extended from the pond across the Curtis Prairie to the highway gutters with buried perforated pipes. Zedler and her students, at the time, were also planting native plants in storm water retention basins and reintroducing native plants that could survive and eventually displace the destructive exotic species.[28] Zedler also suggested that the Arboretum, in the process, could become a demonstration area for improved storm water management, teaching people about the problems of urban runoff. And once the Arboretum had a better storm water management system in place, she suggested, it could "advertise the virtues of infiltration." She also offered the depressing assessment in 1999 that it was probably too late to rid the Arboretum totally of reed canary grass, but that rigorous, continuing experimentation with sustainable, aggressive native plants could someday, hopefully, restore a portion of the damaged ecosystems.[29]

Ten years later, in 2008, however, the pond problem was no closer to being solved. An Arboretum Storm Water Committee was formed to help find a solution. David S. Liebl, chair of the committee, observed in a 2008 *NewsLeaf* article that the collapsed flume in the Curtis Pond was still a problem and that the only pond in the whole system that met storm water management standards for removing sediment and controlling flow was the Marion Dunn Pond off Monroe Street that was rehabilitated in 2003. In 2008 the decision was made to rehabilitate all of the Arboretum's remaining storm water ponds. The first one was Johannsen Pond (Pond 2) in the Arboretum's southeast corner. A wetland basin between the pond and the Beltline Highway was added by summer 2009. Whether it alleviates the problem remains to be seen. In 2008, plans also were drafted for opening a vista to Lake Wingra from the pond off Manitou Way, often referred to as the "secret pond." Plans were also drafted for the rehabilitation of the old pond in the Curtis Prairie. Building on the idea introduced in 1999 by Zedler for combining the restoration work with a teaching focus to help the public better understand problems with unfiltered urban storm runoff, the Arboretum Storm Water Committee planned to demonstrate to interested visitors, when possible, the restoration techniques being used in the field. During the construction period, signs were posted explaining "the importance of each project, the methods . . . being used to rebuild the ponds, and the benefits to the Arboretum and Lake Wingra." The education effort was commendable, but the ultimate object of the rehabilitation was finally to get clean storm water, clear of silt and sediment, draining through the Arboretum and into Lake Wingra.[30]

A Bountiful Time: Increased Staff and New Programs

Gregory Armstrong inherited a lot of problems, but they never outweighed the blessings. The Arboretum was growing in stature, reputation, and bounty, and Dame Fortune was smiling. Endowments between 1983 and 2004, thanks largely to Armstrong's direction, increased from $250,000 to $8,000,000. The growth phase of his tenure also included an increase in the professional staff to full strength for the first time in Arboretum history. He had continuing and growing financial support from the FOA, which, when combined with outside funding, enabled him over the next twenty years to add both staff and a series of pioneering public education programs. The objective all along, in keeping with the "Wisconsin Idea," was the deepening of the Arboretum's public outreach program, a brand-new, university-sanctioned direction that underscored the Arboretum's future teaching and research mission through the 1980s, the 1990s, and into the present.

Evidence of the new emphasis became apparent as early as February 1984 when Armstrong enlisted the voluntary assistance of Jean Rideout, a retired UW librarian, to help set up three McKay Center libraries: a browsing library for the public, a more restricted general reference library, and a staff library on restoration ecology. In 1984, Armstrong also had enough money to hire Donna Thomas, Dame Fortune personified. Thomas, an art history specialist with a background in adult education and a penchant for successful grant writing, was the Arboretum's first education coordinator, responsible for classes, lectures, exhibits, and tours as well as an impressive series of successfully funded grant proposals. In 1991, Thomas was named assistant to the director.

In winter 1984, under Thomas's direction, a series of in-house nature-based public lectures were offered on a trial basis. In fall 1985, Gene Glover enlisted volunteers in an even more ambitious Arboretum-sponsored speaker's bureau that Thomas organized as part of a new public outreach program. The speaker's bureau provided a variety of specialists for public lectures scheduled throughout the year. In 1988, the Arboretum also advertised a "Winter Guide Enrichment Program" in a colorful flyer intended for public circulation. The program, essentially, was a continuation of the instructional program initiated as early as 1966 by Rosemary Fleming, designed for the winter training of guides and naturalists. Twenty-three years later, in 1989, the series, still intact, but now advertised as the "Winter 1989 Enrichment Program for Naturalists," continued the tradition of limiting enrollment to Arboretum volunteers and persons involved in environmental education programs. For the 1989 winter series, however, for the first time, attendance was extended to the general public, but "on a space-permitting basis." An enrollment fee of five dollars was charged to cover administrative costs.

The public response to that 1989 program was impressive. Small wonder, given the appeal of the topics presented. The list for the cold, shadowy winter months of January through March 1989 included "An Introduction to Madison Area Natural History Tour Programs," followed by talks by field specialists on "Ticks and Lyme Disease in Wisconsin," "Peregrine Restoration," "Wisconsin Moundbuilder Indians," "Bat Conservation," and "Wisconsin Butterflies and Their Habitats." In 1990, likely as the result of the unexpected 1989 sellout, the restrictions on attendance at the winter lectures were dropped completely. The 1990 winter program, now partly funded by the FOA, was free and open to the public. The program, with its open admission and its traditional preliminary session on phenology, was the most immediate predecessor to the currently popular Winter Enrichment Lectures, made available to the public annually from January to March for a slight charge.

In fall 1986, a bountiful time indeed for Armstrong and the Arboretum began in earnest. Enough money was available to hire five talented naturalists who were added to the expanding staff of educators, deepening considerably the Arboretum's public outreach commitment. The new naturalists— Sylvia Marek, Ken Wood, Ursula Peterson, Molly Fifield Murray, and Sherrie Gruder-Adams—were assigned to different areas of the Arboretum from 1:00

to 4:00 each weekend afternoon in order to advise visitors and answer questions. They also conducted regularly scheduled tours.[31] By 1987, Gene Glover, now carrying the weighty title of coordinator of the Friends and volunteers of the Arboretum—most of the volunteers were FOA members—would have an assistant, Linda Bishop, who was hired specifically to work on FOA membership.

As time passed, good news even got better. In May 1990, under Glover's supervision, the first FOA Native Plant Sale was held. The plant sale marked the beginning of a twenty-year tradition that continues to expand in volume and has become one of the Arboretum's most successful outreach projects. In July 1993, Glover was succeeded as FOA coordinator by wildlife biologist Sara Minkoff.

In 1987, Gene Moran retired as field supervisor, after thirty-two years of service. In 1989, Steve Glass was hired as operations supervisor, replacing Dick Slavik. Glass, with graduate degrees from the UW, initiated an educational program designed to enhance the field crew's understanding of the broader significance of their work, including the seasonal importance of regular prairie burns and of maintaining the restored plant communities. He also conducted workshops on restoring and maintaining diversity in prairie and savannah remnants. In 1991, he organized a Restoration and Management Workshop for Wisconsin land managers.

A major change in the Arboretum administration also took place in 1989 that quite unexpectedly further enhanced the Arboretum's fortunes. The Arboretum was moved administratively from being a division of the university reporting to the chancellor to being an academic program reporting to the dean of the Graduate School. As Armstrong noted, the move occurred mainly because the Arboretum, which in the past had been considered a facility—a "building" on the campus—was now identified, owing to the new direction the Arboretum had taken since the Bradley years, with its public outreach and educational programs. Chancellor Donna Shalala informed Armstrong of the status change in July 1989. Her rationale was based on the conclusion, drawn from its recent history, that the Arboretum's core activities by 1989 had actually become teaching and research, both of which were fundamentally "academic."[32] Of major concern when the Arboretum became part of the Graduate School was the potential for a damaging cut in annual funding. Initially, Armstrong was skeptical and worried. But a steady flow of grant money throughout the 1990s, beginning with a series of yearly grants acquired by Thomas from the Institute for Museum Services (IMS), enabled the Arboretum not only to maintain the status quo but also to expand its educational programs and to add more positions to the professional staff.[33]

In 1991, a Wisconsin Environmental Education Board (WEEB) grant enabled Jordan, Kathy Morgan, Brock Woods, and Molly Murray to increase the number of naturalist and enrichment programs being offered and to begin work on a teacher training initiative for the promotion of environmental education in schools. The concentration on the process of teaching restoration ecology, first advocated by Armstrong in March 1984, began to produce results. Brian Bader began work on a native plant propagation program, and Rodney Walter began work on an ecological survey of vegetation in the Curtis Prairie. Also, there was enough funding to hire Pat Brown, a professional journalist, who was added to the staff to oversee Arboretum public relations and to assume the editorship of *NewsLeaf,* the Arboretum's in-house publication.[34]

In 1991, Jim Fitzgibbons, ranger-naturalist since 1986, was also added to the staff to assist Brock Woods in a new outdoors work-study program that would evolve, by 1994, into a model training program for Arboretum stewards and outdoor work party volunteers. Fitzgibbons trained over two dozen stewards—Arboretum ambassadors, who patrolled the grounds and educated visitors on the history and mission of the Arboretum as well as on its rules and regulations. The outdoors work-study program also eventually evolved, under the direction of Gina Kline and Sharon Johnson and with the enthusiastic support of Armstrong, into the present Earth Partnership Field Program, involving regular weekend work parties and managed since 2004 by Marian Farrior, the talented and tenacious Arboretum Earth Partnership field manager. The program engages volunteers in regularly scheduled work parties cutting brush, removing invasive plants, collecting seeds, preparing areas for prescribed burns, and restoring plant communities. The work program also provides volunteer naturalists with an opportunity to gain more specialized training in restoration techniques by joining the Arboretum's elite Habitat Restoration Team.

In 1991, Molly Murray, Arboretum naturalist since 1985, was appointed Arboretum tour guide and program planner. Soon after, supported by an IMS grant, she began work overseeing the Ho-nee-um/Wheeler Council Ring oak savanna restoration project on the Monroe Street corridor. The Ho-nee-um Pond area had once been part of a savanna surrounded by a prairie and wetlands. The project plan called for restoring the savanna under the trees and down the bank leading to the shore of Lake Wingra. The area was cleared of invasive plants and shrubs and then replanted with over 360 native plants.

The project, which took five years, required "many hands."[35] Consequently, in 1992, in an ingenious move, a specialized course-oriented training program was instituted in conjunction with the project, called the Wingra Oak Savanna Project, with Murray as the training program consultant. The project was funded by another IMS grant and provided instruction in the conservation of oak savanna communities and savanna restoration techniques, including site analysis, fieldwork safety, and field first aid.

In 1993, Murray, replacing Brock Woods, was named Arboretum education and earth-keeping manager. In 1993, she also published, with the help of a grant from the UW–Madison Center for Biology Education, a book entitled *Prairie Restoration for Wisconsin Schools*. The book, designed primarily for public school teachers, explained, in conjunction with Leopold's "land ethic," how to gain a better understanding of the land by restoring small plots in schools and backyards. In May 1995, as the Wingra Oak Savanna project leader, Murray accepted a Special Citation Award presented by Mutual of Omaha to the Arboretum and to the Dudgeon–Monroe Street Neighborhood Association in recognition of the unique relationship that the Arboretum shared during the project with the neighborhood volunteers who assisted the Arboretum naturalists. By 1995, Murray was managing the Earth Partnership for Schools (EPS) Program. In June 1997, she also assumed the title of Arboretum education coordinator. She currently serves as the Arboretum outreach and education coordinator.

The late 1980s and the 1990s were years highlighted by the development of the Arboretum's now highly successful and internationally known Earth Partnership Programs. Just how the Arboretum got into the business of developing restoration programs in environmental education exclusively for urban public schools is an interesting story. The story begins in 1988, four years after Armstrong's statement of purpose clarifying the Arboretum's mission for the future. In 1988, administrators from Madison's Aldo Leopold Elementary School requested help with the development of an environmental education program. In response, the Arboretum staff developed a model program that included teacher training at the Arboretum and instruction in the curriculum and development of ecological resource material. The effort culminated in the actual creation of a prairie on Leopold School grounds.

The environmental program marked the beginning of an Arboretum-supported effort to promote school gardening and was aided, in the early stages, by a grant from the UW Center for Biology Education. By 1991, the school gardening program was known officially as the Earth Partnership for Schools Program. EPS, currently being managed by Cheryl Bauer-Armstrong, continues to train teachers and other people in nature-based occupations in methods for engaging students and local citizens in urban environmental work aimed at "healing the land," particularly on school sites and in land-damaged local areas. In 1990 and 1992, two grants from WEEB enabled the Arboretum staff to survey public school teachers on their use of natural areas as sites for environmental education and to offer workshops and in-service programs designed to show teachers how to incorporate gardening activities into their courses. In 1994, the EPS program received a third grant from WEEB for the development of additional curriculum materials that would broaden the scope of course activities.[36] Also in 1994, a $500,000 National Science Foundation (NFS) grant provided the EPS program with a solid financial base upon which to expand.

The NSF grant enabled the EPS staff to offer, in 1995, for the first time, a summer Arboretum course. The course enrolled sixteen teachers from eight Madison area schools. The teachers received specialized instruction in prairie gardening efforts. In 1996, the second EPS summer program enrolled teachers "by the pairs" from participating schools. As a course requirement, the paired teachers returned the following summer, in 1997, with four more teachers. The object was to increase the EPS summer course enrollment by multipliers, and it succeeded. By 1998, thirty-five Wisconsin schools were represented and were experimenting with prairie projects.

In 1998, Mark Leach, who had been hired in 1996 as Arboretum ecologist to replace Gina Kline, began the publication of a series of eight articles in *News-Leaf* that explored, in conjunction with the Arboretum's new commitment to public education, "what it means to think of our cities, towns and suburbs as land communities to which we belong." Leach encouraged his readers to assess their neighborhood ecosystems and to decide how to heal damaged land, an effort, he urged, that would help to provide healthy growth opportunities for "native plants, animals and ourselves." Leach's message reinforced ongoing EPS efforts to introduce conservation to urban neighborhoods by encouraging not only school projects but now also backyard projects.[37] It was a novel idea. In 2004, an Environmental Protection Agency grant funded additional professional development for the promotion of scientific inquiry in EPS classrooms statewide. Another WEEB grant supported the adaptation of an EPS backyard rain garden curriculum for Wisconsin schools. In summer 2009, an Arboretum-sponsored EPS Leadership Institute enrolled sixty-seven participants in teams representing natural resource agencies (e.g., U.S. Forest Service, U.S. Fish and Wildlife), nature-based organizations, and public schools from California, Michigan, Maryland, Oklahoma, Puerto Rico, and Wisconsin.[38]

The 1990s for the Arboretum was a period of rapid growth. Increased funding assisted the hiring of additional personnel to staff new programs. In 1992, for instance, a year after the EPS pilot summer program enrolled local teachers, a corresponding increase in funding supported another program, but this time one designed exclusively for children. The program, called Earth Focus Day Camp, initially enrolled children from ages six to ten, with meetings scheduled three days a month during the summer. Eventually, the Earth Focus Program would enroll youngsters from preschool through middle school and would meet three days a week during the summer. Sue Bridson, a pioneer Arboretum guide and naturalist, was the program's first coordinator. She was replaced in 2001 by Nancy Schlimgen, a middle school science teacher. Currently, the program, which provides a basic exploration of Arboretum ecosystems and involves a talented cadre of Arboretum guides, is directed by Jennifer Mitchell and continues to be supported in large part by outside funding, at present, by the Valerie Kerschensteiner Fund.

Grant funding continued throughout the 1990s, and so did the endowments, three of the largest coming in 1993. Ruth Gardner Reese, Louis Gardner's daughter, funded a project in 1993 designed to restore the Gardner Marsh to its original deep water status, complete with native plants and animals. The marsh was also to be used as a research lab for wetland studies. Construction of a boardwalk made it accessible to the public. The boardwalk provided the first foot access to an area where in the 1930s Leopold and McCabe had burned off large sections of marsh to create feeding stations for migratory game birds.[39] Also in 1993, the Teal Pond wetland boardwalk opened. The 387-foot boardwalk, a gift from George W. Icke, was dedicated to the memory of his father, John F. Icke, an engineer who served on the Arboretum's first Advisory Committee in 1933. On September 19, 1999, the Arboretum was gifted with a brand-new council ring, funded by Margaret Van Alstyne, who had been an Arboretum naturalist guide. The ring, dedicated formally as Margaret's Council Ring, is located south of the McKay Center parking lot on the northern edge of the Curtis Prairie.

By 2000, a new wing for the McKay Center and a new native plant garden, financed entirely by donations from foundations, individuals, businesses, and the FOA, were underway. In 2001, a month-long grand opening celebration of the new wing, which included an auditorium, an exhibit hall, a bookstore, a library, an orientation theater, and offices, commenced on September 8. By 2002, the Arboretum Native Plant Gardens project was making noticeable progress. The five-year garden project, when finished, covered 4 acres around the McKay Visitor Center with more than 350 species of native Wisconsin trees, plants, shrubs, and herbs, all designed to show how native plant communities can be adapted to different southern Wisconsin environmental conditions. In 2004, Susan Carpenter, one of Grant Cottam's graduate students, was hired as the Arboretum's Wisconsin native plant gardener. The gardens, designed originally by former UW professor of landscape architecture Darrel Morrison, offer a labeled native plant collection to visitors. In 2005, Susan Kilmer was named Arboretum horticulturist. Working with volunteers and UW students, she continues to manage, among other responsibilities, an extensive seed collecting and planting operation.

In 2002, the Pleasant Company's Fund for Children awarded a grant to the Arboretum's Earth Partnership for Families (EPF) Program for the support of programs designed "to promote a sense of wonder and exploration of the natural world for children." Earlier, in 1994, the Pleasant Company had funded the purchase of books and print materials about prairies and related topics designed to engage children in prairie lore. The primary objective of the EPF Program is the promotion of nature activities that families can experience together. The sessions, supervised through 2010 by the talented Arboretum naturalist and guide Kathy Miner, working with the most imaginative of the Arboretum's educational guide staff, concentrate on providing hands-on opportunities for children and adults to discover and explore the wonders of nature. The sessions also frequently involve storytelling as well as basic lessons on nature study and are currently among the most popular of the Arboretum's educational public outreach offerings.

And so, given the rationale behind the funding of most of the grants and endowments and the historical record of the years between 1980 and 2010, one must conclude that the shaping direction for the past thirty years at the Arboretum has been directed primarily at education and public outreach. What happened to internal scholarly research? It is still considered a viable part of the Arboretum's mission but one that is now often combined with the Earth Partnership programs. Faculty and graduate students, however, continue to be issued permits to conduct research in the Arboretum that involves mainly limited-time projects. Nature-based outside research projects undertaken by Arboretum staff are directed primarily at maintaining the restored and native plant communities that have defined the ecological substance of the Arboretum since at least the 1940s, the time of Leopold and Curtis. Armstrong would say that "because of their historical significance and particularly because of their potential as a research facility of major importance, the restored biotic communities are of the greatest importance. They are the thing that makes the UW Arboretum unique. They represent the first systematic attempt by humans to restore or heal the land. It is this very special relationship with nature, developed here at the UW Arboretum that is the essence of the place: the healing of the land."[40]

One of the most desired results of the restoration ecology initiative, from the beginning of the effort, had been the establishment of an endowed chair. Much to everyone's satisfaction, but especially Armstrong's and Jordan's, the hope became a reality in 1997–98.

Aldo Leopold Chair of Restoration Ecology

From the time of the inception of *R&MN*, William Jordan had the idea that, with support, an endowed restoration ecology chair might be a possibility. As early as April 25, 1980, he had asked members of the AC to consider creating an endowment that would fund an academic chair—an "Aldo Leopold Chair"—in connection with what was now, increasingly, the promotion of the Arboretum as the center for the study of ecological restoration, a concept that Jordan traced directly to Leopold's thoughts on the "land ethic" and on the value of re-creating, as much as possible, plants and habitat that existed before the time of agrarian settlement.

A subcommittee formed to consider the idea in 1980 concluded that the creation of an endowed chair "was a high risk venture because of the possibility of getting a 'lemon,' or someone who would come and then effectively retire: i.e., become inactive."[41] On April 6, 1981, Jordan again pursued the idea with the AC. As a result, Gerald Gerloff, AC chair, appointed yet another subcommittee, which reviewed the possibility and decided to take no action. Three years later, in 1984, an advisory panel on Arboretum research, chaired by botany professor John J. Ewel from the University of Florida, suggested, as its strongest recommendation, that the Arboretum seek support that would enable it "to hire an internationally known scholar for a new position as Director of Research" and that the scholar should be "physically located at the Arboretum." The report also noted that "the Arboretum's re-created ecosystems" had "immeasurable value commensurate with their uniqueness and age" and that these "communities and the research . . . carried out on them during the last five decades are the Arboretum's greatest strengths."[42] Following the advisory panel's 1984 report, Armstrong and Jordan decided to actively promote an endowed professorship of restoration ecology supported by earnings from a fund to be established with the UW Foundation.[43]

In 1985, the Arboretum, represented by Armstrong, Orrin Rongstad, the new AC chair, and Gerloff, the Research Committee chair, applied to the university for a Capital Campaign Endowment for the support of a restoration ecology research program that also included funding for the establishment of a named professorship of restoration ecology.[44] The request was turned down. Efforts to fund the chair, however, continued. In 1987, the Evjue Foundation donated $100,000 to an endowment specifically for the Aldo Leopold Professorship in Restoration Ecology. On February 6, 1989, Lloyd Bostian, AC chair, sent the following letter to Chancellor Donna Shalala:

> The major priority for the Arboretum is to expand and deepen its research program, primarily in restoration ecology. To this end, the Arboretum Committee has proposed establishment of an endowed professorship to attract an outstanding researcher to guide our research program. . . . Because funding this chair is exigent . . . the Committee voted unanimously at its February 3 meeting to ask Gregory Armstrong to pursue the feasibility of selling a 16-acre tract of land to provide a major portion of the funds needed to endow the chair. . . . Now is the time, we believe, to sell the land and to establish this research professorship in restoration ecology.[45]

There is no record in the Arboretum Archives of a written response from Chancellor Shalala. The tract of land that Bostian referred to was known as the Bishop Tract and was separated from Arboretum land in the southeast corner by commercial buildings and land holdings to the east. The tract originally was acquired in the 1950s from the Madison Roman Catholic Diocese and, hence, was called the Bishop Tract. When it was acquired it was farmland, but it was now of little use in terms of the Arboretum's commitment to anything, especially research and teaching.

Thus Armstrong and the AC wanted to sell the tract and to put the money in the endowment. In 1989, the Board of Regents approved the receipt of offers for the sale of the 16 acres and agreed that proceeds from the sale would be included in the endowment supporting the projected Aldo Leopold Professorship of Restoration Ecology. In 1994, the Livesey Company, a Madison development company, paid a million dollars for the tract and, as part of the federally required mitigation regulation, offered to provide $15,000 for the restoration of the Wingra Fen, the wetland located between the Nakoma Golf Course and Lake Wingra.

The proceeds from the sale—$1 million—made certain that there would be sufficient funding for the Leopold Professorship. In 1995, a $68,000 grant from John and Jane Secord increased the endowment. In 1997, the position was advertised. In late 1997, Professor Joy Zedler, the director of the Pacific

Estuarine Research Laboratory at San Diego State University, was offered the professorship. She officially joined the UW faculty in January 1998 as the first holder of the Leopold Chair of Restoration Ecology. Her time would be divided between the Arboretum and the Botany Department. A specialist in the study of wetland conservation, she chaired, from 2000 to 2001, the Committee on Mitigating Wetland Losses for the National Research Council, an extension of the National Academy of Science. Her commitment to the Arboretum and to Leopold's vision of returning the land to a pre-settlement ideal involved, and continues to involve, an active affirmation of the principle that ecological restoration begins with the soil. She would be a firm and vocal advocate of Leopold's "land ethic" and also of the belief that people do benefit in multiple ways from personal involvements in restoration efforts. Her support for the Earth Partnership programs, for the Arboretum's position as a center for the study of restoration ecology, and for the subsequent call for ethical commitments to reclaim, repair, and heal the land globally are duly appreciated by the Arboretum and the international scientific community.

Gregory Armstrong Retires; Kevin McSweeney Named Director

In February 2004 in a *NewsLeaf* letter to the members of the FOA, Gregory Armstrong publicly announced his retirement, effective in early July. He acknowledged that the mission of his tenure as director had focused on promoting restoration ecology and "the development of a positive relationship between humans and nature." He had also refined the Arboretum's mission to include a well-endowed research program and the development of a successful series of outreach programs. He explained that he wanted the FOA members to understand that he would go into his retirement viewing the past twenty years as director as having been "an exciting adventure." What he didn't realize when he wrote the letter, however, was that the adventure was not quite over, and that in his last days as director, during the evening of June 23, an early summer thunderstorm would provide some unwelcome excitement in the form of a tornado that crossed Lake Wingra and ripped through the Arboretum, felling or heavily damaging nearly a hundred oak trees in Wingra Woods. Ironically, the damage, which was extensive but contained, since the path of the tornado was narrow, actually served to hasten a natural transition, already underway, that was replacing the oaks with native, indigenous sugar maples, yellow birch, and hemlock, trees more at home in a north-facing northern hardwood forest than the fallen red oaks, many of which had been weakened by disease over time. Nevertheless, just the thought of a violent,

Kevin McSweeney, current Arboretum director. (Arboretum Photo Collection)

rotating funnel cloud tearing up the surface of the earth as it whirled through the Arboretum provided everyone connected with the Arboretum a pause for reflection on exactly how much damage could have ensued. The damage to the Wingra Woods ecosystem alone, particularly planted hardwood seedlings, would have been irreparable.[46]

Armstrong concluded his retirement letter to the FOA on a positive, uplifting note by assuring them that, "together, we have put the Arboretum in a very strong position to move . . . into the future, continuing to make important contributions to the world." What he did not tell the FOA, however, was that his decision to retire also resulted from his frustration and disappointment, after having been largely responsible for acquiring the endowments, at not being eligible, owing to his lack of faculty status, to vote for representation on the university screening committee for the Leopold Professorship.[47]

On July 1, 2004, Kevin McSweeney, UW professor of soil science and environmental studies and director of the School of Natural Resources, became the interim director of the Arboretum. A native of northwest England, McSweeney came to the United States in 1978 to pursue a doctoral degree in agronomy and soils at the University of Illinois. Much of his graduate research focused on the potential for restoration of disrupted soils in coal-mining areas. He joined the UW–Madison faculty in 1983. In August 2005, he was officially named Arboretum director. He would focus his administrative talents on three particular Arboretum needs: securing solid financial support for the Earth Partnership programs, expanding the size and capabilities of the field staff, and developing a storm water management plan capable of filtering and reducing the threat of over 470 million gallons of storm water annually devastating Arboretum wetlands and prairies.[48] He would also keep his sights on the 2009–2010 academic year and the seventy-fifth anniversary of the Arboretum's dedication, a celebration that would provide opportunities to engage the public and, as he put it, "to rally around a whole range of Arboretum activities," including backing an incentive to have someone write a new, updated history of the UW Arboretum.[49]

Reflections

In 2012 the Arboretum will be eighty years old, and like an enduring octogenarian, its history over the years is measured by some regrettable losses but also by impressive gains. Early dreams of the Arboretum as a potential wildlife refuge and sanctuary for Leopold's migratory game birds faded almost totally after his death in 1948. Early hopes that the Arboretum in conjunction with the U.S. Forest Products Laboratory would become a federally funded experimental forest preserve also never materialized. Bud Jackson's beloved white oak has not survived into the new millennium, although it still stands, bare branches visible and still firmly extended, on the western edge of the Curtis Prairie, a fading memorial to the "white knight"—that tall elderly gentleman who wore a fedora and walked with his cane for thirty years through Arboretum lands.

But other Arboretum dreams and hopes have fared better. Fears that the Arboretum, for the sake of public appeasement, might become an "ordinary park" never were realized. By the late 1930s, the road providing access to the Arboretum's interior from both east and west became a reality, much to Bernie Chapman's relief and to the benefit of the home owners living within the Arboretum in the sole surviving Lake Forest community. The institutional commitment to restoration ecology that originated during the William Jordan years continues, but the focus over the past twenty years has shifted from a promotion of the Arboretum as a center for the promulgation of restoration ecology to a conception of the Arboretum as a center primarily for the promotion of educational techniques in the practical recovery and, where it applies, restoration of the land.

John Nolen and Michael Olbrich surely would see in today's Arboretum the realization of their dream of "a living laboratory in everything that has to do with wild life." Norman Fassett, Aldo Leopold, and John Curtis would approve of the natural design and bounty of the restored plants native to Wisconsin that now cover the Arboretum's tall grass prairies and forests, including recently a Native Plant Garden that continues to grow in variety. Edward Gilbert would have been pleased with the realization of so many of G. William Longenecker's aesthetic landscape design successes. By the late 1950s, most of Leopold's and Robert McCabe's pheasants were gone. The quail and duck populations were diminishing. Yet, the descendants of Bill Feeney's woodcocks continue to appear every spring to play out their aerial mating dance, to the delight of onlookers. And since the late 1970s, the Arboretum has become home to another game bird not included in its early twentieth-century provenance—a growing, perhaps even "exploding," flock of wild turkeys. Since their revival, migrating sandhill cranes now occasionally lay over for a time, sometimes long enough to foster their young, before continuing their migration north.[50] They also often can be seen in the early spring lingering in the prairies and the gardens or staring at their reflections in the Visitor Center's windows before continuing their journey back to the Gulf Coast or to the wetlands of Florida and southern Georgia. Within the past six years, perhaps owing to the population increase in bald eagles to the north, osprey, feeling the competition, have been seen and have nested near the Lake Wingra shoreline.[51]

Longenecker's blooming lilacs, magnolia trees, and blossoming crabapples, first planted in 1935–36, still draw thousands of visitors in the early spring. In winter, sturdy robins and great blue herons can still be found near the open springs of the Lake Wingra shoreline. The pileated woodpeckers still bore very large, noticeable holes in the trees in Wingra Woods. The spring wildflowers still arrive on time in Gallistel Woods, as do the red-winged blackbirds and an occasional green heron in the numerous bordering marshes. Annually, by April, the barn swallows return and nest under the eaves of the McKay Visitor Center and, just as predictably, by early September they disappear. Vestiges of the CCC days are still evident along the old service road. Although nine of the original barracks are gone, victims of time and regretful neglect, one barracks, two garages, and the old CCC bathhouse have survived and are now celebrated as historical centerpieces during Arboretum history tours. Albert Gallistel, the architect who designed the camp in 1934, would be pleased. A commemorative plaque set into a large rock that sits on the edge of the Native Plant Garden now memorializes the hundreds of young CCC enrollees who worked in Camp Madison from 1935 to 1941.

The forest of pines, red maple, and white birch bordering the southern and western edges of the Curtis Prairie, the Arboretum's first serious forestation effort, proposed in 1933 by Ed Gilbert and supervised by Longenecker while he was still a graduate student, continues to form a protective shield against the noise and much of the pollution from roads surrounding the Arboretum, including the offensive Beltline, Grant Cottam's "cement desert." Yet, the natural rhythm of life at the Arboretum beats on in spite of invasive threats, or budget cuts, or crucial staff losses. And all one needs to do to feel that rhythm is to make that turn, either from Mills Street or Seminole Highway, into the Arboretum to where the tree canopies begin. Within seconds, although less than 300 feet from the city, the light drops, the shadows descend, the noise subsides, the quiet envelops. Perhaps the turkeys are walking the road, belligerently slowing down traffic and life. Perhaps a piercing "scream-like" sound signals a red-tailed hawk soaring out of sight high in the air. Perhaps, if one looks deep enough into the woods, deer or a red fox can be spotted. If it's dusk, one might even hear a distant series of "yipping" sounds—coyotes, relatively recent Arboretum residents, on the move.

The past five years have seen the rediscovery, thanks to the Arboretum's new Interactive Map system, of effigy mounds in the Lost City Forest that were originally located by Charles E. Brown and singled out for preservation as parks in the 1920s by Bernie Chapman and the Lake Forest Land Development Company. The location of the mounds was lost during the years of negotiation for the Lake Forest lots. The Lost City Forest itself is now more overgrown than ever. Remnants of concrete from the company's failed housing development—the ill-fated "Venice of the North"—have become harder to spot. Only a few experienced Arboretum guides know where the old abandoned house foundations and concrete steps are found. Ex-residents of the neighborhoods close to the Lost City Forest recall old times when kids rode bikes on the concrete leading into the forest trails and young lovers spooned in parked cars in the abandoned streets. Now, on occasion, a Lost City adventurer bushwhacking through the forest, against all regulations, finds a remnant in the woods—an old open cistern, for example—and triumphantly reports it to the field staff, who must then fill it in. The Lost City Forest continues to draw many hikers and joggers down its trails and into its depths, some of whom, once turned around in the maze of trails, actually do get "lost."

One very noteworthy development that now characterizes the Arboretum, setting it apart in many ways from its early days, is its conscious and acknowl-edged commitment to children. An Arboretum as a place for children to visit, to enjoy, to walk through the woods and the prairies, and to learn about na-ture was what Olbrich and Paul Stark originally had in mind. As early as 1928, Olbrich, in his famous speech to the Madison Rotary Club, was promoting his dream of a natural area on the edge of the city that "would cater to the needs of all," but that "would mean the most . . . to the children." And how could the children, he proposed in that early address, "be more effectively educated in the love of the beautiful than by the creation and preservation of a wonder-land of natural beauty here at the capital of the state."

Because of the Arboretum's Earth Partnership for Families bi-monthly pro-grams that are designed for children, the Earth Focus Day Camp for children in the summer, and the annual tours for hundreds of school children and their teachers, the Arboretum has indeed become Olbrich's "wonderland of natural beauty" for children. For at least the past forty years, thousands of children from the Madison area, many of whom are now adults with children and grandchildren of their own, have had the privilege of meeting the Arboretum's dedicated nature guides under whose tutelage they search for turtles at the Teal Pond, study the formation of snowflakes, learn to identify birds, learn the history of the Arboretum's effigy mounds, ponder how glaciers have shaped the land and why leaves change color, study animal tracks in the mud and the snow, and see up close the plants and wildflowers that herald the changes in the seasons.

But the Arboretum's wonderment and beauty are not just there for children. Adults also come to the Arboretum regularly for weekly guided tours and walks, for regular lectures and nature-based presentations, for cross-country skiing or snowshoeing, for jogging on McCaffrey Drive or on the trails. In the warm seasons, local running events often include the Arboretum road as a link between downtown and West Side bike or running trails. The Arboretum's Earth Partnership for Schools Program, which shows teachers and naturalists how to incorporate restoration activities into a teaching curriculum, offers a two-week adult workshop annually in the summer.

Most of the Arboretum's adult visitors, however, are solitary nature lovers who come alone or with friends to walk the trails, to stroll around Longe-necker Gardens or the Native Plant Garden to see what's blooming or what's coming into season, or to wander through the prairies to check out the height of the grasses or the flowering prairie plants, some eight to ten feet tall by September, or just to marvel at the changes in color. For many visitors who have never seen or walked through a tall grass prairie, the Arboretum prairies, first worked on so many years ago by Fassett, Sperry, Curtis, and Greene, now offer a fulfilling and rich visual treat, particularly during late autumn. Visitors come long distances to see the prairie flowers, including the incredible profusion of fall color in the collection of prairie asters. Some Arboretum hikers are avid birders or just earnest watchers, there to see what they can see that they can't see anywhere else in the city. As Arthur Godfrey, himself an earnest bird-watcher, reported in 1967, "You should see the birds. I counted personally 12 different species that I have never seen before. . . . Sit down and watch. . . . Pretty soon the wildlife starts to come around." Anyone coming to the Arboretum—a bird lover, a budding naturalist, a cross-country skier, a nature photographer, a professional botanist or landscape architect, or just someone who wants to find a quiet spot to relax and to put the miasma of life's taxing problems temporarily out of mind—will soon find what Olbrich saw as the essential appeal of the nature preserve he had in mind in 1928, for it is a place, as the memorial plaque to his memory at the western entrance records, that brings "back into the lives of all confronted by a dismal industrial tangle, whose forces we so little comprehend, something of the grace and beauty which nature intended all to share."

Such a place deserves to be cherished, protected, and supported. And so we are well advised to do whatever is necessary to guarantee that this "treasure in our midst," the University of Wisconsin Arboretum, with its eighty-year legacy, in the words of the city of Boston's 1882 pledge to Harvard's Arnold Arboretum, will also always remain ours "to have and to hold . . . for one thousand years . . . and so on from time to time forever."

Notes

Chapter 1. The Beginnings

1. Hancock, *John Nolen, Landscape Architect*, 16; see also, Mollenhoff, *Madison: A History of the Formative Years*, 324–35.

2. Long, "John Nolen," 26–30.

3. Nolen, *Madison*, 71.

4. UW Archives, 38/4/8 Box 3. The extensive Arboretum archival collection in Steenbock Library on the University of Wisconsin–Madison campus is cataloged under the general heading "Record Group 38." The materials in the collection are contained in boxes listed in series order extending through eighty labeled categories beginning with 38/00/1 (Arboretum Newsletter 1943–1944) and ending with 38/12 (Sachse, Nancy D.—Papers 1960–1976). Items within the individual boxes are not numbered or cataloged. Document citations for this history include the series number and the box number where the documents are located.

5. Ibid., 38/3/2 Box 4.

6. "Michael Balthazar Olbrich," 1–2.

7. Sachse, *A Thousand Ages*, 15–18.

8. UW Archives, 38/4/8 Box 3.

9. Ibid., 38/3/2 Box 1.

10. "Dedicated Arboretum Family."

11. UW Archives, March 23, 1925, 38/4/8 Box 3.

12. Ibid., 38/3/2 Box 3.

13. Ibid., 38/4/8 Box 3.

14. Ibid.

15. Levitan, *Madison*, 178.

16. J. Stephen Tripp, a wealthy banker from Prairie du Sac, Wisconsin, died in 1915 and left a large endowment to the University of Wisconsin. In 1927, $83,000 was officially set aside for the future development of the Arboretum but only after

the university acquired the land, which didn't happen until 1932. See "Excerpts from Minutes of Meetings of the Board of Regents with reference to the Arboretum," sent to A. F. Gallistel on November 2, 1945, UW Archives, 38/1/11 Box 2.

17. "Assurance of Arboretum in OK on Budget."

18. UW Archives, 38/3/1 Box 1.

19. Hubbell, "'U' Botany Students Make Plant Survey on Proposed Arboretum Site"; italics mine.

20. Lapham, "The Forest Trees of Wisconsin."

21. UW Archives, 38/4/3 Box 3.

22. UW Archives, 38/4/8 Box 3.

23. Olbrich, "Speech before the Madison Rotary Club, May 17," UW Archives, 38/4/8 Box 2.

24. The quotation on the Olbrich Memorial Entrance plaque on the Arboretum's west end is an abbreviated version of these now famous lines. The revision was suggested by Aldo Leopold. See Leopold's letter to Joseph Jackson, February 3, 1939, UW Archives, 38/4/8 Box 2.

25. Ibid.

26. Jones, "Jackson, the Man behind the Mighty Oak," 2.

27. Levitan, *Madison*, 243.

Chapter 2. 1930–1932

1. Fred, "Dedication of McCaffrey Drive."

2. UW Archives, 38/3/2 Box 1.

3. The Plough Inn, 3402 Monroe Street, was constructed in 1853.

4. UW Archives, 38/4/8 Box 2.

5. Levitan, *Madison*, 190.

6. Unfortunately, there is to date no extant evidence that Wright designed the stone wall, although Nancy Sachse claims that he did (*A Thousand Ages*, 47); but the wall collapsed and was later restored (Sachse, "Madison's Public Wilderness"). Charlotte Peterson, in an article on the question, emphasizes the lack of documentation confirming that Wright designed the walls inside Spring Trail Park as well as the walls and steps across from the Old Spring Tavern ("Frank Lloyd Wright Walls in the Arboretum").

7. UW Archives, 38/4/8 Box 3.

8. Ibid.

9. UW Archives, 38/3/2 Box 3.

10. Sachse, *A Thousand Ages*, 16.

11. UW Archives, 38/4/1–7 Box 1.

12. Griffin, "Found in Garage," 4.

13. UW Archives, 38/4/3 Box 3.

14. Ibid.

15. Roberts, *Opinions of the Attorney General*, 631–32.

16. UW Archives, 38/4/8 Box 3.

17. Ibid., NPS Memo, UW Archives, 38/7/3 Box 4.

18. The Barlett-Noe property was particularly adaptable to prairie restoration because only the western two-thirds had been plowed and planted since about 1863. The eastern third likely was never plowed. The northern half of the unplowed section was only lightly grazed. The southern half was a mowing meadow. The Bartlett family abandoned cultivation in 1920, and the land remained fallow until 1926 or 1927 when a veterinarian named West leased the land until 1932 as pasture for thirty-five to forty horses.

19. UW Archives, 38/4/1–7 Box 1.

20. Doehlert, "Do Ghosts Walk Arboretum Glades?"

21. UW Archives, 38/3/2 Box 1.

22. Ibid.

23. Ibid., 38/3/2 Box 3.

24. Ibid., 38/3/1 Box 1.

25. Ibid., 38/6/3 Box 1.

26. Ibid., 38/4/1–7 Box 1.

27. Ibid.

28. Ibid., 38/3/2 Box 1.

29. Ibid., 38/4/1–7 Box 1.

30. Ibid., 38/2/12–16 Box 4.

31. Ibid., 38/3/2 Box 2.

32. Ibid., 38/3/2 Box 3.

33. See memo from Leopold, January 13, 1936, UW Archives, 38/3/1 Box 1; 38/7/3 Box 5; Sachse, *A Thousand Ages*, 45–47.

34. Sachse, *A Thousand Ages*, 45.

Chapter 3. 1932–1934

1. Simonds quoted in Olbrich, "Speech before the Madison Rotary Club."

2. "Regents Approve Plan for Wingra Arboretum"; "Announce Establishment of Wisconsin Arboretum," AA.

3. Foss, "The Arboretum," 123, 144.

4. UW Archives, 38/3/2 Box 1.

5. Ibid.

6. "Dedicated Arboretum Family."

7. UW Archives, 38/3/1 Box 1.

8. Ibid.

9. Ibid., 38/3/2 Box 1.

10. Ibid., 38/3/1 Box 1.

11. Sachse, *A Thousand Ages*, 24–26.

12. UW Archives, 38/7/3 Box 6.
13. See Brown's letter to Leopold and Jackson's letter to Nolen, January 29, 1934, UW Archives, 38/3/1 Box 1.
14. UW Archives, 38/4/1–7 Box 1.
15. Ibid., 38/3/1 Box 1.
16. Leopold, "The Chase Journal."
17. UW Archives, 38/3/1 Box 1.
18. Ibid.
19. Ibid.
20. See Jackson's letter to Aaron M. Brayton, August 8, 1933, ibid., 38/3/1 Box 1.
21. See Jackson's letter to M. J. Gillen, March 10, 1933, ibid., 38/3/2 Box 1.
22. Ibid.
23. Ibid.
24. "School Forests."
25. UW Archives, 38/3/1 Box 1.
26. Ibid., 38/1/11 Box 2.
27. Ibid., 38/3/2 Box 1.
28. See Gilbert's letter to Jackson, August 26, 1933, ibid., 38/3/1 Box 1.
29. Gallistel to Gilbert, August 28, 1933, ibid.
30. Ibid., 38/1/11 Box 2.
31. Ibid., 38/3/1 Box 1.
32. Ibid.
33. Hall, *Earth Repair*, 171.

Chapter 4. 1934–1935

1. Hall, *Earth Repair*, 171–72.
2. Ibid., 172.
3. Meine, *Aldo Leopold*, 313–14.
4. Ted Sperry, Arboretum Collection Tape (AC Tape) 020, September 23, 1981.
5. Sperry, AC Tape 109, September 13, 1990.
6. Jackson, "Dedication of the Aldo Leopold Pines."
7. James Hendricks, AC Tape 007, July 1, 1981.
8. Sachse, "G. William Longenecker, Master Planner," [3].
9. Hasselkus, "The Longenecker Horticultural Gardens."
10. Cottam, "Professor G. William Longenecker."
11. Sperry, AC Tape 109.
12. "Arboretum Residence Demolished."
13. See Leopold's letter to Paul V. Brown, NPS Regional Director, February 10, 1940, UW Archives, 38/3/1 Box 1.
14. Lorbiecki, *Aldo Leopold*, 121.
15. Leopold, "Conservation Ethic," 185.

16. Meine, *Aldo Leopold*, 320.

17. Leopold, "Conservation Economics," 196.

18. Foss, "The Arboretum."

19. Tarkow, "In the Beginning . . . Sixty Years Ago."

20. UW Archives, 38/3/2 Box 1.

21. "Owls Help 8 Speakers Dedicate U. Arboretum."

22. For draft of Leopold's speech, "What Is the Arboretum," see UW Archives, 38/3/1 Box 1.

23. Ibid.

24. Leopold, "What Is the Arboretum?"; Leopold, "Conservation Ethic."

25. Leopold, "What Is the Arboretum?"; "Owls Help 8 Speakers."

26. "Owls Help 8 Speakers;" "Dedication Speakers Vision Arboretum."

27. UW Archives, 38/3/1 Box 1.

28. Nolen, "Address. . . at Dedication," UW Archives, 38/3/2 Box 1.

29. Ibid.

30. See Jackson's letters to John Nolen, January 25, February 13, April 30, 1935, UW Archives, 38/3/2 Box 1.

31. UW Archives, 38/3/1 Box 1.

32. Ibid., 38/3/2 Box 1.

33. Ibid..

34. Meine, *Aldo Leopold*, 314–19.

35. UW Archives, 38/3/2 Box 1.

36. "Wingra Marshes Are Designated as Game Refuge."

37. UW Archives, 38/3/1 Box 1.

38. Meine, *Aldo Leopold*, 351–52, 362–63.

39. UW Archives, 38/3/1 Box 1.

40. See Leopold's letter to Howard I. Potter, UW Archives, 38/1/11 Box 1.

41. UW Archives, 38/3/1 Box 1.

42. Newton, *Aldo Leopold's Odyssey*, 2006, 147–48.

43. UW Archives, 38/7/3 Box 9.

44. Harold Madden, AC Tape 016, September 17, 1983.

45. UW Archives, 38/3/2 Box 1.

46. See " Arboretum Argus," 3; see also John Wawrzaszek, AC Tape 023, September 17, 1983.

47. Sachse, *A Thousand Ages*, 31–32.

48. UW Archives, 38/3/1 Box 1.

49. Ibid., 38/2/12–16 Box 4.

50. Ibid., 38/3/1 Box 1.

51. "Arboretum Argus," 3, 5.

52. UW Archives, September 24, 1935, 38/3/2 Box 2.

53. James Hendricks, AC Tape 007, July 1, 1981.

54. Robert Herbert, AC Tape 009, September 17, 1983.

55. Eugene Adler, AC Tape 011, September 17, 1983.

56. Egan, *Worst Hard Time*, 157–58.

57. Theodore Rathje, AC Tape 018, September 17, 1983.

58. Sarah Longenecker, AC Tape 145, July 31, 1986.

59. Sperry, AC Tape 109.

60. Sperry, AC Tape 156, July 6, 1992.

61. Sperry, AC Tape 021, September 23, 1981.

62. Hendricks,, AC Tape 007.

63. UW Archives, 38/1/11 Box 1.

64. Ibid., 38/3/1 Box 1.

65. Ibid., 38/7/3 Box 4.

Chapter 5. 1935

1. Meine, *Aldo Leopold*, 375.

2. UW Archives, 38/3/1 Box 1.

3. Ibid.

4. See Blewett and Cottam, "History of the University of Wisconsin Arboretum Prairies," 131; "Curtis Prairie," 1; Fassett and Thomson, "Establishment of Prairie."

5. UW Archives, 38/3/1 Box 1.

6. "Owls Help 8 Speakers," 1.

7. Fassett and Thomson, "Establishment of Prairie."

8. UW Archives, 38/7/3 Box 10.

9. John Thomson and Grant Cottam, UW Oral History Tape 134, 1978.

10. Blewett and Cottam, "History of the University of Wisconsin Arboretum Prairies," 131.

11. Thomson, UW Oral History Tape 134.

12. Ted Sperry, AC Tape 109, September 13, 1990.

13. Ibid.

14. Sperry, AC Tape 021, September 23, 1981.

15. UW Archives, 38/7/3 Box 5.

16. Letter, Leopold to Ted Sperry, July 18, 1935, Sperry-Gladys Galliger Collection, Pittsburg State University Correspondence, Series July 1935. My thanks to W. R. Jordan III for sending me a copy of the letter.

17. UW Archives, 38/7/3 Box 5.

18. Ibid.

19. Waller, "Prairie 'Father' Takes Pride in His Baby."

20. Sperry, AC Tape 109.

21. Waller, "Prairie 'Father' Takes Pride in His Baby."

22. Sperry, "Reflections on the U.W. Arboretum," 8–9.

23. A series of six interviews with Sperry, taped between 1981 and 1992, are in

the UW Oral History Collection, the UW Archives, and the Arboretum Archival Files.

24. Sperry, AC Tape 020, September 23, 1981.

25. Ibid., AC Tape 021.

26. Ibid., AC Tape 020.

27. Kading, Adler, Hackett, AC Tape 011, September 17, 1983.

28. Sperry, AC Tape 109.

29. Ibid.

30. Ibid., AC Tape 156, July 6, 1992.

31. UW Archives, 38/1/11 Box 1.

32. Sperry, AC Tape 020.

33. Ibid., AC Tape 109.

34. Ibid., AC Tapes 109, 021, 152, 1992.

35. UW Archives, 38/7/3 Box 6.

36. Sperry, AC Tape 109.

Chapter 6. Late 1930s

1. UW Archives, 38/1/11 Box 1.

2. Ibid., 38/4/1–7 Box 1.

3. Sachse, *A Thousand Ages*, 57–58.

4. UW Archives, 38/3/2 Box 3.

5. Ibid.

6. Ibid.

7. Ibid., 38/3/1 Box 1.

8. "Careful with Fire!"

9. "$10,000 Fire Levels CCC Building"

10. UW Archives, 38/3/1 Box 1.

11. *Inland Bird Banding News* 12 (June 1940), [13].

12. See "Wild Life Accomplishments," UW Archives, 38/7/3 Box 1. In 1940, W. S. "Bill" Feeney was appointed by the WCD to survey Wisconsin deer, with an eye toward reducing their number. Feeney was also an early advocate, with Leopold, of the repeal of the state of Wisconsin bounty on wolves, which remained until 1957.

13. UW Archives, 38/1/11 Box 1.

14. Ibid.

15. Ibid.

16. Ibid.

17. Ibid..

18. Ibid., 38/3/2 Box 3.

19. "Twenty Thousand Dollar Gift to Arboretum."

20. UW Archives, 38/7/3 Box 6.
21. Ibid., 38/1/11 Box 1.
22. Curtis, "Information Bearing on Arboretum Policy," 1–2.
23. See Leopold, "Wilderness Values"; see also Leopold's letter to Conrad Wirth, July 17, 1940, UW Archives, 38/7/3 Box 7.
24. Sachse, *A Thousand Ages*, 56.
25. UW Archives, 38/1/11 Box 1
26. "Add 55 More Acres to U.W. Arboretum."
27. UW Archives, 38/4/1–7 Box 1.

Chapter 7. The 1940s

1. Burgess, "John Thomas Curtis," 1–11; Cottam, "The n-Dimensional Niche of John T. Curtis."
2. See Cottam, "The n-Dimensional Niche of John T. Curtis", 47–48; Cottam on Curtis, AC Tape 041, November 13, 1981.
3. Cottam on Curtis, AC Tape 041.
4. See Leopold's letter to Joseph Jackson, May 17, 1937, UW Archives, 38/3/2 Box 3.
5. UW Archives, 38/3/1 Box 1.
6. Meine, *Aldo Leopold*, 408–11.
7. "Story of Wisconsin's Last Remaining Virgin Prairie."
8. Partch, "The Distribution of Plants on the Faville Prairie."
9. Meine, *Aldo Leopold*, 313.
10. Ibid., 410.
11. Leopold, "Exit Orchis."
12. "Story of Wisconsin's Last Remaining Virgin Prairie."
13. "Prairie Tract Preserved by Badger Couple."
14. "Dedicated Arboretum Family."
15. The 92-acre Faville Grove Preserve, the first of the Arboretum's satellite properties, was acquired in 1945. Others, designated as state natural areas, were acquired later: New Observatory Woods, a small oak-hickory woods in Dane County, in 1956; the Finnerud Forest, a 110-acre woods in Oneida County, in 1957; Abraham's Woods, a 40-acre tract in Green County, in 1961; Lodde's Mill Bluff, a cliff in Sauk County, in 1967; and the Brunsweiler Forest, an 882-acre tract of hardwood forest in Ashland County, in 1969. The satellite holdings, amounting to a total of 1,365 acres, also include the Bolz and Pasque Flower Hill Prairies near Madison; the Oliver Prairie in Green County, which includes an undisturbed remnant dry prairie; and three wetland areas: the Hub City Bog in northern Richland County, Anderson's Bottomland, located on the Wisconsin River near the town of Arena, and McKenna Pond, a deepwater marsh, located near Cross Plains (Miner, "We're More Than Madison").

16. See Jackson's letter to the WARF Directors, September 21, 1936, UW Archives, 38/1/11 Box 1.

17. UW Archives, 38/2/1–6 Box 1.

18. Ibid., 38/1/11 Box 2.

19. Ibid., 38/2/1–6 Box 1.

20. Ibid., 38/1/11 Box 2.

21. Ibid., 38/2/1–6 Box 1.

22. Ibid.

23. Ibid., 38/1/11 Box 2.

24. Ibid., 38/3/2 Box 3.

25. Ibid., 38/1/11 Box 2.

26. Ibid., 38/2/1–6 Box 1.

27. Ibid., 38/3/1 Box 1.

28. Curtis, "Information Bearing on Arboretum Policy"; UW Archives, 38/1/11 Box 6.

29. Sachse, *A Thousand Ages*, 115.

30. Burgess, "John Thomas Curtis," 10–13.

31. Backus and Evans, "H. C. Greene," 997–98.

32. Allsup, "Greene Prairie," 3–4.

33. Backus and Evans, "H. C. Greene," 997.

34. "How Greene Is This Prairie?"; "Evolution of a Prairie."

35. Greene, "Notes on Revegetation"; Allsup, "Greene Prairie," 8.

36. "How Greene Is This Prairie?"

37. Ibid.

38. Kline, "Henry Greene's Remarkable Prairie," 36.

39. Sachse, *A Thousand Ages*, 98.

40. "How Greene Is This Prairie?"

41. Greene, "John T. Curtis."

42. Backus and Evans, "H. C. Greene," 994.

43. Hone, "Dr. Henry Campbell Greene."

44. Emlen and McCabe, "In Memoriam: Robert A. McCabe," 674.

45. Robert McCabe, in a 1977 interview, observed that as soon as he arrived on campus, he knew he "wanted to . . . work with Leopold . . . for here was not only a man of science, but a man of literature as well" (Haines, "Familiar Faces . . . McCabe," 5). Although McCabe was hesitant to discuss his relationship with Leopold, nevertheless, in 1987, for the centennial of Leopold's birth, he published his remembrances in *Aldo Leopold: The Professor*.

46. Emlen and McCabe, "In Memoriam: Robert A. McCabe," 675.

47. Haines, "Familiar Faces . . . McCabe," 4.

48. Emlen and McCabe, "In Memoriam: Robert A. McCabe."

49. McCabe, *Aldo Leopold*, 46.

50. UW Archives, 38/7/3 Box 1.

51. *Arboretum News Letter*, November 24, 1943; July 1, 1944.

52. Haines, "Familiar Faces . . . McCabe," 4.

53. McCabe, *Aldo Leopold*, 46, 50.

54. Blewett and Cottam, "History of the University of Wisconsin Arboretum Prairies."

55. Ted Sperry, AC Tape 109, September 13, 1990.

56. Peattie, "Norman Carter Fassett," 239.

57. UW Archives, 38/3/1 Box 1.

58. Hickey, [Tribute to Aldo Leopold].

59. See AC Minutes (ACM), March 13, 1945, UW Archives, 38/1/11. See also *Arboretum News Letter*, November 24, 1943.

60. *Arboretum News Letter*, November 24, 1943, and July 1, 1944.

61. Cronon and Jenkins, *University of Wisconsin*, 679.

62. *Arboretum News Letter*, July 1, 1944.

63. UW Archives, 38/2/7–10 Box 2.

64. Lenehan, "Aldo Leopold and the Lakeshore Nature Preserve."

65. Hasler, "Teaching Exhibits Which Should Be Installed."

66. Leopold, "Wildlife in the Picnic Point Program."

67. University Bay Committee, "Preliminary Detailed Development Program."

68. Lenehan, "Aldo Leopold and the Lakeshore Nature Preserve," 3–4.

69. October 4, 1940, AA.

70. UW Archives, 38/1/11 Box 2.

71. Ibid., 38/2/7–10 Box 2.

72. "U. Arboretum Acquisition of Nakoma Course Is Discussed."

73. UW Archives, 38/2/7–10 Box 2.

74. See Longenecker's letter to Maurice McCaffrey, August 20, 1945, AA.

75. See A. L. Masley's letter to McCaffrey, August 18, 1945, AA.

76. UW Archives, 38/1/11 Box 2.

77. See Sachse, *A Thousand Ages*, 105; see also "Annual Technical Committee Report," July–August 1947, UW Archives, 38/00/3 Box 2.

78. See "Arboretum Technical Committee minutes," 1947–48, AA.

79. See Curtis, *Arboretum Master Development Plan*, 1; ACM, February 26, 1948, May 23. 1949, June 27, 1949, UW Archives, 38/1/11.

80. Max Partch, AC Tape 095, September 16, 1980.

81. Meine, *Aldo Leopold*, 512, 506-20.

82. Cronon, "A Voice in the Wilderness"; Meine, *Aldo Leopold*, 520.

83. Buss, "The Passenger Pigeon."

84. Meine, *Aldo Leopold*, 522.

85. Errington, "Appreciation of Aldo Leopold."

86. UW Archives, 38/1/11 Box 3.

87. Ibid.

88. Ibid.

89. Ibid., 38/1/11 Box 5.

90. Jackson, "Dedication of the Aldo Leopold Pines," 3–5; Schorger, "Dedication of the Aldo Leopold Pines."

91. Leopold, *Sand County Almanac*, 70.

92. Fassett, "An Appreciation of Aldo Leopold"; see also Fassett's letter to Estella Leopold, April 28, 1948, Aldo Leopold Papers, 1903–1948, UW Archives, 10–8, 3.

93. Hickey, [Tribute to Aldo Leopold].

Chapter 8. The 1950s

1. Curtis, *Arboretum Master Development Plan*, 4.

2. Ibid., 9–12, 16–17, [30].

3. UW Archives, 38/7/3 Box 1.

4. Ibid., 38/1/11 Box 2.

5. Ibid., 38/1/11 Box 3.

6. Ibid.

7. Ibid.

8. Ibid.

9. Greene, "Retirement of J. R. Jacobson."

10. Sachse, *A Thousand Ages*, 101.

11. UW Archives, 38/1/11 Box 5.

12. Ibid., 38/1/11 Box 4.

13. AC Minutes (ACM), April 18, 1951, AA; UW Archives, 38/1/11 Box 6.

14. UW Archives, 38/1/11 Box 6.

15. Peattie, "Norman Carter Fassett," 241–42.

16. Ibid., 233–34.

17. See Peattie, "Norman Carter Fassett" and Thomson, "Norman C. Fassett."

18. Peattie, "Norman Carter Fassett," 238.

19. ACM, July 16, November 19, 1953, AA.

20. UW Archives, 38/1/11 Box 6.

21. Ibid.

22. ACM, October 15, 1953, AA.

23. ACM, November 25, 1953, in UW Archives, 38/1/11 Box 6.

24. Nelson, "Letter to the Editor—the Tragedy at Redwing Marsh"; see also Hale, "Redwing Marsh Site of Fatal Crash 39 Years Ago."

25. See Hale, "Redwing Marsh Site of Fatal Crash"; see also "Jet Crashes in Arboretum Marsh"; Dieckmann, "2 Truax Men Killed in Jet's Crash Here."

26. "Another Desiccated Autumn."

27. See "Fire in the Grady Tract"; see also ACM, March 11, 1954, in UW Archives, 38/1/11 Box 6.

28. UW Archives, 38/1/11 Box 6.

29. Ibid.

30. Ibid., 38/1/11 Box 7.
31. Ibid.
32. Ibid., 38/1/11 Box 6.
33. Cottam, "Dr. James H. Zimmerman."
34. "General Use of the Arboretum."
35. UW Archives, 38/1/11 Box 10.
36. Ibid., 38/1/11 Box 7.
37. [Greene], "Widening of Beltline Highway through Arboretum."
38. UW Archives, 38/1/11 Box 8.
39. Ibid.
40. Ibid., 38/1/11 Boxes 9 and 10.
41. Sachse, "Salute to an Old Friend."
42. "Dinner Honoring A. F. Gallistel."

Chapter 9. The 1960s

1. UW Archives, 38/7/3 Box 4.
2. Ibid., 38/1/11 Box 10.
3. Kline, "John Curtis and the University of Wisconsin Arboretum," 53.
4. Cottam, "The n-Dimensional Niche of John T. Curtis," 49.
5. UW Archives, 38/1/11 Box 10.
6. Cottam, "The n-Dimensional Niche of John T. Curtis," 50.
7. "Arboretum Seed Exchange in 1961."
8. Greene, "John T. Curtis"; UW Archives, 38/7/1 Box 1.
9. Howell and Stearns, "Preservation, Management, and Restoration of Wisconsin Plant Communities," 63.
10. "Grant Cottam."
11. See "'Great' Treasure Hunt"; Sachse, *A Thousand Ages*, 102.
12. Kline, "John Curtis and the University of Wisconsin Arboretum," 53.
13. UW Archives, 38/7/1 Box 1.
14. Ibid., 38/00/6 Box 1.
15. "Introducing the Friends of the Arboretum."
16. "Dedication of John T. Curtis Prairie."
17. "Friends of the Arboretum Meeting."
18. Sachse, "Town and Gown."
19. Ibid., 8.
20. "Spring Meeting of the Friends of the Arboretum"; Sachse, "Town and Gown."
21. "New Brochure, Map Details Highlights of U. Arboretum."
22. The Arboretum movie was originally the idea of David Archbald, newly appointed managing director (1962), working closely with Walter J. Meives, director, Department of Photography, UW Extension, and James Larsen, science editor,

University News and Publications. Larsen wrote the script. At Archbald's urging, the Madison City Council donated $1,500 (UW Archives, 38/00/06 Box 1; 38/7/1 Box 1).

23. "Report on the Friends of the Arboretum."

24. Krapfel, "Movie Is First 'Living Record' of Sounds and Sights of Arboretum."

25. "Arboretum Movie."

26. Sachse, "Town and Gown," 10.

27. UW Archives, 38/1/11 Box 10.

28. Ibid., 38/7/3 Box 4.

29. David Archbald, AC Tape 036, February, 27, 1989, AA.

30. The Arboretum botanist position was a Botany Department tradition dating from 1943 when Robert McCabe, while still a graduate student, was appointed at Leopold's urging (see chapter 7) to the position of Arboretum biologist. In 1946, after McCabe gave up the position, the title was changed to Arboretum botanist, owing to the Botany Department's decision to fund the position. After 1946, the Arboretum botanists were botany graduate students, listed as teaching assistants, with duties that involved ensuring a technical liaison between the Botany Department and the Arboretum. The first Arboretum botanist, appointed in 1946, was Max Partch, who worked closely with Curtis. He also drew one of the Arboretum's first vegetation maps (AC Tape 095, September 16, 1980). Two years after Partch, in 1951, David Archbald was appointed to the position and remained in it until 1955. From 1962 to 1964, F. Glenn Goff held the position. Goff also conducted tours, designed leaf prints, and wrote much of the text for Arboretum trail guides. Paul H. Zedler followed Goff, and during his time as Arboretum botanist, he coauthored, with Goff, a picture key to Wisconsin trees for elementary botany students and a tree identification guide that Cottam, who directed Zedler's Ph.D. work, cited as "one of the most useful" he had ever seen (Cottam, "Arboretum Botanist").

In 1962, the Arboretum "overseer's house," built in 1932, was torn down. From 1935 to 1941, it served as the Officers' Quarters for Camp Madison. After 1941, it became the caretaker-watchman's residence until it was razed in 1962. But the need for a live-in caretaker remained a concern. As a temporary measure, the university bought a forty-six-foot trailer and parked it on the site of the old caretaker-watchman's house. A precedent was established when the Botany Department permitted two graduate students, Bob Riehm and his wife, Kathy, to live in the trailer rent free in return for keeping watch. Botany Department graduate students lived in the trailer until summer 1968, when Paul Zedler and his wife, Joy, both doctoral students who had occupied the trailer since 1964, left the campus for positions at the University of Missouri. After the Zedlers left, the Arboretum removed the trailer. In 1971 a small house with offices in the back was built on the spot. With the opening of the house, the official Arboretum address changed from

329 Birge Hall to its present address—1207 Seminole Highway (Cottam, "Personnel Changes in the Arboretum"; "Building in the Arboretum").

The Zedlers returned to the UW and to the Arboretum in 1998 from California's San Diego State University. Paul would divide his time between the Institute for Environmental Studies and the Arboretum; Joy would become the first holder of the Leopold Chair of Restoration Ecology and currently also serves as the Arboretum director of research.

31. Archbald, Tape AC 036, February 27, 1989.

32. "Arboretum Office."

33. UW Archives, 38/7/1 Box 1.

34. From 1947 onward, efforts, mainly Jackson's, to acquire individually owned lots in the Lake Forest development created a myriad of problems. Two of the most aggravating instances involved efforts to acquire Block 25 and a parcel of land known as the Herling property. Block 25 (containing twenty-two lots) was owned by a group of Norwegian families. They had originally paid $1,000 per lot. In fall 1947, Jackson learned that the Norwegians, in frustration, had sold the block for $750 to two daughters of one of the original owners. John McKenna, a former Lake Forest Company sales manager who originally sold Block 25 to the Norwegians, felt duty bound to help the daughters get back as much of their original investment as possible. He had plans to build houses on the lots. The possibility left Jackson numb. Houses on the lots would interfere with Arboretum plans to close off streets.

Jackson contacted the two women, a Mrs. Johnson from La Crosse, Wisconsin, and a Mrs. Logan from Westby, Wisconsin, and told them he would try to get them $1,000 for the block. They wanted $3,000. Jackson, dismayed, told one of the University Regents that Mrs. Johnson, in particular, was "what one might call a bit cagey and quite certain of herself and her viewpoints." He raised the offer to $1,500. The sisters still were not selling. Finally, in early 1949, two years into the negotiations, the sisters agreed to sell the twenty-two lots for $2,200—$100 a lot. Jackson and the university held out at $50 a lot, but finally gave up. Jackson, the original old fox, had been "outfoxed" this time by the woman who turned out to be "a bit cagey" after all (UW Archives, 38/1/11 Box 2).

But the biggest aggravation, by far, was the Herling property. George Herling owned land adjacent to the Arboretum's southeastern boundary. When the Arboretum began acquiring Lake Forest land, it fenced off large portions of acreage. In September 1951, the AC was told that because of the fencing of property and streets in the southeast, Herling no longer had access to his land and demanded a right-of-way. The conflict initiated a series of university offers to purchase the land. On April 16, 1953, Gallistel reported to the AC that the Herlings were not selling. In May 1953, they threatened a lawsuit over a right-of-way. Their lawyer, Randolph R. Connors, told the university that his clients would trade the property if the university provided them with a "piece of land comparable to the size" of the current property "within the vicinity of Madison . . . in a nice neighborhood" with land of comparable beauty (UW Archives, 38/1/11 Box 5). In July 1953, the university stretched a cable with a gate across Draper Street (the street in question), gave Herling a key, and asked him to keep the gate locked. Connors advised Herling to leave the gate unlocked, which he did. In 1954, Herling threatened to build on the property. In 1956, Conners told the university that Herling would sell his 8 acres for $2,000 an acre (38/1/11 Box 8). The university was not interested. In 1959, eight years into what some members of the AC referred to "as the perennial dispute," Jackson reported that Herling threatened to cut the cable stretching across Draper Street. The AC considered denying him access by tearing up Draper Street. The dispute waged on through 1959. Finally, in 1960, Herling sold his property, and the new owner sold it to the university. Eventually Draper Street was torn up. Within three years, in 1963, the last Lost City lots were acquired, putting a welcome end to any more threats of litigation (38/1/11 Box 10).

In October 1972, the three-quarter acre Kipen property on the Arboretum's northeastern fringe was purchased by The Nature Conservancy for $5,300 with the understanding that it would be resold to the FOA for $700, TNC's additional costs. The Kipen property actually was the last of the Lake Forest lots ("Kipen Property File," AA).

35. See Groy, "Lake Forest, the Lost City," 115–20, "Men and the Marsh: Lake Forest I," and "Men and the Marsh: Lake Forest II"; Sachse, "The Lost City," 6; Newhouse, "Madison's 'Lost City'"; Bordsen, "Madison's Lost City," 67; Cronon and Jenkins, *University of Wisconsin*, 702–5.

36. Newhouse, "Madison's 'Lost City,'" 1.

37. Ibid.

38. See *Lake Forester*, UW Archives, 38/6/5 Box 1; "Introducing the Lake Forest Golf Club."

39. Alden, "Legend of the Lost City."

40. Newhouse, "Madison's 'Lost City,'" 1.

41. "Concrete Pavement to New House of Lake Forest Homes Company," 2.

42. See " Growing Community of Lake Forest"; "New Lake Forest Home Receives Roof"; Smith, "Truth and the Lost City."

43. UW Archives, 38/6/5 Box 1.

44. My thanks to Kathy Miner, Arboretum naturalist and guide, for researching the details and doing the footwork to identify the location of the houses in the Carver-Martin Streets area.

45. Smith, "Truth and the Lost City."

46. Groy, "Men and the Marsh: Lake Forest II," 5–6.

47. Jones, "Dream Suburb Is Bad News for Developers," 2; "Arnold Still in Jail."

48. "Banker in Sermon Calls Poor to Him."

49. "Town Hall's Banker-Preacher Arrested."

50. "Arnold Still in Jail"; Jones, "Dream Suburb Is Bad News for Developers," 2.

51. UW Archives, 38/4/1–7 Box 1.

52. Gesteland, "The Lost City."

53. See memorandum, C.J. Stathas to Grant Cottam, March 6, 1963, Lake Forest File 1, AA.

54. See letter to Jackson, October 27, 1961, Lake Forest File 1, AA.

55. UW Archives, 38/1/11 Box 2.

56. Sachse, "Salute to an Old Friend."

57. Cottam, "Dr. James H. Zimmerman."

58. [Fleming], "Arboretum Tours on the Grow."

59. Jones, "Guide Program Completes 30 Years."

60. Bradley, "Rosemary Fleming Retires."

61. [Fleming], "Arboretum Tours on the Grow."

62. Newhouse, "Pied Piper of Madison."

63. Ibid.

64. Zimmerman and Zimmerman, "Arboretum Requires Special Knowledge."

65. James Zimmerman, AC Tape 036, February 27, 1989, AA.

66. Leslie, "In Arboretum Friends Eyes, Beltline Is a 'Cement Desert,'" 1.

67. Horton, "Recreation Department–Community Center Project."

68. "What We Did in 1972," 10.

69. Fleming, "Overview '72 Guide Training Program."

70. [Fleming], "Arboretum Tours on the Grow"; Jones, "Guide Program Completes 30 Years."

71. Katharine Bradley, AC Tape 141, October 15, 1991, AA.

72. "Arboretum Guides," 4; Kline, "John Curtis and the University of Wisconsin Arboretum," 54.

73. [Fleming], "Arboretum Tours on the Grow."

74. Hone, "Appointment of Roger Anderson."

75. Robbins, "Godfrey Tells the World He Liked It Here."

76. "University in an Arboretum."

77. Zimmerman and Cottam, "Protecting the Arboretum."

78. Leslie, "In Arboretum Friends' Eyes"; Zimmerman and Cottam, "Protecting the Arboretum"; Cottam, "Arboretum and the Beltline."

79. Leslie, "In Arboretum Friends' Eyes."

80. Archbald, AC Tape 036, February 27, 1989, AA.

81. Zimmerman and Zimmerman, "Arboretum Requires Special Knowledge."

82. Newhouse, "We Are on a Collision Course with Nature."

83. Kienitz, "UW Charts Describe Dangers of Pollution."

84. See AC Minutes for January 14, February 18, March 11, 1970, AA.

85. David Archbald remained in Madison working with the Man–Environment–Communications Center until 1981 and from 1986 to 1987 as director of Unified Learning Approach, a project involving UW faculty and area K–12 teachers in the design of problem-based curriculum units. In 1987, he moved back

to Buffalo, New York, his hometown, where he directed Project ThinkWell, a systems problem-solving and communications networking project, operating in conjunction with the Department of Catholic Education for the Buffalo Roman Catholic Diocese and the Research Foundation of Buffalo State College. He directed Project ThinkWell until his death on October 27, 2009.

In an email follow-up to a telephone interview, his daughter, Lynne Archbald, observed that her father's "life mission as an ecologist and environmentalist was to provide schools with a unified learning approach to environmental education using a systems model for teaching natural and human systems" (October 15, 2010).

86. "New Arboretum Director."

87. "Managing Directors Report, 1972–73," AA.

Chapter 10. The 1970s

1. See minutes of March 16, 1971, AC Minutes (ACM) 1967–1974, AA; see also "New Arboretum Director"

2. In the early 1970s, some members of the AC wanted a ceiling on the number of guided tours. There was also a feeling among some members that, in order to alleviate pedestrian traffic, the guide program should be encouraged to utilize non-Arboretum sites. In addition, there was concern that the tour program was straining the FOA budget. Some AC members wanted to exclude the public or, at least, prohibit access to critical research areas. Despite the objections, Anderson expanded the guide program to include A Beginning Nature Guide Training Program (December 9, 1970, and January 13, 1971, ACM 1967–1974, AA).

3. Waixel, "New Arboretum Chief Links People to Nature"; Hone, "Appointment of Roger Anderson."

4. Gould, "Urban Crush Is Closing In on the Arboretum."

5. In 1958, the AC refused WHA's request for a tower and advised WHA representatives that, contrary to their claim, the Arboretum's Curtis Prairie was not "grown to weeds . . . nor [was] it idle" (UW Archives, 38/1/11 Box 9). In 1958, the WHA tower in use was located on private land that WHA was paying $4,500 per year to lease. Arboretum property, under UW control, was rent free. By September 1971, WHA again asked for and this time received AC approval to erect a tower—but in an Arboretum marsh, not the middle of the Curtis Prairie. Anderson argued against giving up any land, even marsh land, noting that the image of the AC and Arboretum staff as defenders of green space could suffer publicly, an opinion not shared at the time by a majority of AC members (September 10, 1971, ACM, AA). Anderson, however, was right. By January 1972, rumor of a clandestine conspiracy between WHA and the Arboretum administration was given a sympathetic airing in a February 1, 1972, article in the *Daily Cardinal*, the UW student newspaper. The construction of the tower had been approved, the article asserted, after "secret

meetings" between WHA representatives and FOA members. Just as Anderson had feared, the Arboretum administration was depicted as phony "defenders of green space" (Carman, "Arboretum Jeopardized by WHA Tower").

Occasional grumbling about the ugly tower on the eastern Arboretum boundary continued throughout the 1970s and became a sustained growl in August 1981 when Katharine Bradley, the director who succeeded Anderson, took WHA to task for not honoring its 1972 agreement. Bradley's assault began in late summer 1981 with a report to the AC on a Lake Forest water control project that had potential for flooding the tower area and that included a detailed historical record of events leading up to 1972, including information on an agreement that WHA had not honored that limited its occupancy to ten years. Bradley urged a renegotiation of the agreement. Otherwise, she claimed, "WHA may . . . remain in the Arboretum for all eternity—an undesirable possibility." She added: "WHA has a very interesting agreement, apparently forced on the Arboretum by various influential officers of the UW System, in which WHA gets everything and the Arboretum gets nothing. It is high time the Arboretum stopped being the University's fall guy" (memo from Bradley to Gerald Gerloff, August 21, 1981, AA).

In spring 1988, with the tower still in place, WHA installed a satellite dish and assured Gregory Armstrong, Arboretum director at that time, that the tower would be located close to the building and that granite or washed rocks (not limestone) would be placed around the pad at the tower's base (letters, March 8, 15; April 6, 28; AA). Twenty years later, in December 2008, WHA had two towers on Arboretum property, one brand new and the other waiting to be removed. Concerns about soil compaction in the marsh caused by heavy equipment were ignored. Steve Glass, Arboretum field manager, likened the irreparable soil damage to "trying to turn" a "mound of hamburger . . . back into a cow." The compaction meant slow death for that piece of wetland. In response, Steve Johnston, Wisconsin Public Radio director of engineering and operations, claimed that the site "was far more disturbed years ago when it was a farmer's field than by anything we've done." Johnston was baffled by the Arboretum's desire to protect the site since, years ago, he maintained, "it was chosen to be the sacrificial one" (Lueders, "Tower Project Scars the Arboretum").

6. Gould, "Urban Crush Is Closing In on the Arboretum."

7. Gould, "Young Says UW's 'Bulldozer Era' Is Over."

8. "Fermilab Prairie," http://ed.fnal.gov/samplers/prairie/fnal_prairie.html; Schulenberg, "Summary of Morton Arboretum Prairie Restoration Work."

9. Hone, "U.W. Prairies—Where Man-Tall Grasses, Brilliant Flowers Abound."

10. Wilford, "Prairie Partisans Move to Save Grasslands," 72.

11. Ibid.

12. Ibid.

13. Wade, "Small Show Prairie Extends Education"; Ode, "Some Aspects of Establishing Prairie Species by Direct Seeding," 52.

14. The properties Anderson was eying included the Fisher tract, the Pahl tract, the Bewick property, the Kipen property (see chapter 9, note 34), and property offered for sale by Strassburger Realty. A full list of available properties was distributed by Anderson to the AC at the September 10, 1971, meeting. On February 18, 1972, the committee recommended acquiring the Bewick property and that the FOA make the purchase.

15. [Anderson 1972?], undated, unsigned manuscript, AA; see also Catherine Manson, untitled news release, UW University News and Publication Service, March 21, 1972.

16. See "New Jobs, New Faces on Arboretum Committee," 6; see also memo from Secretary of the Faculty, September 2, 1971, AA.

17. See February 18, 1972, ACM 1967–74, AA.

18. February 2, 1973, ACM, AA.

19. March 2, 1973, ACM, AA.

20. See March 1, 1974, ACM, AA; see also "McKay Center Site Selection Committee Report," AA.

21. Chairman Robert J. Dicke's report to the AC took place on November 3, 1972. See ACM 1967–1974, AA.

22. May 4, 1973, ACM, AA.

23. Letter to the Chancellor, undated; "Managing Director Reports, 1972–73," AA.

24. January 8, 1974, ACM, AA.

25. Gould, "Explosion of Ducks at Arboretum Pond."

26. December 1, 1972, January 12, 1973, ACM, AA.

27. Katharine Bradley, AC Tape 141, October 15, 1991.

28. September 7, 1973, ACM, AA.

29. "Arboretum Welcomes New Managing Director"; Hine, "Familiar Faces . . . Katharine Bradley"; Bradley, AC Tape 141.

30. Bradley, AC Tape 141.

31. *Friends of the Arboretum* was published separately from *Arboretum News* from 1975 until 1983, when *Friends of the Arboretum* ceased publication and was fully integrated with *Arboretum News* under the title *Arboretum News and Friends Newsletter*. In 1987, the journal's name was changed to *NewsLeaf: Friends of the Arboretum Newsletter*, and since 1987 it has been published monthly.

32. Bradley, AC Tape 141.

33. Bradley, "A Question of Balance," 2.

34. Matheson, "'Housewife' Finds Career," 7.

35. Bradley, "A Question of Balance," 1–2.

36. Irwin, "A Natural History of East Marsh."

37. "Arboretum Loses a Friend."

38. Bradley, "A Question of Balance," 4.

39. Ibid., 1; Bradley, AC Tape 141.

40. Bradley, AC Tape 141; Haines, "Familiar Faces . . . Kline."

41. Haines, "Familiar Faces . . . Kline."

42. Bradley, AC Tape 141.

43. "Director's Report," January 1977, AA.

44. "Arboretum Ecologist Virginia Kline Retires."

45. Ground was broken for the McKay Center on December 12, 1975. Robert Ellarson, AC chair, served as master of ceremonies. By summer 1977, the center was ready for occupancy. Funding came from the McKay Foundation, the James Olin Reed bequest, and the University of Wisconsin Foundation. In 1976, the State Building Commission approved $30,000 for a solar heating system, the amount to be matched by the Arboretum Trust Fund. The solar system was expected to provide approximately 40 percent of the heat for the new building.

46. Bradley, AC Tape 141; see also "Director's Report," January 1977, AA.

47. Sachse, "A Not-So-New Face."

48. Reports, April 1977, AA.

49. Monthly Report, June 1977, AA.

50. Monthly Report, January 1978, AA.

51. Monthly Report, March 1977, AA.

52. Zimmerman and Zimmerman, "UW Arboretum Tackles Ecosystem Restoration."

53. See DeKalb County Forest Preserve District, http://www.dekalbcounty .org/forest/forest_preserve.html.

54. Allison, "When Is a Restoration Successful?" 10, 12; Howell and Jordan, "Tallgrass Prairie Restoration in the North American Midwest."

Chapter 11. The 1980s and Beyond

1. Morrison, "Native Plants for Man-Made Moonscapes."

2. See Monthly Report, February–March 1979, AA.

3. See July 18, 1980, AC Minutes (ACM), AA.

4. Grant Cottam,, AC Tape 040, October, 1980.

5. Jordan, "Restoration and Management Notes."

6. Katharine Bradley, AC Tape 141, October 15, 1991.

7. See December 2, 1983, ACM, AA; see also "Director's Report," July–August 1984, AA.

8. In 2011, the editorial side of the publication moved to the Center for Urban Restoration Ecology at Cook College, School of Agriculture, Marine and Environmental Sciences, at Rutgers University, New Brunswick, New Jersey. *Ecological Restoration*, however, continues to be published by the University of Wisconsin Press.

9. Leopold, "Ecological Conscience," 345.

10. Bradley, "Goodbye."

11. Jordan, "Adding Strip of Marshland."

12. The availability of $47,000 in federal LAWCON funds (Federal Land and Water Conservation Fund) guaranteed the purchase of the Selix property. The LAWCON grant covered half the purchase price. The remainder was made up by state land acquisitions funds, money from the FOA, and a fiftieth anniversary gift from the UW Class of 1927. The Sinaiko money was the gift of the surviving five of eight children of Alex and Rachel Sinaiko, immigrants to the United States from Russia. An engraved boulder honoring the Sinaikos is located at the western end of the prairie overlook on McCaffrey Drive. The Gardner money was designated for the acquisition of additional land.

13. Hauda, "Couper Seeks Safety for Arboretum Runners."

14. Bradley, AC Tape 141.

15. Ibid.

16. November 5, 1982, ACM, AA; Bradley, AC Tape 141.

17. Bradley, AC Tape 141.

18. Ibid.

19. "A Look at the Bradley Decade," adapted from remarks by Bill Hilsenhoff at an FOA breakfast and Emily Earley at a farewell banquet in honor of Bradley.

20. Bradley, "Goodbye."

21. "Our Purpose," in "Director's Report," March 1984.

22. Leopold, "What Is the Arboretum?"

23. See Rongstad, Gerloff, and Armstrong, "UW–Madison Capital Campaign Proposal."

24. Ibid.

25. "Pond Experiment Shows Control."

26. Fixmer, "Storm Sewers Damage 30 Acres of Arboretum."

27. "Workers Clear Way for Arboretum Pond."

28. Zedler, "What Has Happened to the Arboretum Wetlands?"

29. Zedler, "What Can Be Done to Minimize Negative Impacts."

30. Liebl, "Ponds of the Arboretum."

31. "Five Naturalists Hired."

32. See letter from Donna Shalala to Gregory Armstrong, July 13, 1989, "Directors Reports, 1983 to 89," AA.

33. Three grants designated for funding new education projects were acquired by Donna Thomas in 1991, 1992, and 1993 from the Institute of Museum Services (IMS). The 1993 grant was earmarked especially for research programs and field staff projects under Steve Glass's supervision. In 1993 and 1995, the Arboretum received two generous donations from the Neckerman estate. In 1994 and 1996, the National Science Foundation (NSF) awarded the Arboretum grants for support of the Earth Partnership for Schools (EPS) program. In 1995, the Secord family donated funds supporting the Leopold Professorship endowment; generous grants were also received from Jack and Marion Goetz, the Eisenhower Founda-

tion, Mildred Gill, Mary A. Muckenhirn, the estate of Susanne Ames, the Evjue Foundation, and the Oscar Mayer Foundation. In 1996, the Eisenhower Foundation provided another generous teacher training grant; two grants were received from Lenore Landry for the land restoration program. In 1997, Wisconsin Capital Campaign Funds funded two major capital projects: a new visitor center wing and a Wisconsin Native Plant Garden; IMS provided a grant for a two-year study of aerial photography in conjunction with ArborLIS in order to monitor vegetation. In 1999, Sally Mead Hands donated $850,000 toward the cost of the new McKay Center wing and auditorium. In 2002, the Howard Hughes Medical Institute provided $500,000 in support of the EPS training institutes.

34. Wilger-Gaskell, "Meet Pat Brown, Next Editor."

35. Murray, "Clearing for Savanna Restoration."

36. Teuke, "WEEB Grant Funds K–12 Curriculum Materials."

37. Leach, "Urban Ecology."

38. Currently, the EPS program is funded by the Institute of Museum and Library Services, the U.S. Environmental Agency's Office of Environmental Education, the USDA Forest Service's More Kids in the Woods Project, the Wisconsin Environmental Education Board, and the National Oceanic Atmospheric Administration Wisconsin Sea Grant Earth Partnership.

39. Jacobs, "Gardner Marsh Targeted for Restoration."

40. See Director's Report, December 1984, AA.

41. April 25, 1980, ACM, AA.

42. Ewel, "Summary and Synthesis of Advisory Panel Discussion."

43. See Director's Reports, 1983–89, AA.

44. Rongstad, Gerloff, and Armstrong, "UW–Madison Capital Campaign Proposal."

45. Lloyd Bostian to Chancellor Donna Shalala, February 6, 1989, AA.

46. Armstrong, letter announcing retirement to "Friends"; see also "Recent Tornado Damages Arboretum Trees, Closes Trails"; Jacky Kelley, "Wingra Woods in Transition."

47. Gregory Armstrong, UW Oral History interview, 2009, author's notes.

48. McSweeney, "Interim Director Sees Challenges and Opportunities Ahead"; see also Basu, "Kevin McSweeney Named Arboretum Director."

49. Nathans, "McSweeney Is Arboretum Chief."

50. White, "Arboretum Welcomes New Residents." White notes that sandhill cranes were first spotted in the Arboretum in 1993.

51. Miner, "Osprey Spotted in the Arboretum."

Bibliography

"Add 55 More Acres to U.W. Arboretum." *Capital Times*, 1 August 1937.

Alden, Sharyn. "Legend of the Lost City." *Wisconsin State Journal* (Madison), 24 March 2007.

Allison, Stuart K. "When Is a Restoration Successful? Results from a 45-Year-Old Tallgrass Prairie Restoration." *Ecological Restoration* 20, no. 1 (March 2002): 10–17.

Allsup, Mark Steven. "Greene Prairie: A Model for Prairie Restoration." MS thesis, University of Wisconsin–Madison, 1977.

"Another Desiccated Autumn." *Arboretum News* 3, no. 1 (January 1954).

"The Arboretum Argus." Vol. 1. (Mimeographed newsletter of Camp Madison, 2670 Company, CCC). April 1936. Uncataloged Personal Collection.

"Arboretum Ecologist Virginia Kline Retires." *NewsLeaf: Friends of the Arboretum Newsletter* 10, no. 5 (May 1996).

"Arboretum Guides: A Visitor's Key to Understanding and Appreciating Restoration Ecology." *NewsLeaf* 7, no. 1 (January 1993): 4–5.

"Arboretum Loses a Friend." *NewsLeaf* 6, nos. 11–12 (November–December 1992): 2.

"The Arboretum Movie." *Arboretum News* 13, no. 4 (October 1964).

"Arboretum Office." *Arboretum News* 11, no. 1 (January 1962).

"Arboretum Residence Demolished." *Arboretum News* 11, no. 1 (January 1962).

"The Arboretum Seed Exchange in 1961." *Arboretum News* 10, nos. 3–4 (July–October 1961).

"Arboretum Welcomes New Managing Director." *Arboretum News* 23, no. 1 (Winter 1974): 2.

Armstrong, Greg. Letter announcing retirement to "Friends." *NewsLeaf* 19, no. 2 (February 2004).

"Arnold Still in Jail; Finances in Chaos." *New York Times*, 28 December 1922.

"Assurance of Arboretum in OK on Budget." *Capital Times*, 1 January 1928.

Backus, M. P., and R. I. Evans. "H. C. Greene (1904–1967)." *Mycologia* 60, no. 5 (September–October 1968): 994–98.

"Banker in Sermon Calls Poor to Him." *New York Times*, 6 November 1922.

Basu, Paroma. "Kevin McSweeney Named Arboretum Director." *NewsLeaf* 20, no. 9 (September 2005).

Blewett, Thomas J., and Grant Cottam. "History of the University of Wisconsin Arboretum Prairies." *Wisconsin Academy of Sciences, Arts and Letters* 72 (1984): 130–44.

Bordsen, John. "Madison's Lost City." *Wisconsin Regional* (January 1979): 67–68.

Bradley, Katharine. "Goodbye." *Arboretum News and Friends Newsletter* 32, no. 3 (Summer 1983): 4.

———. "A Question of Balance." *Arboretum News* 23, no. 3 (Fall 1974): 1–5.

———. "Rosemary Fleming Retires." *Friends of the Arboretum* 7, no. 1 (Winter 1981).

"Building in the Arboretum." *Arboretum News* 17, no. 3 (Late Summer 1968).

Burgess, Robert L. "John Thomas Curtis: Botanist, Ecologist, Conservationist." In Fralish, McIntosh, and Loucks, *John T. Curtis*, 1–43.

Buss, Irven O. "The Passenger Pigeon." *Wisconsin Society for Ornithology* 11 (January 1949).

"Careful with Fire! Is Arboretum Lesson." *Wisconsin State Journal*, 18 August 1937.

Carman, Diane. "Arboretum Jeopardized by WHA Tower." *The Daily Cardinal*, 1 February 1972.

Chesterton, G. K. "Victorian Age in Literature." In *Collected Works*. Vol. 15. San Francisco: Ignatius Press, 1989.

"Concrete Pavement to New House of Lake Forest Homes Company." *Lake Forester* 2, no. 18 (15 September 1921): 1–2.

Cottam, Grant. "Arboretum Botanist." *Arboretum News* 13, no. 4 (October 1964).

———. "The Arboretum and the Beltline." *Arboretum News* 19, no 2 (Summer 1970): 12–14.

———. "Dr. James H. Zimmerman—Our Ranger-Naturalist." *Arboretum News* 14, nos. 2–3 (April–July 1965).

———. "The n-Dimensional Niche of John T. Curtis." In Fralish, McIntosh, and Loucks, *John T. Curtis*, 45–50.

———. "Personnel Changes in the Arboretum." *Arboretum News* 17, no. 3 (Late Summer 1968).

———. "Professor G. William Longenecker." *Arboretum News* 18 (Spring 1969).

Cronon, Edmund David, and John W. Jenkins. *The University of Wisconsin: A History*. Vol. 3. Madison: University of Wisconsin Press, 1994.

Cronon, William. "A Voice in the Wilderness." *Wilderness* (1998).

Curtis, John T. *Arboretum Master Development Plan—I, The Prairie*. Madison: University of Wisconsin Arboretum (March 1951); AA.

————. "Information Bearing on Arboretum Policy." 1 February 1948, AA.

————. *The Vegetation of Wisconsin: An Ordination of Plant Communities.* Madison: University of Wisconsin Press, 1959.

"Curtis Prairie: 75-Year Old Restoration Research Site." *Arboretum Leaflets* 16 (August 2008): 1–4.

"A Dedicated Arboretum Family—The Gilberts." *Arboretum News* 16, no. 1 (Winter 1967).

"Dedication of John T. Curtis Prairie." *Arboretum News* 11, no. 4 (October 1962).

"Dedication Speakers Vision Arboretum as Realization of 'Consciousness of Nature.'" *Capital Times,* 18 June 1934.

Dieckmann, June. "2 Truax Men Killed in Jet's Crash Here." *Wisconsin State Journal,* 24 November 1953.

"Dinner Honoring A. F. Gallistel." *Arboretum News* 8, no. 4 (October 1959).

Doehlert, Betsy. "Do Ghosts Walk Arboretum Glades?" *Capital Times,* 31 October 1977.

Egan, Timothy. *The Worst Hard Time: The Untold Story of Those Who Survived the Great American Dust Bowl.* Boston: Houghton Mifflin, 2006.

Emlen, John T., Jr., and Thomas R. McCabe. "In Memoriam: Robert A. McCabe, 1914–1995." *Auk: A Quarterly Journal of Ornithology* 113, no. 3 (July 1996): 674–77.

Errington, Paul L. "Appreciation of Aldo Leopold." *Journal of Wildlife Management* 12, no. 4 (October 1948): 341–50.

"Evolution of a Prairie—The Grady Tract." *Arboretum News* 15, nos. 2–3 (Spring–Summer 1966).

Ewel, John J. "Summary and Synthesis of Advisory Panel Discussion on Arboretum Research at the University of Wisconsin." 1984. Uncataloged manuscript, AA.

Fassett, Norman, John Thomson, et al. "Establishment of Prairie—UW Arboretum." 1935. Uncataloged manuscript, AA; reprinted as John W. Thomson, *Experiments with Prairie Plants.* 1937. Uncataloged manuscript, AA.

"Fire in the Grady Tract." *Arboretum News* 3, no. 2 (April 1954).

"Five Naturalists Hired." *NewsLeaf* 1, no. 1 (January 1987).

Fixmer, Rob. "Storm Sewers Damage 30 Acres of Arboretum." *Capital Times,* 24 August 1981.

Flader, Susan L., and J. Baird Callicott, eds. *The River of the Mother of God and Other Essays.* Madison: University of Wisconsin Press, 1991.

[Fleming, Rosemary]. "Arboretum Tours on the Grow." *Arboretum News* 17, no. 1 (Winter 1968).

Fleming, Rosemary. "Overview '72 Guide Training Program." *Arboretum News* 21, no. 2 (Summer 1972): 6–7.

Foss, Robert. "The Arboretum." *Wisconsin Alumni Magazine* (February 1934): 123, 144.

Fralish, James S., Robert P. McIntosh, and Orie L. Loucks, eds. *John T. Curtis: Fifty Years of Wisconsin Plant Ecology.* Madison: Wisconsin Academy of Sciences, Arts and Letters, 1993.

Fred, E. B. "Dedication of McCaffrey Drive." *Arboretum News* 2 (1953): 1–3.

"Friends of the Arboretum Meeting." *Arboretum News* 11, no. 4 (October 1962).

"General Use of the Arboretum." *Arboretum News* 4, no. 4 (October 1955).

Gesteland, Bernice Elver. "The Lost City: Wrecked Dream of a Few of Madison's Leading Citizens." Mrs. Gesteland's handwritten reminiscences of her Lake Forest experience. Manuscript copyedited by Kathy Miner (UW Arboretum). July 2007.

Gould, Whitney. "Explosion of Ducks at Arboretum Pond Poses Threat to Land, Man and Fowl." *Capital Times*, 10 January 1973.

———. "Urban Crush Is Closing In on the Arboretum." *Capital Times*, 19 March 1971.

———. "Young Says UW's 'Bulldozer Era' Is Over." *Capital Times*, 15 March 1972.

"Grant Cottam." *Deseret News* (Salt Lake City), 20 May 2009.

"The 'Great' Treasure Hunt." *Arboretum News* 10, nos. 3–4 (July–October 1961).

[Greene, Henry]. "Widening of Beltline Highway through Arboretum." *Arboretum News* 5, no. 2 (April 1956).

Greene, Henry C. "John T. Curtis." *Arboretum News* 10, no. 2 (June 1961).

———. "Notes on Revegetation of a Wisconsin Sandy Oak Opening" [1943–1949]. Unpublished manuscript, 1949. UW Archives, 38/7/3 Box 9.

———. "Retirement of J. R. Jacobson." *Arboretum News* 11, nos. 2–3 (April–July 1962).

Griffin, Harold M. "Found in Garage . . . Nelson Turns Gun on Self." *Wisconsin State Journal*, 7 March 1927.

"The Growing Community of Lake Forest." *Lake Forester* 1, no. 13 (15 November 1920): 1, 4.

Groy, Jeffrey B. "Lake Forest, the Lost City (Madison, Wisconsin): One of Wisconsin's First Totally Planned Communities." Bachelor of Science thesis, University of Wisconsin–Madison, 1981.

———. "Men and the Marsh: Lake Forest I." *Arboretum News* 30, no. 1 (Winter 1981): 1–4

———. "Men and the Marsh: Lake Forest II." *Arboretum News* 30, no. 2 (Spring 1981): 1–6.

Haines, Dorothy. "Familiar Faces . . . #9, Professor Robert McCabe." *Friends of the Arboretum* 3, no. 3 (Summer 1977): 4–5, 10.

———. "Familiar Faces . . . #10, Virginia Kline." *Friends of the Arboretum* 3, no. 4 (Autumn 1977): 5, 8.

Hale, Jim. "Redwing Marsh Site of Fatal Crash 39 Years Ago." *NewsLeaf* 6, nos. 11–12 (November–December 1992): 4–5.

Hall, Marcus. *Earth Repair: A Transatlantic History of Environmental Restoration.* Charlottesville: University of Virginia Press, 2005.

Hancock, John L. *John Nolen, Landscape Architect, Town, City, and Regional Planner: A Bibliographical Record of Achievement.* Ithaca, N.Y.: Cornell University, Program in Urban and Regional Studies, 1976.

Hasler, A. D. "Teaching Exhibits Which Should Be Installed in the University Bay Area." University Bay Committee Publication, University of Wisconsin–Madison, 1944.

Hasselkus, Edward R. "The Longenecker Horticultural Gardens." In *Our First 50 Years: The University of Wisconsin Arboretum, 1934–1984,* edited by William R. Jordan III. Madison: University of Wisconsin Arboretum, 1984.

Hauda, Bill. "Couper Seeks Safety for Arboretum Runners." *Capital Times,* 22 December 1978.

Hickey, Joseph J. [Tribute to Aldo Leopold.] In *Aldo Leopold: Mentor,* edited by Richard E. McCabe. Madison: University of Wisconsin, Department of Wildlife Ecology, 1989.

Hine, Ruth L. "Familiar Faces . . . Katharine Bradley: Four Years Later." *Friends of the Arboretum* 4, no. 2 (Spring 1978): 3, 6–7.

Hone, Vivien M. "Appointment of Roger Anderson" (news release). University of Wisconsin News and Publications Service, 16 September 1970.

———. "Dr. Henry Campbell Greene" (obituary). *Capital Times,* 10 May 1967; reprinted in *Arboretum News* 16, no 2 (Spring 1967).

———. "U.W. Prairies—Where Man-Tall Grasses, Brilliant Flowers Abound." *Capital Times.* 19 September 1970.

Horton, Meredith. "Recreation Department–Community Center Project: Children Learn Wonders of Nature in Arboretum Tours." *Capital Times,* 6 August 1969.

Howell, E. A., and W. R. Jordan III. "Tallgrass Prairie Restoration in the North American Midwest." In *The Scientific Management of Temperate Communities for Conservation,* edited by I. F. Spellerberg, 395–414. London: Blackwell Scientific Publications, 1991.

Howell, Evelyn, and Forest Stearns. "The Preservation, Management, and Restoration of Wisconsin Plant Communities: The Influence of John Curtis and His Students." In Fralish, McIntosh, and Loucks, *John T. Curtis,* 57–66.

"How Greene Is This Prairie?" *University of Wisconsin–Madison News.* 25 March 1998. http://www.news.wisc.edu/4233.

Hubbell, Margaret S. "'U' Botany Students Make Plant Survey on Proposed Arboretum Site." *Capital Times,* 1 June 1928.

"Introducing the Friends of the Arboretum." *Arboretum News* 11, nos. 2–3 (April–July 1962).

"Introducing the Lake Forest Golf Club." *Lake Forester* 2, no. 9 (1 May 1921).

Irwin, Harriet. "A Natural History of East Marsh of the University of Wisconsin Arboretum." Master's thesis, University of Wisconsin–Madison, 1973.

Jackson, J. W. "Dedication of the Aldo Leopold Pines." *Arboretum News* 2, no. 3 (July 1953): 3–5.

Jacobs, Jane. "Gardner Marsh Targeted for Restoration." *NewsLeaf* 7, no. 3 (March 1993): 1–2.

"Jet Crashes in Arboretum Marsh." *Capital Times*, 23 November 1953.

Jones, Vera. "Dream Suburb Is Bad News for Developers, Good News for Arboretum." *NewsLeaf* 7, no. 2 (February 1993): 1–2.

———. "Guide Program Completes 30 Years." *NewsLeaf* 11, no. 1 (January 1997): 1–2.

———. "Jackson, the Man behind the Mighty Oak." *NewsLeaf* 10 (March 1996): 2–3.

Jordan, William III. "Adding Strip of Marshland Means a Lot to Arboretum." *Wisconsin State Journal*, 30 December 1979.

———. "Restoration and Management Notes: A Beginning." *Restoration and Management Notes* 1 (June 1981): 2.

Kelley, Jacky. "Wingra Woods in Transition." *NewsLeaf* 20, no. 10 (October 2005).

Kienitz, Richard C. "UW Charts Describe Dangers of Pollution." *Milwaukee Journal*, 6 April 1996.

Kline, Virginia M. "Henry Greene's Remarkable Prairie." *Restoration and Management Notes* 10 (Summer 1992): 36–37.

———. "John Curtis and the University of Wisconsin Arboretum." In Fralish, McIntosh, and Loucks, *John T. Curtis*, 51–56.

Krapfel, Joan. "Movie Is First 'Living Record' of Sounds and Sights of Arboretum." *Capital Times*, 28 September 1964.

Lapham, Increase. "The Forest Trees of Wisconsin." *Wisconsin State Agricultural Society* 4 (1853).

Leach, Mark. "Urban Ecology." *NewsLeaf* 13, no. 6 (1999).

Lenehan, Roma. "Aldo Leopold and the Lakeshore Nature Preserve." *Friends of the Lakeshore Nature Preserve*, n.d. http://waa.uwalumni.com/lakeshorepreserve /leopoldandpreserve.html.

Leopold, Aldo. "The Chase Journal: An Early Record of Wisconsin Wildlife." *Transactions of the Wisconsin Academy of Science, Arts and Letters* 30 (1937): 69–76.

———. "Conservation Economics." *Journal of Forestry* 32 (May 1934): 537–44; reprinted in Flader and Callicott, *River of the Mother of God*, 193–202.

———. "The Conservation Ethic." *Journal of Forestry* 31 (October 1933): 634–43; reprinted in Flader and Callicott, *River of the Mother of God*, 181–92.

———. "The Ecological Conscience." *Bulletin of the Garden Club of America* (September 1947): 45–53; reprinted in Flader and Callicott, *River of the Mother of God*, 338–46.

———. "Exit Orchis" (with Leopold's March 1941 update). *Faville Grove Prairie-*

Notes. Robert W. Freckmann Herbarium, University of Wisconsin–Stevens Point, 1940. http://wisplants.uwsp.edu/Favilleprairie.html.

——. *A Sand County Almanac*. New York: Oxford University Press, 1949.

—— "What Is the Arboretum?" With note to Jackson, undated. UW Archives, 38/3/1 Box 1.

——. "Wilderness Values." 6-page manuscript. 16 July 1940. UW Archives, 38/7/3 Box 7. First published in *Yearbook, Park and Recreation Progress* (NPS, 1941), 27–29.

——. "Wildlife in the Picnic Point Program." University Bay Committee, University of Wisconsin–Madison, 15 May 1944. http://waa.uwalumni.com/lake shorepreserve/historicaldocuments/wildlife1944.html.

Leslie, Gay. "In Arboretum Friends' Eyes, Beltline Is a 'Cement Desert.'" *Wisconsin State Journal*, 8 June 1969.

Levitan, Stuart D. *Madison: The Illustrated Sesquicentennial History. Vol. 1, 1856–1931*. Madison: University of Wisconsin Press, 2006.

Liebl, David S. "The Ponds of the Arboretum." *NewsLeaf* 23, no. 1 (January 2008): 4–5.

Long, Barbara Jo. "John Nolen: The Wisconsin Activities of an American Landscape Architectural and Planning Pioneer, 1908–1937." MA thesis, University of Wisconsin–Madison, 1978.

"A Look at the Bradley Decade." *Arboretum News and Friends Newsletter* 32, no. 3 (Summer 1983): 5.

Lorbiecki, Marybeth. *Aldo Leopold: A Fierce Green Fire*. Guilford, Conn.: Falcon, 2005.

Lueders, Bill. "Tower Project Scars the Arboretum." *Isthmus*, 12 December 2008.

Matheson, Helen. "'Housewife' Finds Career." *Wisconsin State Journal*, 27 January 1974.

McCabe, Robert A. *Aldo Leopold: The Professor*. Madison: Rusty Rock Press, 1987.

McSweeney, Kevin. "Interim Director Sees Challenges and Opportunities Ahead." *NewsLeaf* 20, no. 1 (January 2005): 3.

Meine, Curt. *Aldo Leopold: His Life and Work*. Madison: University of Wisconsin Press, 1988.

"Michael Balthazar Olbrich." *Olbrich Botanical Gardens' History*. Madison: Olbrich Botanical Gardens, n.d.

Miner, Kathy. "Osprey Spotted in the Arboretum." *NewsLeaf* 20, no. 11 (November 2005): 3.

——. "We're More Than Madison . . . Arboretum Has 12 Outlying Properties." *NewsLeaf* 20, no. 7 (July 2005): 6.

Mollenhoff, David V. *Madison: A History of the Formative Years*. Madison: University of Wisconsin Press, 2003.

Morrison, Darrel. "Native Plants for Man-Made Moonscapes." *Arboretum News* 27, no. 3 (Summer 1978): 1–3.

Murray, Molly. "Clearing for Savanna Restoration Will Begin Soon in Monroe Street Area." *NewsLeaf* 5, no. 12 (November 1991): 2.

———. *Prairie Restoration for Wisconsin Schools.* University of Wisconsin–Madison Arboretum, 1993.

Nathans, Aaron. "McSweeney Is Arboretum Chief." *Capital Times*, 8 August 2005.

Nelson, Les. "Letter to the Editor—the Tragedy at Redwing Marsh." *NewsLeaf* 7, no. 1 (January 1993).

"New Arboretum Director." *Arboretum News* 19, no. 3 (Fall 1970): 5.

"New Brochure, Map Details Highlights of U. Arboretum." *Capital Times*, 31 October 1963.

Newhouse, John. "Madison's 'Lost City.'" *Wisconsin State Journal*, 31 December 1967.

———. "Pied Piper of Madison: 'Jim Zim.'" *Wisconsin State Journal*, 20 October 1968.

———. "We Are on a Collision Course with Nature." *Wisconsin State Journal*, 30 March 1969.

"New Jobs, New Faces on the Arboretum Committee." *Arboretum News* 20, no. 3 (Fall 1971): 6–7.

"New Lake Forest Home Receives Roof." *Lake Forester* 2 (15 July 1921).

Newton, Julianne Lutz. *Aldo Leopold's Odyssey.* Washington, D.C.: Island Press, 2006.

Nolen, John. *Madison: A Model City.* Boston, 1911.

———. "Address . . . at Dedication of the Arboretum and Wild Life Refuge of the University of Wisconsin. . . June 17, 1934." UW Archives, 38/3/2 Box 1.

Ode, Arthur H. "Some Aspects of Establishing Prairie Species by Direct Seeding." In *Proceedings of a Symposium on Prairie and Prairie Restoration*, edited by P. Schramm, 52–60. Special publication no. 3. Galesburg, Ill.: Knox College Biological Field Station, 1968.

Olbrich, Michael. "Speech before the Madison Rotary Club, May 17, 1928." UW Archives, 38/4/8 Box 2.

"Owls Help 8 Speakers Dedicate U. Arboretum." *Wisconsin State Journal*, 18 June 1934.

Partch, Max. "The Distribution of Plants on the Faville Prairie from Records in the 1940s and 1970s." 1997. Unpublished manuscript, AA.

Peattie, Donald C. "Norman Carter Fassett, 1900–1954." *Rhodora: Journal of the New England Botanical Club* 56, no. 671 (November 1954): 233–42.

Peterson, Charlotte. "Frank Lloyd Wright Walls in the Arboretum." *NewsLeaf* 2 (November 1988).

"Pond Experiment Shows Control of Pesky Lake Weeds." *Capital Times*, 31 May 1978.

"Prairie Tract Preserved by Badger Couple." *Milwaukee Journal*, 23 March 1941.

"Recent Tornado Damages Arboretum Trees, Closes Trails." *NewsLeaf* 19, no. 8 (August 2004).

"Regents Approve Plan for Wingra Arboretum." *Wisconsin State Journal*, 27 April 1932.

"A Report on the Friends of the Arboretum." *Arboretum News* 12, nos. 1–2 (January–April 1963).

Robbins, William C. "Godfrey Tells the World He Liked It Here." *Wisconsin State Journal*, 30 July 1967.

Roberts, Glenn D. *Opinions of the Attorney General of the State of Wisconsin.* Vol. 17, January 1, 1928–December 31, 1928. Madison: Department of Justice, 1928.

Rongstad, Orrin, Gerald Gerloff, and Gregory D. Armstrong. "UW–Madison Capital Campaign Proposal: Restoration Ecology Research Program at the Arboretum." 1985. Uncataloged manuscript, AA.

Ruskin, John. "Lamp of Memory." In *Seven Lamps of Architecture.* London: G. Allen, 1880.

Sachse, Nancy D. "Excerpts from a Journal, May 18, 1963." *Friends of the Arboretum* 3, no. 3 (Summer 1977): 3, 8.

———. "G. William Longenecker, Master Planner." *Arboretum News* 15 (Spring–Summer 1966).

———. "The Lost City. . . ." *Friends of the Arboretum* 2, no. 4 (Autumn 1976): 6–10.

———. "Madison's Public Wilderness: The University of Wisconsin Arboretum." *Wisconsin Magazine of History* 44 (Winter 1960–61).

———. "A Not-So-New Face." *Friends of the Arboretum* 5, no. 3 (Summer 1979): 3, 7.

———. "Salute to an Old Friend." *Arboretum News* 13, no. 1 (January 1964): 1–3.

———. *A Thousand Ages: The University of Wisconsin Arboretum.* Rev. ed. Madison: University of Wisconsin Arboretum, 1974.

———. "Town and Gown: How We Work Together." *Friends of the Arboretum* 3, no. 2 (Spring 1977): 5, 9–10.

"School Forests: Their Origin in Wisconsin." Madison Metropolitan School District, 2008. https://envedweb.madison.k12.wi.us/forest/edwischf.htm.

Schorger, A. W. "Dedication of the Aldo Leopold Pines." *Arboretum News* 2, no. 3 (July 1953): 5–6.

Schulenberg, Ray. "Summary of Morton Arboretum Prairie Restoration Work, 1963 to 1968." *Proceedings of a Symposium on Prairie and Prairie Restoration* 3 (1970): 45–46.

Smith, Susan Lampert. "Truth and the Lost City." *Wisconsin State Journal*, 3 May 1998.

Sperry, Theodore M. "Reflections on the U.W. Arboretum—45 Years Later." 10 February 1982. Uncataloged manuscript, AA.

"Spring Meeting of the Friends of the Arboretum—Dedication of the Jackson Oak." *Arboretum News* 12, no. 3 (July 1963).

"Story of Wisconsin's Last Remaining Virgin Prairie Linked Inescapably with Tale of a State Patriarch." *Milwaukee Journal*, 29 June 1941.

Tarkow, Harold. "In the Beginning . . . Sixty Years Ago, the Arboretum Was Born." *NewsLeaf* 8, no. 6 (June 1994): 2.

"$10,000 Fire Levels CCC Building; Blaze Threatens Camp Arboretum." *Capital Times*, 17 March 1937, 1, 6.

Teuke, Molly Rose. "WEEB Grant Funds K–12 Curriculum Materials." *NewsLeaf* 8, no. 2 (February 1994): 3.

Thomson, John W. "Norman C. Fassett, 1900–1954." *Taxon* 4, no. 3 (May 1954): 49–51.

"Town Hall's Banker-Preacher Arrested; Is Accused of Selling Worthless Bonds." *New York Times*, 27 December 1922.

"Twenty Thousand Dollar Gift to Arboretum." *Madison and Wisconsin Foundation* 20, no. 40 (July 1937).

University Bay Committee. "Preliminary Detailed Development Program for the Picnic Point–University Bay Preserve" (May 1944).

"A University in an Arboretum." *Wisconsin State Journal*, 5 January 1968.

Wade, Douglas E. "Small Show Prairie Extends Education." In *Proceedings of the Second Midwest Prairie Conference*, edited by James H. Zimmerman, 200–201. Madison, Wis.: Published by the editor, 1970.

Waixel, Vivian. "New Arboretum Chief Links People to Nature." *Wisconsin State Journal*, 13 September 1970.

Waller, Thomas M. "Prairie 'Father' Takes Pride in His Baby." *Wisconsin State Journal*, 27 June 1982.

"What We Did in 1972." *Arboretum News* 21, no. 1 (Spring 1972): 9–10.

White, Maury. "Arboretum Welcomes New Residents." *NewsLeaf* 10, no. 2 (February 1996): 1–2.

Wilford, John Noble. "Prairie Partisans Move to Save Grasslands." *New York Times*, 18 October 1970.

Wilger-Gaskell, Judith. "Meet Pat Brown, Next Editor." *NewsLeaf* 5, no. 12 (December 1991).

"Wingra Marshes Are Designated as Game Refuge." *Wisconsin State Journal*, 19 June 1934.

"Workers Clear Way for Arboretum Pond." *Wisconsin State Journal*, 11 January 1983.

Zedler, Joy. "What Can Be Done to Minimize Negative Impacts on Arboretum Wetlands." *NewsLeaf* 13, no. 7 (July 1999): 5.

———. "What Has Happened to the Arboretum Wetlands?" *NewsLeaf* 13, no. 6 (June 1999): 4–5.

Zimmerman, Jim, and Grant Cottam. "Protecting the Arboretum." *Arboretum News* 17, no. 2 (Spring 1968).
Zimmerman, James, and Elizabeth Zimmerman. "Arboretum Requires Special Knowledge." *Wisconsin State Journal*, 17 October 1971.
———. "UW Arboretum Tackles Ecosystem Restoration." *Wisconsin State Journal* 21 January 1973.

Index

Page numbers in italics indicate illustrations.

AAC (Advisory Committee), 48–49, 52–53, 96, 134, 154, 180, 254

AC (Arboretum Committee): overview of, 9, 37–38, 180; Anderson and, 221; Beltline subway underpass and, 175–76; Bishop Tract and, 257; Bradley and, 227–28; buffer zone and, 224; Burr Jones Memorial Park and, 125; environmentalism and, 153, 212–13; Faville Grove Preserve and, 131; film on Arboretum and, 166; forestation project and, 52–53, 262; McKay Center and, 222–23; Nakoma Country Club acquisition and, 151; Native American effigy mounds and, 49–52; newsletter publications and, 146; Picnic Point and, 149–50; politics of governance and, 9, 47–49, 62–63, 222–23; rabbit shoots and, 176–77; radio tower and, 224, 281n5; reparations for tree damages and, 170–72; research programs and, 121; Special Committee on Arboretum Planning and, 97; Technical Committee title for, 47, 49, 152; tour guide programs and, 215, 281n2; UBC and, 149–50; UW Buildings and Grounds department and, 58; wildlife refuge conflicts and, 145; WISM's "The Great Treasure Hunt" and, 184–85

Advisory Committee (AAC), 48–49, 52–53, 96, 134, 154, 180, 254

Aldo Leopold Pines memorial, 31, 154, 155–56, 177, 231

American Conservation Movement, xi–xii, 3–4. *See also* environmentalism (conservation)

Ames, Susanne estate, 285n33

Anderson, Roger C.: AC and, 221; biographical information about, 214; buffer zone and, 220–21, 224, 283n14; buildings for classrooms and research and, 214, 221; environmental ecology education and, 214–17, 223–24; legacy of, 224–25; McKay Center and, 223; managing directorship and, 213–14; outside threats and, 216–17; politics of governance and resignation of, 222–24; public outreach and, 224; radio tower and, 167, 216, 224, 281n5; research and, 214, 215–16; tour guide programs and, 215, 216, 281n2; UW's partnership with Arboretum and, 217

aquatic plants: research and, 24, 95, 110, 112, 121, 163–64; Stevens Memorial Aquatic Gardens and, 25, 92, 123. *See also* fish population; plant research

Arboretum (UW's Arboretum): overview and attractions at, 123–25, 158, 260–64; acreage and satellite properties of, xii, 149–50, 244, 272n15; archival collection of, 265n4; dedication of, 40, 74–79;

Arboretum (UW's Arboretum) (*continued*) envisioned description of, 10–11, 14–17, 266n24; federal government's partnerships with, 12–13, 286n38; McKay Center and, 222–23, 232–34, 242, 254, 285n33; mission of, 15, 45, 46, 75–77, 166, 180–81, 213, 244–45; original six land parcels for, 20–30; skepticism about, 46; title for, xiii–xiv, 46, 74, 166; UW's partnership with, 217, 250. *See also* directorships; educational programs; environmentalism (conservation); funding for Arboretum; land acquisition; McCaffrey Drive (formerly Arboretum Drive); migratory game bird refuge; newsletters; personnel staff; planning and beginnings of Arboretum; politics of governance; prairie restoration; problems and issues at Arboretum; research; tour guide programs; *specific committees, ecological features, gardens, memorials, and people involved in Arboretum*

Arboretum Committee (AC). *See* AC (Arboretum Committee)

Arboretum Drive (now McCaffrey Drive). *See* McCaffrey Drive (formerly Arboretum Drive)

Arboretum Executive Committee, 51, 63, 69, 80, 87, 96, 184

Arboretum News, 141, 161, 232, 234, 283n31. *See also* newsletters

Arboretum News Letter, 141, 145, 146. *See also* newsletters

Archbald, David: biographical information about, 153, 191–92, 213, 280n85; botanist and, 153, 191, 277n30; Curtis's relationship with, 191; Dane County Board and, 204–6, 210; environmental activism and, 192, 210–13, 280n85; film about Arboretum and, 276n22; FOA and, 187; on Leopold's legacy, 192; Lost City and, 192; as managing director, 153, 183, 190–91; Midwest Prairie Conference and, 217; politics of governance and resignation of, 223–24, 228; prairie restoration and, 191; resignation of, 192, 210–13, 223–24, 228, 280n85; soil erosion and, 209; tour guide programs and, 204, 210

Armstrong, Gregory: biographical information about, 243–44; Bishop Tract and, 257; as director, 242, 243; funding for Arboretum and, 248–50, 285n33; legacy of, 252; Leopold Chair of Restoration Ecology and, 256–58; McCaffrey Drive traffic issues and, 242; mission statement and, 244–46; photograph of, *244*; pre-settlement ecosystems and, 245; restoration ecology and, 245–46, 248, 258; storm water runoff and, 246; tour guide programs and, 206; UW's partnership with Arboretum and, 250

Arnold, Victor H. (Frank Custer), 197–99

Arnold Arboretum, Harvard University, vi, 3, 5, 8, 11, 13–16, 45, 78, 264

Aust, Franz A., 24, 48, 97, 100–102, 148

Bader, Brian, 251

Bartlett-Noe property, 22, *22*, 30–32, 53, 267n18. *See also* land acquisition

Bauer-Armstrong, Cheryl, 252–53

BBS (Bureau of Biological Survey), 41, 79–81

Beltline Highway, 148, 153, 174–76, 207–8, 216, 231, 247, 262

Betz, Robert F., 217–18

Beuscher, Jacob H., 186, 187

Bewick property, 283n14

biologists, 142, 144–46, 277n30. *See also* botanists

bird populations: banding and record keeping for migratory, 115, 118–19, 143, *144*, 147, 167; crow shoots and, 177; pheasants and, 72, 82, 115, 118, 143, *144*, 145, 147, 152; wildlife management research and, 147–48. *See also* migratory game bird refuge

Bishop, Linda, 250

Bishop Tract, 257

Blewett, Thomas J., 99–100

Bostian, Lloyd, 257

botanical gardens (natural landscape design). *See* natural landscape design (botanical gardens)

botanists, 153, 191, 277n30. *See also* biologists

Bradley, Katharine T.: AC and, 227–28; *Arboretum News* and, 232; biographical information about, 225–26; Cottam's relationship with, 227; as director, 213; FOA and, 227; Jordan's relationship with, 234, 237–38; journal proposal by Jordan for restoration ecology and, 234, 240; land acquisition and, 240–41; legacy of, 225, 240, 242–43; McCaffrey Drive traffic issues and, 241–42; McKay Center and, 232–34; photograph of, *226*; politics of governance and, 226–27; radio tower and, 281n5; record keeping and, 229–30; tour guide programs and, 206, 232–33, 242

Bradley, Nina Leopold, 245

Bridson, Sue, 234, 254

Brink, Royal A., 150, 163

Brittingham Foundation, 232

Brown, Charles E., 49–52, 97–98, 262

Brown, Charles N., 25–26

Brown, Pat, 251

Brown, Paul V., 102–3

buffer zone, 220–21, 224, 283n14. *See also* land acquisition

Buildings and Grounds department (Committee on Constructional Development), UW, 49, 58, 63, 70, 161

buildings for classrooms and research, 214, 220, 221

Bureau of Biological Survey (BBS), 41, 79–81

Burr Jones Memorial Park, 125

Camp Arboretum, 83–87, 98–99

Camp Madison, xii–xiii, 29, 86, 88–94, *89, 91,* 112–14, *114,* 277n30. *See also* CCC (Civilian Conservation Corps)

Carpenter, Susan, 255

carp population, 110–12, 148, 163. *See also* fish population

Catenhusen, John, 142, 159

Cather, Willa, 219

CCC (Civilian Conservation Corps): overview and dates for, xii–xiii, 29, 72, 83–84, 88, 93, 261–62; bathhouse and, 113, 152, 165, 220–21, 261; bird banding and record keeping for migratory population and, 115; Camp Madison and, xii–xiii, 29, 86, 88–94, *89, 91,* 112–14, *114,* 277n30; commemorative plaque and, 261–62; Gallistel and, 86–87; Leopold's relations with enrollees in, 68; McCaffrey Drive and, 40–43, 92; NPS and, 84, 87–93, 99, 102; personnel for research

and, 116; prairie restoration and, 93, 98, 99, 104–5, *105,* 107, 140; projects completed by, 25, 52, 92–93; public recreation areas and, 125; U.S. Army and, 84, 86, 88–91, 93, 99, 114

Chapman, Chandler B. "Bernie": East Marsh and, 41–44, 82; Island plot of land, 36, *37,* 38, 40–42; Lake Forest Land Development Company and, 23, 30, 31, 278n34; McCaffrey Drive and, 35–36, 38–39, 40–41. *See also* Lost City

Chase, Wilfred E., 54

Chicago and North Western (C&NW) Railway, 168–72, 240–41

children as visitors, 16, 123, 174, 181, 204–5, 224, 254–55, 263

Chiwaukee Prairie (UW Parkside campus), 207, 218

Citizens Committee (Arboretum planning committee), 18–19. *See also* Madison Parks Foundation

Civilian Conservation Corps (CCC). *See* CCC (Civilian Conservation Corps)

Civil Works Administration (CWA), 52, 84, 97–99

C&NW (Chicago and North Western) Railway, 168–72, 240–41

College of Agriculture, UW, 15, 19, 54–56, 60, 62–63, 67, 149–50

Committee on Constructional Development (Buildings and Grounds department), UW, 49, 58, 63, 70, 161

conservation (environmentalism). *See* environmentalism (conservation)

Cottam, Grant: AC and, 163, 183; Beltline expansion and, 208–9; biographical information about, 183; Bradley's relationship with, 227; Curtis's relationship with, 126–27; on Fassett's planting experiments, 99–100; FOA and, 184, 186–88; Greene's relationship with, 142; Jackson Oak dedication ceremony and, 188–89, 260; on Longenecker, 70; on outside threats, 207–8; photograph of, *184;* politics of governance and, 226–27; prairie restoration and, 220; soil erosion and, 209; Wilford on relic prairies and, 219–20; WISM's "The Great Treasure Hunt" and, 184–85

Couper, David C., 241–42

crow shoots, 177
Curtis, Jane, 182, 187, 192
Curtis, John T.: overview of, xii; AC and, 180; Archbald's relationship with, 191; biographical information about, 126, 138, 181–82; controlled burns and, 144; Cottam's relationship with, 126–27; diversity of Arboretum and, 158; environmentalism and, 138; Faville Grove Preserve and, 130–33; funding for research and, 166; Grady Tract and, 162–63; Greene's relationship with, 139–40, 141; legacy of, 182–83; managing directorship and, 181–82, 226; native plants and, 128, 131–32, 162–63; Partch's research with, 153, 277n30; photograph of, *127*; plant research and, 127–29, 138; plant research directorship proposal by Leopold and, 128–29, 163; prairie restoration and, xiv, 95, 117, 138–39, 153, 277n30; pre-settlement ecosystems and, 139; record keeping and, 159; as research coordinator, 158–59, 163; on research mission of Arboretum, 180–81; seed collection and, 152, 182; Sperry's research with, 129; storm runoff and, 159; student programs and, 163, 169, 183; superintendent and, 160; on tour guide programs, 174, 177; TRC and, 138, 159
Curtis Prairie: overview of, 31, 71, 95; CCC and, 140; dedication ceremony for, 187–88; forestation project along edges of, 52–53, 262; migratory game bird refuge and, 92; oak trees and, 175; photograph of, *105*; prairie restoration and, 31, 66, 71, 92–93, *105*, 236; reputation of, 218, 219. *See also* prairie restoration
Custer, Frank (Victor H. Arnold), 197–99
CWA (Civil Works Administration), 52, 84, 97–99

Dane County Board, 204–6, 210
Darling, J. N. "Ding," 41, 80–81
Dawson, Gar, 234
deer population, 129–30, 160, 178, 271n12
DeKalb County, Illinois, 236
Dennis, Clifford, 234
Dicke, Robert J., 153, 163, 183–84, 221–22, 227, 228
Dickson, James G., 48, 97, 149–50, 155

directorships: AC and, 62–63; animal research director and, 128–29; assistant director and, 249; directors and, 213, 242, 243; executive director and, 49, 63–64, 68, 70, 75, 129, 180, 184; FOA directors and, 227, 250; funding for, 54, 56, 63; managing directors and, 153, 181–82, 183, 213–14, 226; plant research director and, 128–29, 163; politics of governance and, 53–57, 59–64; Regents and, 63–64; research directors and, 63–64, 71, 119, 128–29, 158, 181, 277n30. *See also* personnel staff; politics of governance; *specific directors*
Disney Corporation, 234
DNR ([Wisconsin] Department of Natural Resources), 218, 224
Doetz, Jack and Marion, 285n33
domestic animal populations, 131–32, 147, 188, 267n18
Duck Pond (Spring Trail Pond), 119, 123, 222, 224

Earley, Emily H., 221, 243
Earth Focus Program, 254, 263
Earth Partnership for Families Program (EPF), 255, 260, 263
Earth Partnership for Schools Program (EPS), 252–54, 260, 263, 285n33, 286n38
East Marsh (Gardner Marsh), 41–44, 82, 147–48, 163, 167–68, 240, 247, 254. *See also* Island plot of land
Ecological Restoration [formerly *Restoration and Management Notes (R&MN)*], xiv, 234–35, 237–39, 242, 284n8
ecosystem restoration, 235–36. *See also* restoration ecology
ecosystems: pre-settlement, 139, 240, 245, 258. *See also* prairie restoration; restoration ecology
educational programs: overview of, 253; buildings for classrooms and, 214, 220, 221; children as visitors and, 16, 123, 174, 181, 204–5, 224, 254–55, 263; classroom space and, 152–53; EPF and, 255; EPS and, 252–54, 285n33, 286n38; funding for, 249–50, 251, 285n33, 286n38; Jordan and, 251, 256; naturalist and, 215, 225, 229, 249–51; outreach programs and, 153; Picnic Point and, 149–50; public

lectures and, 249; restoration ecology and, 214–17, 223–24, 251–52; school gardening program and, 252–53; student programs and, 161, 163, 173–74, 181, 183; superintendent and, 160, 161, 173; teacher training programs and, 251–53, 285n33. *See also* ranger-naturalists

effigy mounds, Native American, 29, 30, 49–52, 262, 263

Eisenhower Foundation, 285n33

Ellarson, Robert, 228, 284n45

Emlen, John T., Jr., 143, 152, 163

environmentalism (conservation): overview of, xi, xii, 65, 153; AC and, 153, 212–13; activists and, 192, 210–12, 210–13, 280n85; American Conservation Movement and, xi–xii, 3–4; Beltline effects and, 207–8, 216; conservation chair and, 56, 59, 60–61; Curtis and, 138; educational programs and, xiv, 214–17, 223–24; environmental effects of Beltline and, 216; game management link with, 15; land management and, 54–55; Leopold and, 56–57, 66–68, 72–73, 192; Longenecker and, 66; prairie restoration and, 217–20; wildlife management and, 216; Zimmerman and, 207–8, 210–11, 212–13. *See also* restoration ecology

EPF (Earth Partnership for Families Program), 255, 260, 263

EPS (Earth Partnership for Schools Program), 252–54, 260, 263, 285n33, 286n38

E. Ray Stevens Memorial Aquatic Gardens, 25, 92, 123. *See also* aquatic plants

Evjue Foundation, 242, 257

"Exit Orchis" (Leopold), 130–31, 156

experimental forests: overview of, xii, xiv, 8, 15, 166; federal government and, 15, 18–19, 27, 74; Grady Tract and, 133, 136, 137–38. *See also* forestation projects

farm practices for game cropping, 153, 239–40, 245–46, 252–53, 255–56, 258

Farrior, Marian, 251

Fassett, Norman C.: aquatic plants and, 95, 163–64; Arboretum Executive Committee and, 96; biographical information about, 95, 163–65; CCC and, 99; Faville Grove Preserve and, 130–33; on Jacobson as superintendent, 160; legacy of, 165; on

Leopold's legacy, 156–57; natural landscape design and, 13; photograph of, *96*; planning the Arboretum and, 13; prairie restoration experiments and, xii–xiii, 13, 96–100, 104, 105; Sperry's relationship with, 101–2; UBC and, 149–50; writings of, 157

Faville Grove Preserve, 122, 130–33, 153, 272n15

Federal Emergency Relief Administration (FERA), 39, 40, 98

federal government and agencies: Arboretum partnerships with, 12–13, 286n38; Fermilab and, 217–18; fire station and, 55–56; Jackson and, 39–42, 46–47, 57, 80, 81, 111; McCaffrey Drive assistance from, 39–42; Madison Parks Foundation and, 19; migratory game bird refuge and, 15, 46–47, 55–57, 61, 67, 79–81. *See also* CCC (Civilian Conservation Corps); *specific federal agencies*

Feeney, W. S. "Bill," 72, 115, 129, 261, 271n12

FERA (Federal Emergency Relief Administration), 39, 40, 98

Fermilab, 217–18. *See also* prairie restoration

films, and Arboretum, 166, 189–90, 218, 276n22

fire/s: C&NW Railway and, 168–72; controlled burns and, 107, 144–45, 146, 231; fire station proposal and, 55–56; Grady Tract and, 148; Greene Prairie and, 152, 171; threats of, 112–14, *114*, 168; Winslow neighborhood, 169

Fisher property, 240–41, 283n14

fish population: carp and, 110–12, 148, 163; minnows and, 111, 148, 163; trout and, 111–12; wildlife management research and, 148. *See also* aquatic plants

Fitzgibbons, Jim, 251

Fleming, Rosemary: personnel staff issues and, 225, 229, 232–33; photograph of, *202*; tour guide programs and, 202–6, 210, 215, 232–33; training for tour guide programs and, 204–5, 215, 249

FOA (Friends of the Arboretum): overview of, 155–56, 184, 186–88; activities of, 188–90; Archbald and, 187; Bradley and, 227; Curtis Prairie dedication

FOA (Friends of the Arboretum) (*continued*)
ceremony and, 187–88; director of, 227,
250; film about Arboretum and, 189–90,
276n22; funding and, 190, 192, 240,
285n12; Kipen property and, 278n34;
Lost City lots acquisition and, 190, 192,
200; membership and, 188, 189, 250;
newsletters and, 227, 283n31; tour guide
program funding and, 190, 205, 281n2
Folley, Susan, 242
forestation projects: along Curtis Prairie,
52–53, 262; Beltline and, 231, 262; Gardner Marsh and, 148; Grady Tract and,
148, 152, 162, 163, 210; politics of governance and, 52–53, 262; reforestation
and, 46, 54–55, 57, 75, 86, 235–36, 258;
reparations for fire damage and, 169–72;
research and, 163; Teal Pond and, 92,
143. *See also* experimental forests; oak
trees; pines
Forest Preserve Arboretum and Wild Life
Refuge, xi–xii, 3, 5, 12, 15, 20–21. *See also*
Arboretum (UW's Arboretum); land
acquisition
Forest Products Laboratory, 15, 19, 49, 55,
133–34, 163, 260
Frautschi, Lowell, 187, 188
Friends of the Arboretum (FOA). *See* FOA
(Friends of the Arboretum)
funding for Arboretum: Armstrong and,
248–50, 285n33; directorships and, 54,
56, 63; Gay-Way tract and, 119, 120–21;
land acquisition funding and, xv, 8–9,
12, 18, 19–20; Leopold on lack of,
115–18; natural landscape design and, 61;
personnel for research issue and, 117–18;
Regents and, 12, 18, 117–18; research
and, 121–22, 166; soil survey and, 117;
Tripp Estate endowment and, 12, 30,
55, 265n16. *See also* Arboretum (UW's
Arboretum); *specific donors*

Gallistel, Albert F.: AC and, 48, 62, 184;
Beltline and, 174, 176; biographical
information about, 58–59, 178–79, 200;
Camp Madison design and, 86; CCC
and, 86–87; ceremonial dinner for, 179,
188; federal funding and, 57; FOA and,
184, 187; Lake Wingra fish population
and, 112; legacy of, 179; Longenecker's

relationship with, 49, 53–54; Nakoma
Country Club acquisition and, 151, 152;
photograph of, *59*; Picnic Point and, 150;
politics of governance and, 58; superin-
tendent and, 160; UW Buildings and
Grounds department and, 49, 58, 63
Gallistel Woods, 29, 179, 185, 189, 210, 261
game cropping in farm practices, 153,
239–40, 245–46, 252–53, 255–56, 258
Gardner, Louis, 42–43, *43*, 119, 120, 136,
190, 241, 285n12
Gardner Marsh (East Marsh), 41–44, 82,
147–48, 163, 167–68, 240, 247, 254
Gay, Leonard W., 21, 23, 119–21, 124
Gay-Way tract, 119–21, 124
Gerloff, Gerald C., 221, 224, 228, 238, 256
Gilbert, Edward M.: AAC and, 48–49;
AC and, 9, 36, 47–48, 61; botanical
gardens and, 61; on Camp Arboretum,
84; Faville Grove Preserve transfer and,
133; federal funding and, 57; forestation
project and, 52–53; land acquisition
funding and, 12; on McCaffrey Drive,
38; migratory game bird refuge and, 57;
native plants and, 8; natural landscape
design and, 57–58; Olbrich's relationship
with, 8–9, 12; photograph of, *48*; planning the Arboretum and, 8–9; politics
of governance and, 53–54, 57, 58, 61–62;
prairie restoration and, 96–97; on Yellow Thunder's Arboretum dedication
speech, 77
Gill, Mildred, 285n33
Glass, Steve, 253, 285n33
Glover, Gene, 227, 242, 249, 250
Godfrey, Arthur, xv, 206–7, 264
Goff, F. Glenn, 277n30
Gorham Farm (parcel H), 21, 22, 24–25
Grady family, 133–37
Grady Tract: acquisition of, 43, 133–38;
Curtis and, 162–63; experimental forests
and, 133, 136, 137–38; fire and, 138,
148, 168–72; forestation projects and,
148, 152, 162, 163, 210; Greene Prairie
and, 138–42, 152, 171, 218, 246; Haen
property acquisition and, 162; Jackson's
funding efforts for, 134–37; legacy of,
140–41, 142, 218, 236; Master Development Plan of 1991 and, 231–32; prairie
restoration and, 138–42; rabbit shoots

and, 177; reparations for tree damages and, 169–72; reputation of, 218; research and, 162–63; storm runoff and, 142, 246; walkway under road to, 176
Grady Tract prairie (now Greene Prairie), 138–42, 152, 171, 218, 246. *See also* Grady Tract
Greene, Henry Campbell: AC and, 141, 180; *Arboretum News* and, 141, 161; on Beltline project, 176; biographical information about, 139, 141–42; Cottam's relationship with, 142; Curtis's relationship with, 139–40, 141; Greene Prairie and, 138–42, 152; on Jacobson as superintendent, 161; native plants and, 139–42; prairie restoration and, xiv, 138–42, 152, 159, 182; pre-settlement ecosystems and, 139; storm runoff and, 176
Greene Prairie (formerly Grady Tract prairie), 138–42, 152, 171, 218, 246. *See also* Grady Tract
Gruder-Adams, Sherrie, 249–50

Haen property, 162
Hale, Jim, *144*
Hall, Marcus, 65–66
Hands, Sally Mead, 285n33
Hanson, Walter F., 71
Harvard's Arnold Arboretum (Arnold Arboretum), vi, 3, 5, 8, 11, 13–16, 45, 78, 264
Hasler, Arthur D., 112, 148–50, 163, 167, 177, 180, 182
Hasselkus, Edward R., 69–70, 223, 243
Henderson murders, 25–30. *See also* Nelson Farm (parcel G)
Herling property, 278n34
Hickey, Joseph J., 147, 155, 157
Hilsenhoff, Bill, 243
Ho-Chunk tribe, 52, 74, 77
Ho-nee-um Pond, 68, 124, 148, 203, 224, 251. *See also* Wheeler Council Ring
Howard Hughes Medical Institute, 285n33
Hoyt transfer, 39

Icke, George W., 254
Icke, John F., 254
IMS (Institute of Museum Sciences), 250–52, 285n33
Irwin, Harriet, 229

Island plot of land, 36, *37*, 38, 40–42, 82. *See also* East Marsh (Gardner Marsh); McCaffrey Drive (formerly Arboretum Drive)

Jackson, Colonel Joseph W. "Bud": AC and, 62; Barlett-Noe property and, 31; biographical information about, 32–33, 200; East Marsh acquisition and, 41–42; effigy mounds and, 50; on E. Ray Stevens Memorial Aquatic Gardens, 25; experimental forests and, 133; federal support and, 39–42, 46–47, 57, 80, 81, 111; forestation projects and, 52–53, 262; Forest Products Laboratory collaboration and, 55; as fundraiser, 34; on Gallistel's legacy, 179; Gay-Way tract and, 120; Grady Tract acquisition and, 43, 134–37; Herling property and, 278n34; Jackson Oak dedication ceremony and, 188–89, 260; Jackson School Forest and, 205, 231; Lake Forest land acquisition and, 193, 199–200, 278n34; Lake Wingra fish population and, 111–12; land acquisition and, 20; legacy of, 188, 200; McCaffrey Drive and, 38–39; Madison and Wisconsin Foundation and, 48, 119; on Marston Estate property, 23; migratory game bird refuge and, 47, 55, 80, 81; Nolen's relationship with, 32–35; on Olbrich's arboretum interests, 8; photograph of, *33*; planning the Arboretum and, 13–14, 16–17; politics of governance and, 47, 55; prairie restoration and, 97; public recreation areas and, 123–25; scientific research and, 125; Stark's relationship with, 10
Jackson School Forest (Madison School Forest), 205, 231
Jacobson, Jacob R. "Jake," 160–62, 173, 184, 185, 190, 228
James Olin Reed bequest, 284n45
Jensen, Jens, xii, 65–66, 124
Johannsen Pond (Pond 2), 248
Johnson, Sharon, 251
Jones, Burr W., 125
Jordahl, Harold C., 251
Jordan, William R., III: biographical information about, 233; Bradley's relationship with, 234, 237–38; educational

Jordan, William R., III (*continued*)
  programs and, 251, 256; Gardner Marsh
  and, 240; Leopold Chair of Restoration
  Ecology and, 256–58; McKay Center
  and, 233–34; photograph of, *238*; public
  service coordinator and, xiv, 233; restoration ecology and, 235, 236, 239–40, 242,
  245; *R&MN* (now *Ecological Restoration*)
  and, xiv, 234–35, 237–39, 242, 284n8;
  tour guide programs and, 245
journals for restoration ecology: Bradley
  and, 234, 240; *Ecological Restoration* and,
  xiv, 234–35, 237–39, 242, 284n8. *See also*
  newsletters

Kenneth Jensen Wheeler Council Ring,
  92, 124, 251
Kilmer, Susan, 255
Kipen property, 278n34, 283n14
Kline, Virginia "Gina": biographical information about, 232; controlled burns
  and, 231; ecologist and, 230–31; forestation as shield against Beltline and, 231;
  on Greene's prairie restoration success
  and, 140–41, 142; legacy of, 232; McKay
  Center and, 234; Marion Dunn Pond
  design by, 24; Master Development Plan
  of 1991 and, 231–32; photograph of, *230*;
  pre-settlement ecosystems and, 240;
  record keeping and, 230; as research
  director, 181; tour guide programs and,
  202, 205, 230–31; writings of, 253
Knox College, 217, 236

laborers for projects: Camp Arboretum
  and, 83–87, 98–99; city of Madison
  and, 121; CWA and, 84, 97; funding
  for, 170–71, 182; Transfer Building and,
  85; WERA and, 83–87, 98–99. *See also*
  Camp Madison
Lake Forest Land Company: acquisition
  of land from, 81; bankruptcy and, 194,
  197–99; Hoyt transfer and, 39; Jackson's
  role in land acquisition from, 193,
  199–200, 278n34; McCaffrey Drive and,
  35–36, 38–39, 40–41; parcel F and, 21,
  22, 29–30, 35–36, 38; partners in, 193. *See
  also* Chapman, Chandler B. "Bernie";
  Gay, Leonard W.; Lost City
Lake Forest Land Development Company,

23, 30, 31, 278n34. *See also* Lake Forest
  Land Company; Lost City
Lake Wingra: overview of, xii, xv, 7; carp
  problem in, 110–12; McCaffrey Drive
  and, xi–xii, 9–12, 18, 35–36, *37*, 38–43,
  82, 92; storm water runoff and, 246,
  248; trout population and, 111–12. *See
  also* McCaffrey Drive (formerly Arboretum Drive)
Lake Wingra Game Refuge, 81–82, 84, 85,
  177–78. *See also* migratory game bird
  refuge
land acquisition: overview of, xi–xii, 18, 21,
  22, 44; Bartlett-Noe property and, 22,
  22, 30–32, 53, 267n18; Bewick property
  and, 283n14; Bradley and, 240–41;
  buffer zone and, 220–21, 224, 283n14;
  Fisher property and, 240–41, 283n14;
  funding for, xv, 8–9, 12, 18, 19–20;
  Gardner Marsh and, 41–44, 82, 147–48,
  163, 167–68, 240, 247, 254; Gorham
  Farm parcel H and, 21, 22, 24–25; Haen
  property and, 162; Herling property
  and, 278n34; Island plot of land and,
  36, *37*, 38, 40–42, 82; Kipen property
  and, 278n34, 283n14; Lake Forest Land
  Company parcel F land and, 21, 22,
  29–30, 35–36, 38; Lake Wingra and,
  xv, 7; Madison Parks Foundation and,
  12, 20–21, 22, 24–26, 29–30; Nakoma
  Country Club location and, 21, 22, 24,
  30, 31; Nelson Farm parcel G and, 21, 22,
  25–30, 31; Olbrich and, xv, 8–9; parcel I
  and, 21, 22, 24; satellite properties and,
  149–50, 272n15; Silex property and,
  240–41, 285n12; Spring Trail Park parcel
  J and, 21, 22, 23–24, 266n6; Stark and,
  xv, 23–24, 26–27, 29–30; Wingra Marsh
  property and Marston Estate parcel K
  and, 21, 22, 23; Winslow property and,
  149, 169. *See also* Lost City
"land ethic," and farm practices for game
  cropping, 153, 239–40, 245–46, 252–53,
  255–56, 258
Landry, Lenore, 285n33
Lapham, Increase A., vi, 13
LAWCON funds (Federal Land and Water
  Conservation Fund), 285n12
Leach, Mark, 253
Leopold, Aldo: AAC and, 49; Aldo Leop-

old Pines memorial and, 31, 154, 155–56, 177, 231; animal research directorship and, 128–29; BBS and, 80; biographical information about, 14–15, 49, 56, 57, 133, 154–55; bird banding and record keeping for migratory population and, 115, 117–18; CCC and, 68, 84, 87–88; conservation chair proposal and, 56; Coon Valley project and, 67; Curtis's plant research directorship and, 128–29, 163; dedication of Arboretum and, 74; directorship and, 55–56, 59–64; effigy mounds tablet markers and, 51–52; environmentalism and, 56–57, 66–68, 192; experimental forests and, 133; farm practices for game cropping, and "land ethic" of, 153, 239–40, 245–46, 252–53, 255–56, 258; Faville Grove Preserve and, 130–33, 153; Forest Products Laboratory and, 55; on funding for research, 121–22; Lake Wingra fish population and, 110–11; legacy of, 155–58, 192; Leopold Chair of Restoration Ecology and, 246, 256–58, 277n30, 285n33; Longenecker's relationship with, 64, 66; McCabe's relationship with, 143–44, 273n45, 277n30; migratory game bird refuge and, 15, 42, 55–57, 61, 64, 67, 71–72, 79–82, 260; mission statement and, 75–77, 245; native plants and, 128; natural landscape designers and, 65–66; NPS policies and, 72; Olbrich Memorial Entrance plaque and, 266n24; personnel for research issue and, 115–19; photograph of, 56; planning the Arboretum and, 14–15, 16; prairie restoration and, 13, 68, 100, 102–4; on predator control, 118–19; professional style of, 65, 66–68; public recreation areas versus research and, 72, 73, 116, 121–22, 125, 145–46; reforestation and, 57; research and, 37–38, 65, 66, 67–68, 71–74; research directorship of, 63–64, 71, 119, 129, 158; soil erosion and, 67, 73, 76; soil survey and, 117; Sperry's appointment and, 100, 102–4; on tour guide programs, 153; TRC and, 138; UBC and, 149–50; wildlife research management and, 71–74, 82–83, 145; writings of, 118, 130–31, 133, 154–55, 154–56

Liebl, David S., 247
Livesey Company, 257
Longenecker, G. William "Bill": AC and, 180, 184; Arboretum Executive Committee and, 69, 184; biographical information about, 49, 70, 210; Buildings and Grounds department and, 49, 63, 70; on Camp Arboretum, 84; on carp population, 112; Curtis Prairie and, 71; dedication of Arboretum and, 75; effigy mounds tablet markers and, 51–52; environmentalism and, 66; as executive director, 49, 63–64, 68, 70, 75, 129, 180, 184; FOA and, 187; forestation and, 148; Gallistel's relationship with, 49, 53–54; Gay-Way tract and, 120; Ho-nee-um Pond and, 68; on Jacobson as superintendent, 160; Leopold's relationship with, 64, 66; Nakoma Country Club acquisition and, 151; native plants and, 148; natural landscape design and, 64, 66, 68–70; photograph of, 69; politics of governance and, 53–54, 57, 62–64; prairie restoration and, 68, 71; professional style of, 65, 68–69, 70; superintendent and, 160; UBC and, 149–50
Longenecker Gardens, 69–70, 234, 263. See also Longenecker, G. William "Bill"
Lorbiecki, Marybeth, 73
Lost City: overview and history of, 192, 193–200, 262; FOA funds for acquisition of last lots of, 190, 192, 200; Halloween tour guide programs and, 196; Jackson's central role in acquisition of, 193, 199–200, 278n34; Master Development Plan of 1991 and, 231, 241; photograph of abandoned foundation in, 202; "The Great Treasure Hunt" and, 185. See also land acquisition

Madden, Harold, 85–87, 104
Madison, Wisconsin: as authority for Lake Monona project, 7; civic beautification efforts in, 5–6; Forest Products Laboratory and, 55, 133–34, 163; Godfrey's relationship with, 207; land acquisition for parks and, 6, 7–8; Nolen's relationship with, 3–5
Madison: A Model City (Nolen), 4, 6, 10–11, 12

Madison and Wisconsin Foundation, 48, 119. *See also* Jackson, Colonel Joseph W. "Bud"

Madison Bond Company, 194, 197–99

Madison Parks Foundation, 7–8, 12, 18–21, 22, 24–26, 29–30, 35, 39, 46. *See also* land acquisition

Madison Pleasure Drive Association, 3, 9, 25, 36, 43–44, 120

Madison School Forest (Jackson School Forest), 57, 205, 231

Marek, Sylvia, 249–50

Margaret's Council Ring, 254

Marion Dunn Prairie and Pond, 24, 247–48

Marston Estate and Wingra Marsh property (parcel K), 21, 22, 23, 167. *See also* land acquisition

Master Development Plan of 1991, 231–32

MATC (Madison Area Technical College), 201, 231

McCabe, Robert A.: AC and, 157, 180; *Arboretum News Letter* and, 141, 145, 146; biographical information about, 142–43; biologists and, 144–46, 277n30; bird refuge and, 143–46, *144*; controlled burns and, 144–45, 146; Faville Grove Preserve and, 131; Leopold memorial plaque and, 155; Leopold's relationship with, 143–44, 273n45, 277n30; migratory game bird refuge and, 143; native plants and, 143–44, 162; newsletter publication and, 146; pheasant population and, 143, *144*, 145; photograph of, *144*; prairie restoration and, 146, 159; public recreation areas versus research and, 145–46; rabbit shoots and, 176–77; record keeping and, 143

McCaffrey, Maurice Erve, 19–20, 27, 38, 58, 62, 156

McCaffrey Drive (formerly Arboretum Drive), xi–xii, 9–12, 18, 35–36, *37*, 38–43, 82, 92; overview of, 10, 263; dedication of, 19; land parcels acquisition and, 30, 200; memorial to McCaffery and, 156; Sinaikos donation commemoration and, 285n12; traffic issues and, 216, 224–25, 241–42

McKay Foundation, 221–22, 284n45

McKay Nature Awareness Center, 222–23, 232–34, 242, 254, 285n33

McSweeney, Kevin, *259*, 259–60

Mead, George W., 30–31, 57–58, 61, 75, 81

Midwest Prairie Conferences (North American Prairie Conferences), 217–20

migratory game bird refuge: bird banding and record keeping for migratory population and, 115, 118–19, 143, *144*, 147, 167; caretaker for feeding station protection and, 71; Curtis Prairie and, 92; environmental effects of Beltline and, 216; federal government and, 15, 46–47, 55–57, 61, 67, 79–81; forestation and, 124–25; Gilbert and, 57; Leopold and, 15, 42, 55–57, 61, 64, 67, 71–74, 79–82, 260; McCabe and, 143–44, 143–46; outside threats and, 216; politics of governance and, 55–57, 60–61; research on wildlife management and, 71–74; state game refuge status and, 81–82, 84, 177–78. *See also* wildlife management; *specific marshes*

Miles, Mr. and Mrs. Philip E., 132–33

Miner, Kathy, 255

Minkoff, Sara, 250

minnows, 111, 148, 163. *See also* fish population

Mitchell, Jennifer, 254

Moran, Eugene D. "Gene," 190, 229, 241, 242, 250

Morgan, Kathy, 251

Morrison, Darrel, 237, 255

Morton, Joy, 14

Morton Arboretum in Illinois, 14–15, 217

mounds, Native American effigy, 29, 30, 49–52, 262, 263

Muckenhirn, Mary A., 285n33

Muckenhirn, Robert J., 163

Muir, John, xi, 7, 77, 132, 136, 215

Murray, Molly Fifield, 249–52

Nakoma Country Club: acquisition of, 150–52; land parcels for Arboretum locations and, 21, 22, 24, 30, 31

National Park Service (NPS). *See* NPS (National Park Service)

national prairie restoration efforts, 218–20

National Science Foundation (NSF), 220, 253, 285n33

Native Americans: overview and history of, 50–51, 131, 188, 205, 249; effigy mounds of, 29, 30, 49–52, 262, 263; Ho-Chunk tribe and statue and, 52, 74, 77; Ho-nee-um Pond and, 68, 124, 148, 203, 224, 251; Wheeler Council Ring and, 92, 124, 251

native plants (wildflowers): overview of, xii, 8, 13, 45; Curtis and, 128, 131–32, 162–63; Gardner Marsh and, 44, 254; Gilbert and, 8; Longenecker and, 148; McCabe and, 143–44, 162; Native Plant Gardens project and, 254–55, 261, 263, 285n33; Olbrich and, 6, 8, 10, 45; plant sales and, 250; propagation program for, 251; public recreation areas and, 145; research and, 162–63; seed collection and exchange program and, xiii, 152, 182, 255. *See also* prairie restoration

natural landscape design (botanical gardens): overview of, 60–61, 65–66; Fassett and, 13; funding for, 61; Gay-Way tract and, 120, 124; Gilbert and, 57–58, 61; landscape architecture chair and, 4; Longenecker and, 64, 66, 68–70; politics of governance and, 57–58. *See also* public recreation areas (parks)

nature center (McKay Nature Awareness Center), 222–23, 232–34, 242, 254, 285n33

The Nature Conservancy (TNC), 219, 278n34

Neckerman estate donations, 285n33

Nelson Farm (parcel G), 21, *22*, 25–30, 31

newsletters: *Arboretum News* and, 141, 161, 232, 234, 283n31; *Arboretum News Letter* and, 141, 145, 146; *NewsLeaf: Friends of the Arboretum Newsletter* and, 246–47, 251, 253, 283n31. *See also* journals for restoration ecology

Noe Woods, 31, 68, 143–44, 148, 149, 169, 175

Nolen, John: Arboretum and, 78–79; dedication of Arboretum and, 74, 77–78; on federal transient camps, 84; Forest Preserve Arboretum and Wild Life Refuge and, xi–xii, 3, 5; Jackson's relationship with, 32–35; *Madison: A Model City*, 4, 6, 10–11, 12; Madison's relationship with, 3–5; natural landscape designers and, 65–66; Olbrich's correspondence with, 9, 10–12; Olbrich's relationship

with, 3, 5–6, 79; Olmsted's relationship with, 3–4; photograph of, *5*; planning the Arboretum and, xi–xii, 6; Wingra Project correspondence between Olbrich and, 3, 9, 33

North American Prairie Conferences (Midwest Prairie Conferences), 217–18

NPS (National Park Service): Camp Madison and, 112; CCC and, 84, 87–93, 99, 102; Feeney and, 115; Gay-Way tract and, 120, 121; Leopold and, 72; McCaffrey Drive and, 41; personnel for research and, 116; public recreation areas and, 122–23, 125; Sperry and, xii, 103–4, 109; USFS versus, 102–4

NSF (National Science Foundation), 220, 253, 285n33

oak trees: Beltline expansion and, 175; Curtis Prairie and, 175; fire destruction and, 170–72; Grady Tract and, 138, 170–72; infestations and, 163; Noe Woods and, 31, 175; pre-settlement ecology and, 76, 240; rabbit populations and, 147; reforestation and, 235–36, 258; Sperry and, 108; tornado damage and, 258. *See also* forestation projects

Ode, Arthur H., 220

Olbrich, Michael Balthazar: Arboretum described by, 10–11; biographical information about, 6–7, 17; civic beautification efforts by, 5–6; experimental forests and, 133; Forest Products Laboratory collaboration and, 55; Gilbert's relationship with, 8–9, 12; Harvard's Arnold Arboretum visit by, 8; land parcels acquisition and, xv, 8–9, 20; land parcels acquisition for parks and, 6, 7–8; McCaffrey Drive and, 23–31, 36; Madison Parks Foundation investments and, 7–8; mission statement and, 15, 45; native plants and, 6, 8, 10, 45; Nolen's correspondence with, 9, 10–12; Nolen's relationship with, 3, 5–6; planning for Arboretum and, xi–xii, 7–9; prairie plants and, 6, 10; Rotary Club address on Arboretum and, 14–17, 266n24; University Board of Regents and, 8–9, 12; Wingra Project correspondence between Nolen and, 3, 9, 33

Olbrich Memorial Entrance plaque, 25, 266n24
Olmsted, Frederick Law, Jr., 3–4, 5, 65–66
original parcels of land for Arboretum, 20–30. *See also* land acquisition
Oscar Mayer Foundation, 285n33
outside threats, 167–69, 186, 216–17, 262, 282. *See also* Beltline Highway; problems and issues at Arboretum

parcels of land. *See* land acquisition
parks (public recreation areas), 72, 73, 116, 121–25, 123–25, 145–46. *See also* natural landscape design (botanical gardens)
Partch, Max, 153, 277n30
Pauly, Wayne, 229
Peattie, Donald C., 164–65
personnel staff: biologist and, 142, 144–46, 277n30; botanist and, 153, 191, 277n30; ecologist and, 230–31, 253; horticultur-ist and, 255; native plant gardeners and, 255; naturalist and, 215, 225, 229, 249–50; office staff and, 220, 222, 225, 229; operations manager and, 190, 229, 241, 250, 282; public service coordina-tor and, xiv, 233; research and, 115–19, 146; research coordinator and, 158–59, 163; security and, 161, 162, 173, 189, 229; superintendent and, 160–62, 173, 184, 185, 190, 228. *See also* directorships; ranger-naturalists
Peterson, A. W., 149–50, 169–72, 187
Peterson, Ursula, 249–50
pheasant population, 72, 82, 115, 118, 143, *144*, 145, 147, 152
Picnic Point, 149–50
pines: Aldo Leopold Pines memorial and, 31, 154, 155–56, 177, 231; Beltline, 174, 176, 208, 231; Curtis Prairie, 262; fire destruction and, 112–13; forestation projects and, 52–53, 68, 124; Grady Tract and, 31, 148, 152, 154, 155–56, 162, 177, 231. *See also* forestation projects
plane crash into Redwing Marsh, 167–68
planning and beginnings of Arboretum: overview of, xi–xiii, 6–18; Forest Preserve Arboretum and Wild Life Refuge as original title and, xi–xii, 3, 5, 12, 15, 20–21; Wingra Project as original name and, 9, 15, 18, 20, 25, 33. *See also* Arboretum (UW's Arboretum); land acquisition

plant research: biologist and, 142, 144–46, 277n30; botanist and, 153, 191, 277n30; Curtis and, 127–29, 138; directorships and, 128–29, 163, 277n30. *See also* aquatic plants
Pleasant Company's Fund for Children, 255
politics of governance: overview of, xii, 45–47, 213; AAC and, 48–49, 52–53; AC and, 9, 47–48, 62–63, 222–23; Anderson's resignation and, 222–24; Archbald's resignation and, 223–24, 228; Bradley and, 226–27; conservation chair and, 56, 59, 60–61; Cottam and, 226–27; directorship and, 53–57, 59–64; effigy mounds tablet markers and, 49–52; forestation projects and, 52–53, 262; Gallistel and, 58; Gilbert and, 53–54, 57, 58, 61–62; Leopold and, 55–56, 59–64; Longenecker and, 53–54, 57, 62–64; McCaffrey and, 58; migratory game bird refuge and, 55–57, 60–61; mission of Arboretum and, 46; naming the Arbo-retum and, 46; natural landscape design and, 57–58; Regents and, 63–64. *See also* directorships; problems and issues at Arboretum
ponds: Duck Pond (Spring Trail Pond), 119, 123, 222, 224; Ho-nee-um Pond, 68, 124, 148, 203, 224, 251; Johannsen Pond (Pond 2), 248; Marion Dunn Pond, 24; Teal Pond, 92, 143. *See also* aquatic plants; fish population
Potter, Ken, 247
prairie restoration: overview of, xii–xiii, xiv, 95, 162, 235–36; Archbald and, 191; Bartlett-Noe property and, 31, 267n18; CCC and, 93, 98, 99, 104–5, *105*, 107, 140; controlled burns and, 107, 144–45, 146, 231; Cottam and, 220; Curtis and, xiv, 95, 117, 138–39, 153, 277n30; Curtis Prairie and, 31, 66, 71, 92–93, *105*, 236; environmentalism and, 217–20; Fassett and, 101, 104–7; Fermilab and, 217–18; Gilbert and, 96–97; Greene and, xiv, 138–42, 152, 159, 182; Greene Prairie and, 138–42, 152; Jackson and, 97; legacy of, 108–9; Leopold and, 13, 68, 100, 102–4; Longenecker and, 68, 71; McCabe and,

146, 159; oak trees and, 108; personnel for research and, 116–17; planning for Arboretum and, xii–xiii, 10–11; planting experiments and, 13, 96–100, 101, 104–9; pre-settlement ecosystems and, 139, 240, 245, 258; relic prairies and, 217–20; reparations for fire damage and, 169–72; reputation of Arboretum and, 217–20; Sperry and, 13, 101, 104–9; wildlife management and, 159. See also restoration ecology; specific prairies

predator control, 118–19

pre-settlement ecosystems, 139, 240, 245, 258

problems and issues at Arboretum: overview of, 167; Beltline and, 148, 153, 174–76, 207–8, 216, 231, 247, 262; bird banding and record keeping for migratory population and, 115, 144; carp population and, 110–12, 148, 163; classroom space and, 152–53; fire danger and, 112–14, 114; funding for research and, 121–22, 166; laboratory space and, 152, 165, 220–21; outside threats and, 167–69, 186, 216–17, 262, 282; personnel for research and, 115–19, 146; plane crash into Redwing Marsh and, 167–68; rabbit population and, 147, 176–77; radio tower and, 167, 216, 224, 281n5; record keeping and, 115–18, 229–30, 243; research versus public recreation areas and, 72, 116, 121–22, 125; sewage and, 166, 167, 211; soil erosion and, 67, 73, 76, 174, 208–9; storm water runoff and, 142, 159, 176, 246–48, 260; "The Great Treasure Hunt" of WISM and, 184–85; tornado damage and, 258–59; trespassers and, 72, 148. See also Arboretum (UW's Arboretum); politics of governance

public outreach, 173, 224, 255. See also educational programs; FOA (Friends of the Arboretum); tour guide programs

public recreation areas (parks), 65, 72, 73, 116, 121–25, 145–46. See also natural landscape design (botanical gardens)

public service coordinator, xiv, 233

rabbit population, 147, 176–77

radio station WISM, and "The Great Treasure Hunt," 184–85

radio tower, 167, 216, 224, 281n5

ranger-naturalists: Fitzgibbons and, 251; funding for, 173, 204, 224, 229; Wendt and, 241; Zimmerman and, 152, 173, 189, 201–4. See also educational programs

"Reading the Landscape" (Zimmerman), 201–2, 204–5, 230–31

record keeping: overview of, 115–18, 143, 159, 229–30, 243; Bradley and, 229–30; Kline and, 230; migratory game bird refuge and, 115, 118–19, 143, 144, 147, 167; superintendent and, 160, 161, 184; for tour guide programs, 232–33

Redington, Paul, 79–81. See also Bureau of Biological Survey (BBS)

Redwing Marsh, 167–68

Reese, Ruth Gardner, 254

Reeve, Roger, 97, 100

reforestation, 46, 54–55, 57, 75, 86, 235–36, 258. See also forestation projects

Regents (University Board of Regents): as authority for Arboretum, 7, 47; Beltline subway underpass and, 174–75; dedication of Arboretum and, 75; endowments for land parcels acquisition and, 12, 18; Forest Preserve Arboretum and Wild Life Refuge and, xi, 12, 20–21, 45; Forest Products Laboratory and, 55; Gay-Way tract and, 121; Grady Tract and, 136–37; land acquisition funding and, 12, 18, 192, 200, 257; landscape architecture chair and, 4; Nakoma Country Club acquisition and, 150–52; Olbrich and, 8–9, 12; personnel for research funding and, 117–18; Picnic Point and, 150; politics of governance and, 63–64; reparations for fire damage and, 172; UW Buildings and Grounds and, 58

Reihm, Robert and Kathy, 189, 277

relic prairie restoration, 217–20. See also prairie restoration

Rennebohm Foundation, 229

reputation of Arboretum, xi, 152, 206–7, 217–20, 248

research: overview of, 162–63, 213, 215–16; Anderson and, 214, 215–16; aquatic plants and, 24, 95, 110, 112, 121, 163–64; buildings for classrooms and, 214, 220, 221; Curtis as coordinator of, 158–59, 163; directors of, 63–64, 71, 119, 128–29,

research (*continued*)
158, 181, 277n30; forestation projects
and, 163; funding for, 117–18, 121–22,
166; laboratory space and, 152, 165,
220–21; Leopold and, 37–38, 65, 66,
67–68, 71–74; Leopold as director
of, 63–64, 71, 119, 129; as mission of
Arboretum, 180–81; native plants and,
162–63; personnel for, 115–19; plant,
127–29, 128–29, 138, 162–63, 277n30;
plant research directorship for Curtis
and, 128–29, 163, 277n30; public recreation areas versus, 72, 116, 121–22, 125,
145–46; record keeping and, 115–18;
wildlife management and, 71–74, 82–83
*Restoration and Management Notes
(R&MN)* [now *Ecological Restoration*],
xiv, 234–35, 237–39, 242, 284n8
restoration ecology: overview of, xi, xiv,
245, 256–57, 260–61; Armstrong and,
245–46, 248, 258; Bradley on journals
for, 234, 240; *Ecological Restoration* and,
xiv, 234–35, 237–39, 242, 284n8; ecosystem restoration and, 235–36; education
and, 214–17, 223–24, 251–52; Jordan
and, 235, 236, 239–40, 242, 245; Leopold
Chair of Restoration Ecology and, 246,
256–58, 277n30, 285n33; pre-settlement
ecosystems and, 139, 240, 245, 258;
teacher training and, 251–53, 285n33;
training programs and, 251–52; Zedler
and, 246; Zimmerman and, 257–58. *See
also* environmentalism (conservation);
prairie restoration
Richards, C. A., 163
Rideout, Jean, 248
Riker, A. J., 163
*R&MN (Restoration and Management
Notes)* [now *Ecological Restoration*], xiv,
234–35, 237–39, 242, 284n8
Rongstad, Orrin, 256–57
Roosevelt, Franklin, 19, 55, 67, 84, 86, 97
Ruskin, John, vi
Russell, Harry, L., 38, 49, 54–57, 60, 77,
133, 137, 156

Sachse, Nancy D., 44, 161, 178–79, 189,
200, 266n6
Sand County Almanac (Leopold), 118, 130,
133, 154–56

Sargent, Charles Sprague, 14, 65–66, 78
satellite properties, 149–50, 272n15. *See also*
Arboretum (UW's Arboretum)
Sauthoff, Harry, 87
Schlimgen, Nancy, 254
School Forest Movement, 57. *See also*
reforestation
Schulenberg, Ray, 217–18
Secord, John and Jane, 257, 285n33
security issues, 161, 162, 173, 189, 229
seed collection and exchange program, xiii,
152, 182, 255
SER (Society for Ecological Restoration),
232, 239
sewage problems, 166, 167, 211
Shalala, Donna, 250, 257
Shepard, Paul, 236
Silcox, F. A. "Gus," 81
Silex property, 240–41, 285n12
Simonds, Ossian Cole, 14, 15, 45
Sinaiko family gift, 241, 285n12
six original parcels of land for Arboretum,
20–30. *See also* land acquisition
Society for Ecological Restoration (SER),
232, 239
soil erosion, 55, 67, 73, 76, 174, 208–9
Sperry, Ted: biographical information
about, 100–101, 109; bird banding and
record keeping for migratory population
and, 115; bureaucratic issues over hiring
of, 101–3; CCC and, 93, 98, 104–5, *105*,
107; controlled burns and, 107; Curtis's
research with, 129; Fassett's relationship with, 101–2; legacy of, 107–9; on
Leopold's professional style, 67–68; on
Longenecker, 71; on McCabe, 146; NPS
and, xii, 103–4, 109; NPS biologist and,
115, 142; oak trees in curtis Prairie and,
108; photograph of, *105*; prairie restoration experiments in Curtis Prairie and,
xii–xiii, xiv, 93, 101, 104–7, 159; USFS
and, 100–104
Spring Trail Park (parcel J), 21, *22*, 23–24,
266n6. *See also* land acquisition
Spring Trail Pond (Duck Pond), 119, 123,
222, 224
Stark, Paul E.: on aquatic plant research,
24; Grady Tract funding efforts by, 135;
Jackson's relationship with, 10; land
acquisition and, xv, 20; land parcels

acquisition and, xv, 23–24, 26–27,
29–30; McCaffrey Drive and, 9–12;
Nakoma Country Club acquisition and,
151; photograph of, *10*; planning the
Arboretum and, xi–xii, 18
Stevens Memorial Aquatic Gardens, 25, 92,
123. *See also* aquatic plants
storm water runoff, 142, 159, 176, 209,
246–48, 260
Stoughton Faville Grove Prairie Preserve,
122, 130–33, 153, 272n15
subway underpass, and Beltline, 174–76
superintendents, 160–62, 173, 184, 185,
190, 228
Sutherland, Marion, 205

teacher training programs, and restoration
ecology, 251–53, 285n33. *See also* educational programs
Teal Pond and Teal Pond wetlands, 31,
73–74, 92, 143, 254, 263
Technical Committee, 47, 49, 152. *See also*
AC (Arboretum Committee)
Technical Research Committee (TRC),
138, 159, 166, 174
Thomas, Donna, 248–51, 285n33
Thomson, John W., 97, 99–100, 159, 164,
221
Thoreau, Henry David, 16, 78, 156
threats from the outside, 167–69, 186,
216–17, 262, 282. *See also* problems and
issues at Arboretum
TNC (The Nature Conservancy), 219,
278n34
tour guide programs: overview and history
of, xiv, 205–6, 263, 281n2; AC and, 215,
281n2; Anderson and, 215, 216, 281n2;
Archbald and, 204, 210; Armstrong and,
206; Bradley and, 206, 232–33; children
as visitors and, 16, 123, 174, 181, 204–5,
224, 254–55, 263; Curtis, Jane on, 174,
177; Dane County Board and, 204–6,
210; Fleming and, 202–6, 215, 232–33;
FOA and, 190, 205, 281n2; funding for,
190, 205, 242, 251, 281n2; Halloween
and, 196; Jacobson and, 161, 173; Jordan
and, 245; Kline and, 202; Leopold on,
153; personnel for, 173–74, 204–5, 206;
record keeping for, 232–33; statistics for,
205–6; Thomas and, 249–50; training programs and, 201–2, 204–6, 215,
230–33, 249, 281n2; Zimmerman and,
189, 201–3
training programs: restoration ecology and,
251–53, 285n33; teacher training and,
251–53, 285n33; tour guide programs
and, 201–2, 204–6, 215, 230–33, 249,
281n2. *See also* educational programs
Transfer Building, 85
TRC (Technical Research Committee),
138, 159, 166, 174
trespassers, 72, 148
Tripp Estate endowment, 12, 30, 55, 265n16
trout populations, 111–12. *See also* fish
population
Twenhofel, W. H., 38, 195
Twenhofel Hill, 38, 40, 195

UBC (University Bay Committee), 149–50
University Board of Regents (Regents). *See*
Regents (University Board of Regents)
University of Wisconsin (UW): AAC and,
48–49, 52–53, 96, 134, 154, 180, 254;
Buildings and Grounds department and,
49, 58, 63, 70, 161; College of Agricul-
ture and, 15, 19, 54–56, 60, 62–63, 67,
149–50; conservation chair and, 56, 59,
60–61; game management chair and, 59;
Leopold Chair of Restoration Ecol-
ogy and, 246, 256–58, 277n30, 285n33;
Midwest Prairie Conferences and, 218;
Parkside campus of, 207, 218; partner-
ship between Arboretum and, 217, 250.
See also AC (Arboretum Committee);
Arboretum (UW's Arboretum)
U.S. Army, 84, 86, 88–91, 93, 99, 114. *See
also* CCC (Civilian Conservation Corps)
U.S. Army Reserve, 88
USDA (U.S. Department of Agriculture),
12–13, 286n38
U.S. Forest Products Laboratory, 15, 19, 49,
55, 133–34, 163, 260
USFS (U.S. Forest Service), 53, 55, 81,
102–4, 134
U.S. Soil Erosion Service, 67
UW (University of Wisconsin). *See* AC
(Arboretum Committee); Arboretum
(UW's Arboretum); Forest Preserve
Arboretum and Wild Life Refuge; Uni-
versity of Wisconsin (UW)

UW Foundation, 256, 284n45, 285nn33
UW Parkside campus (Chiwaukee Prairie), 207, 218
UW's Arboretum (Arboretum). *See* Arboretum (UW's Arboretum)

Valerie Kerschensteiner Fund, 254
Van Alstyne, Margaret, 254
The Vanishing Prairie (documentary film), 234
Viburnum Garden, 25

Wade, Douglas E., 220
Wallace, Tom, 118–19
Walter, Rodney, 251
Ward, George, 236
WARF (Wisconsin Alumni Research Foundation), 54–57, 59, 63, 77, 116, 137
Way, Lulu A., 119, 121, 124. *See also* Gay-Way tract
WCD (Wisconsin Conservation Department), 55, 80–82, 111–12, 134, 147–48, 155, 177–78, 271n15. *See also* migratory game bird refuge
WEEB (Wisconsin Environmental Education Board), 251, 253
Wendt, Keith, 241
WERA (Wisconsin Emergency Relief Administration), 83–87, 98–99
W. G. McKay Foundation, 221–22, 284n45
WHA radio tower, 167, 216, 224, 281n5
Wheeler Council Ring, 92, 124, 251. *See also* Ho-nee-um Pond
Wilde, John H., 189
Wilde, Sergei A., 163
wildflowers (native plants). *See* native plants (wildflowers)
wildlife management: overview of, 76, 271n12; animal research director and, 128–29; bird populations and, 147–48; deer population and, 129–30, 160, 178, 271n12; environmental effects of Beltline and, 216; farm practices for game cropping and, 153, 239–40, 245–46, 252–53, 255–56, 258; fish population and, 148; pheasant population and, 72, 82, 115, 118, 143, *144*, 145, 147, 152; prairie restoration and, 159; predator control and, 118–19; research on, 71–74, 82–83, 145. *See also* migratory game bird refuge

Wilford, John Noble, 219–20
Wingra Fen, 257
Wingra Marsh property and Marston Estate (parcel K), 21, *22*, 23, 167. *See also* land acquisition
Wingra Oak Savanna Project, 252
Wingra Project, 3, 9, 15, 18, 20, 25, 33. *See also* Arboretum (UW's Arboretum)
Wingra Woods: forestation and, 148; Native American effigy mounds in, 29, 30, 49–52, 262, 263; oak tree damage and, 163, 258; tornado damage and, 258–59
Winslow property, 149, 169
Wisconsin Alumni Research Foundation (WARF), 54–57, 59, 63, 77, 116, 137
Wisconsin Capital Campaign funds, 285n33
Wisconsin Conservation Department (WCD), 55, 80–82, 111–12, 134, 147–48, 155, 177–78, 271n15. *See also* migratory game bird refuge
Wisconsin Department of Natural Resources (DNR), 218, 224
Wisconsin Emergency Relief Administration (WERA), 83–87, 98–99
Wisconsin Environmental Education Board (WEEB), 251, 253
Wisconsin State Highway Commission (WSHC), 108, 148, 174–75, 186, 208–9
WISM radio station, and "The Great Treasure Hunt," 184–85
Wood, Ken, 234, 249–50
Woods, Brock, 251–52
Wright, Frank Lloyd, 24, 266n6
WSHC (Wisconsin State Highway Commission), 108, 148, 174–75, 186, 208–9

Yellow Thunder, Chief Albert, 74, 77

Zedler, Joy, 246–48, 257–58, 277n30
Zedler, Paul H., 277n30
Zimmerman, James H. "Jim Zim": biographical information about, 229; ecosystem restoration and, 235–36; environmentalism and, 207–8, 210–11, 212–13; naturalist and, 215, 225, 229; ranger-naturalists and, 173, 189, 201–4; "Reading the Landscape," 201–2, 204–5, 230–31; restoration ecology and, 257–58; seed collection and, 152

WISCONSIN LAND AND LIFE

Arnold Alanen
Series Editor

Spirits of Earth: The Effigy Mound Landscape of Madison and the Four Lakes
Robert A. Birmingham

*Pioneers of Ecological Restoration: The People and Legacy of the University of
Wisconsin Arboretum*
Franklin E. Court

A Thousand Pieces of Paradise: Landscape and Property in the Kickapoo Valley
Lynne Heasley

*Environmental Politics and the Creation of a Dream:
Establishing the Apostle Islands National Lakeshore*
Harold C. Jordahl Jr., with Annie L. Booth

A Mind of Her Own: Helen Connor Laird and Family, 1888–1982
Helen L. Laird

When Horses Pulled the Plow: Life of a Wisconsin Farm Boy, 1910–1929
Olaf F. Larson

North Woods River: The St. Croix River in Upper Midwest History
Eileen M. McMahon and Theodore J. Karamanski

Buried Indians: Digging Up the Past in a Midwestern Town
Laurie Hovell McMillin

Wisconsin Land and Life: A Portrait of the State
Edited by Robert C. Ostergren and Thomas R. Vale

Condos in the Woods: The Growth of Seasonal and Retirement Homes in Northern Wisconsin
Rebecca L. Schewe, Donald R. Field, Deborah J. Frosch,
Gregory Clendenning, and Dana Jensen

Door County's Emerald Treasure: A History of Peninsula State Park
William H. Tishler